QUALITY OF LIFE IN OLD AGE

Social Indicators Research Series

Volume 31

This new series aims to provide a public forum for single treatises and collections of papers on social indicators research that are too long to be published in our journal *Social Indicators Research*. Like the journal, the book series deals with statistical assessments of the quality of life from a broad perspective. It welcomes the research on wide variety of substantive areas, including health, crime, housing, education, family life, leisure activities, transportation, mobility, economics, work, religion and environmental issues. These areas of research will focus on the impact of key issues such as health on the overall quality of life and vice versa. An international review board, consisting of Ruut Veenhoven, Joachim Vogel, Ed Diener, Torbjorn Moum, Mirjam A.G. Sprangers and Wolfgang Glatzer, will ensure the high quality of the series as a whole.

The titles published in this series are listed at the end of this volume.

QUALITY OF LIFE IN OLD AGE
International and Multi-Disciplinary Perspectives

Editors

Heidrun Mollenkopf
Senior Research Scientist (retired),
Formerly German Center for Research on Ageing
at the University of Heidelberg
Department of Social and Environmental Gerontology,
Germany

and

Alan Walker
Professor of Social Policy and Social Gerontology,
Department of Sociological Studies,
University of Sheffield,
UK

 Springer

A C.I.P. Catalogue record for this book is available from the Library of Congress.

ISBN 978-1-4020-5681-9 (HB)
ISBN 978-1-4020-5682-6 (e-book)

Published by Springer,
P.O. Box 17, 3300 AA Dordrecht, The Netherlands.

www.springer.com

Printed on acid-free paper

*Heidrun Mollenkopf would like to dedicate this book
to her grandchildren Anne Sophie, Marie Claire, Mathieu,
and Joscha, and Alan Walker to his children Alison and Christopher.
We hope that their later lives will be high-quality ones.*

TABLE OF CONTENTS

Preface   ix

Section I – Understanding Quality of Life in Old Age

1. International and Multi-Disciplinary Perspectives on
    Quality of Life in Old Age: Conceptual Issues   3
    Alan Walker and Heidrun Mollenkopf

2. Quality of Life in Older Age: What Older People Say   15
    Ann Bowling

Section II – Key Aspects of Quality of Life

3. Well-being, Control and Ageing: An Empirical Assessment   33
    Svein Olav Daatland and Thomas Hansen

4. Social Resources   49
    Toni C. Antonucci and Kristine J. Ajrouch

5. Economic Resources and Subjective Well-Being in Old Age   65
    Manuela Weidekamp-Maicher and Gerhard Naegele

6. Quality of Life in Old Age, Inequality and Welfare State Reform:
    A Comparison Between Norway, Germany, and England   85
    Andreas Motel-Klingebiel

7. Environmental Aspects of Quality of Life in Old Age:
    Conceptual and Empirical Issues   101
    *Hans-Werner Wahl, Heidrun Mollenkopf, Frank Oswald,
    and Christiane Claus*

8. The Environments of Ageing in the Context of the Global
    Quality of Life among Older People Living in Family Housing   123
    *Fermina Rojo-Pérez, Gloria Fernández-Mayoralas,
    Vicente Rodríguez-Rodríguez, and José-Manuel Rojo-Abuín*

9. Perceived Environmental Stress, Depression, and Quality
    of Life in Older, Low Income, Minority Urban Adults   151
    *William B. Disch, Jean J. Schensul, Kim E. Radda,
    and Julie T. Robison*

10. Ageing and Quality of Life in Asia and Europe:
    A Comparative Sociological Appraisal   167
    Mohammad Taghi Sheykhi

11. Ethnicity and Quality of Life   179
    Neena L. Chappell

12. Health and Quality of Life   195
    Dorly J.H. Deeg

13. Care-related Quality of Life: Conceptual and Empirical
    Exploration   215
    Marja Vaarama, Richard Pieper, and Andrew Sixsmith

Section III – Conclusions and Outlook

14. Quality of Life in Old Age: Synthesis and Future
    Perspectives   235
    Heidrun Mollenkopf and Alan Walker

List of Contributors   249

Index   253

PREFACE

This book started its life during a symposium we organised on Quality of Life in Old Age at the Fifth Conference of the International Society for Quality of Life Studies (ISQOLS) in Frankfurt in July 2003. We are extremely grateful to Alex Michalos for sponsoring that session, encouraging us to use the symposium presentations as the basis for a book and for his support during the commissioning and editing process. We are also very grateful to the authors of this volume for the prompt delivery of their manuscripts and responses to our editorial comments. Finally our special thanks to Marg Walker for her expert and efficient preparation of the manuscript for publication.

Heidrun Mollenkopf   *Alan Walker*
Heidelberg   *Sheffield*

SECTION I

UNDERSTANDING QUALITY OF LIFE IN OLD AGE

1. INTERNATIONAL AND MULTIDISCIPLINARY PERSPECTIVES ON QUALITY OF LIFE IN OLD AGE

Conceptual issues

Quality of life (QoL) is a multidimensional, holistic construct assessed from many different perspectives and by many disciplines. Moreover, the concept of QoL can be applied to practically all important domains of life. Thus, QoL research has to include social, environmental, structural, and health-related aspects, and be approached from an interdisciplinary perspective. This holds even more when QoL in old age is the focus because ageing itself is a multidimensional process. General QoL studies have used age for many years as a social category like gender or social class, but apart from a few exceptions (e.g. Diener and Suh, 1997; Michalos, 1986; Michalos *et al.*, 2001) they have largely neglected older people.

Recent research in gerontology has begun to systematically study QoL – following the World Health Organization (WHO) dictum 'years have been added to life and now the challenge is to add life to years'. However, there are very few overarching texts available on this topic and none of an international and multidisciplinary nature. Given the size and growth of this population, it is time to publish a volume on this topic that systematically pursues a comprehensive perspective and includes theoretical approaches and empirical findings with respect to the most important components of QoL in old age.

This volume brings together leading researchers on QoL in old age and summarises, on the one hand, what we know and, on the other, what further research is needed. It consists of three main parts with an extended introduction, the main chapters on the various aspects of what contributes to ageing people's QoL, and finally a concluding chapter pointing to knowledge gaps and necessary further developments in theory and methodology.

The introductory part emphasises the amorphous, multidimensional and complex nature of QoL as well as the high level of inconsistency between scientists in their approach to this subject. Drawing on an extensive literature review (Brown *et al.*, 2004), eight different models of QoL are distinguished. These range from objective social indicators, subjective indicators of life satisfaction and well-being, health and functioning, to interpretative approaches emphasising the individual values and theories held by older people. Moreover, this chapter summarises the main areas of consensus about QoL in old age: its dynamic multifaceted nature, the combination of life course and immediate influences, the similarities and differences in the factors determining QoL between younger and older people, the most common associations with QoL and the likely variations between groups, and the powerful role of subjective self-assessment.

The main part of the book spans the whole range of the most important issues in ageing people's QoL: their subjective evaluations (Chapter 2), personal control

H. Mollenkopf and A. Walker (eds.), Quality of Life in Old Age, 3–13.

beliefs (Chapter 3), economic resources (Chapter 5), and social relations and net-works (Chapter 4). The impact of diverging national welfare systems and social policies is investigated (Chapter 6) and environmental conditions are explored to detect their supporting or hindering potential with respect to older people's well-being (Chapters 8 and 9). Differences in the conditions of ageing between Asia and Europe are highlighted (Chapter 10) as is the diverging conditions of ethnic groups ageing in different host countries (Chapter 11). Last but not least, QoL in the case of decreasing health (Chapter 12) and the challenge of care (Chapter 13) are considered.

Not unexpectedly in view of the various topics and the empirical and scientific backgrounds, the contributions differ in approach, style, and degree of differentia-tion. Some of them provide a comprehensive overview on the available knowledge in the domain they deal with while others focus on a specific study. Some throw light on the micro cosmos of the individual, investigating psychological aspects and their role for well-being with increasing age, while others locate individual QoL in the meso and macro contexts of family, networks, cultural habits, societal structures, and national or regional conditions.

We did not try to level out these differences. More important in our view, as editors, was that the authors explained carefully their theoretical frame of reference and methodological approach and that their specific contributions deepened our knowledge about what makes up a good QoL in old age in different parts of the world. That said, we have simultaneously touched a limitation to this volume: it was not possible to consider, in fact, all parts of the world. However, our aim was not to establish a global map of older people's living conditions. Instead, this volume pro-vides a comprehensive perspective on what we know – and what we do not know – about the most important components of QoL in old age from as many national and disciplinary perspectives as possible.

Finally, the main research priorities and gaps in knowledge are outlined together with the key theoretical and methodological issues that must be tackled if compara-tive, interdisciplinary research on QoL is to develop further. That part draws on the conclusions stated by the authors of this volume and charts, as an outlook, the recent evolution of a new perspective on ageing.

THE SCOPE OF RESEARCH ON QoL IN OLD AGE[1]

QoL is a rather amorphous, multilayered, and complex concept with a wide range of components – objective, subjective, macro societal, micro individual, positive, and negative – which interact (Lawton, 1991; Tesch-Römer et al., 2001). It is a con-cept that is very difficult to pin down scientifically and there are competing disci-plinary paradigms. Three central limitations of QoL are its apparent open-ended nature, its individualistic orientation, and its lack of theoretical foundations (Walker and van der Maesen, 2004). The widely acknowledged complexity of the concept, however, has not inhibited scientific inquiry. As Fernández-Ballesteros (1998a) has shown, in the final third of the last century, there was a substantial increase in citations of QoL across five different disciplinary databases. While the growth was significant in the psychological and sociological fields, in the biomedical one, starting from a lower point, it was 'exponential' (e.g. increasing from 1 citation in 1969 to 2,424 in 1995 in the 'Medline' database). This reflects the fact that in many countries recent discussions of QoL have been dominated by health issues, and a subfield, health-related quality of life (HRQoL), has been created which emphasises the longstanding pre-eminence of medicine in gerontology (Bowling, 1997; Walker, 2005b).

Another key factor behind this growth in scientific inquiry is the concern among policymakers about the consequences of population ageing, particularly for spending on health and social care services, which has prompted a search for ways to enable older people to maintain their mobility and independence, and so avoid costly and dependency-enhancing institutional care. These policy concerns are not peculiar to Europe but are global (World Bank, 1994); nor are they necessarily negative because the new policy paradigms such as 'a society for all ages' and 'active ageing', both of which are prominent in the 2002 Madrid International Plan of Action on Ageing, offer the potential to create a new positive perspective on ageing and a major role for older people as active agents in their own QoL. A significant part of the impetus for this positive approach comes from within Europe (Walker, 2002).

MODELS OF QoL

Given the complexity of the concept and the existence of different disciplinary perspectives, it is not surprising that there is no agreement on how to define and measure QoL and no theory of QoL in old age. Indeed, it is arguable whether a theory of QoL is possible because, in practice, it operates as a meta-level construct, which encompasses different dimensions of a person's life. Nonetheless, a theory would not only lend coherence and consistency but also strengthen the potential of QoL measures in the policy arena (Noll, 2002). As part of the European FORUM project, Brown and colleagues (2004) prepared a taxonomy and systematic review of the English literature on the topic of QoL. In this, Bowling (2004) distinguishes between macro (societal, objective) and micro (individual, subjective) definitions of QoL. Among the former, she includes the roles of income, employment, housing, education, and other living and environmental circumstances; among the latter, she includes perceptions of overall QoL, individuals' experiences and values, and related proxy indicators such as well-being, happiness and life satisfaction. Bowling also notes that models of QoL are extremely wide-ranging, including potentially everything from Maslow's (1954) hierarchy of human needs to classic models based solely on psychological well-being, happiness, morale, life satisfaction (Andrews, 1986; Andrews and Withey, 1976; Larson, 1978), social expectations (Calman, 1984), or the individual's unique perceptions (O'Boyle, 1997; Brown et al., 2004, p.4).

She distinguishes eight different models of QoL which may be applied, in the adapted form here, to the gerontological literature:

1. Objective social indicators of standard of living, health, and longevity typically with reference to data on income, wealth, morbidity, and mortality. Scandinavian countries have a long tradition of collecting such national data (Hornquist, 1982; Andersson, 2005). Recently, attempts have been made to develop a coherent set of European social indicators (Noll, 2002; Walker and van der Maesen, 2004) but, as yet, these have not been applied to subgroups of the population.
2. Satisfaction of human needs (Maslow, 1954), usually measured by reference to the individual's subjective satisfaction with the extent to which these have been met (Bigelow et al., 1991).
3. Subjective social indicators of life satisfaction and psychological well-being, morale, esteem, individual fulfilment, and happiness usually measured by the use of standardised, psychometric scales and tests (Bradburn, 1969; Lawton, 1983; Mayring, 1987; Roos and Havens, 1991; Suzman et al., 1992; Veenhoven, 1999; Clarke et al., 2000).
4. Social capital in the form of personal resources, measured by indicators of social networks, support, participation in activities and community integration (Wenger, 1989, 1996; Bowling, 1994; Knipscheer et al., 1995; see also Chapter 4).
5. Ecological and neighbourhood resources covering objective indicators such as levels of crime, quality of housing and services, and access to transport, as well as subjective indicators such as satisfaction with residence, local amenities and transport, technological competence, and perceptions of neighbourliness and personal safety (Cooper et al., 1999; Kellaher et al., 2004; Mollenkopf et al., 2004; Scharf et al., 2004). Recently, this approach to QoL has become a distinct subfield of ecological or architectural gerontology, with German researchers playing a prominent role (Mollenkopf and Kaspar, 2005; Wahl and Mollenkopf, 2003; Wahl et al., 2004; Weidekamp-Maicher and Reichert, 2005).
6. Health and functioning focussing on physical and mental capacity and incapacity (e.g. activities of daily living and depression) and broader health status (Verbrugge, 1995; Deeg et al., 2000; Beaumont and Kenealy, 2004; see also Chapter 12).
7. Psychological models of factors such as cognitive competence, autonomy, self-efficacy, control, adaptation, and coping (Brandtstädter and Renner, 1990; Filipp and Ferring, 1998; Grundy and Bowling, 1999; see also Chapters 3 and 9).
8. Hermeneutic approaches emphasising the individual's values, interpretations, and perceptions usually explored via qualitative or semi-structured quantitative techniques (WHOQoL Group, 1993; O'Boyle, 1997; Bowling and Windsor, 2001; Gabriel and Bowling, 2004a). This model, which is growing in its research applications, includes reference to the implicit theories that older people themselves hold about QoL (Fernández-Ballesteros et al., 1996, 2001). Such implicit theories and definitions may be of significance in making cross-national comparisons by providing the basis for a universal understanding of QoL (and will be revisited later).

A common feature of all of these models identified by Brown et al. (2004) is that concepts of QoL have invariably been based on expert opinions rather than on those of older people themselves (or, more generally, those of any age group). This limitation has been recognised only recently in social gerontology but has already led to a rich vein of research (Farquhar, 1995; Grundy and Bowling, 1999; Gabriel and Bowling, 2004a, b). This does not mean, however, that QoL can be regarded as a purely subjective matter, especially when it is being used in a policy context. The apparent paradox revealed by the positive subjective evaluations expressed by many older people living in objectively adverse conditions, such as poverty and poor housing conditions, is a longstanding observation in gerontology (Walker, 1980, 1993). The processes of adjustment involved in this 'satisfaction paradox' have been the focus of interest in recent research (Mollenkopf *et al.*, 2004; Staudinger and Freund, 1998), and this is emphasised in Chapter 5. As Bowling (2004, p.6) notes, there may be a significant age-cohort effect behind the paradox, as older people's rating of their own QoL is likely to reflect the lowered expectations of this generation, and they may therefore rate their lives as having better quality than a person in the next generation of older people in similar circumstances would do (Schilling, 2006).

Empirical research is required to test whether or not the satisfaction paradox is a function of age-cohort but, nonetheless, the caution concerning subjective data on older people's QoL is particularly apposite in a comparative European context where expectations may differ markedly on the north/south and east/west axes (Mollenkopf *et al.*, 2004; Polverini and Lamura, 2005; Weidekamp-Maicher and Reichert, 2005). For example, there are substantial variations in standards of living between older people in different European countries: in the 'old' EU 15 the at-risk poverty rate among those aged 65 and over varied, in 2001, from 4% in the Netherlands to more than 30% in Greece, Ireland, and Portugal (European Commission, 2003).

A recent review of QoL in old age in five European countries found a fairly widespread national expert consensus about the range of indicators that constitute the concept, particularly in the two countries with the most developed systems of social reporting, the Netherlands and Sweden, but with a dominance of objective measures (Walker, 2005b). The southern European representative, Italy, does not consistently distinguish older people's QoL from the general population and frequently does not differentiate among the older age group. In all five countries health-related QoL is the most prevalent approach in gerontology. Also, while there is no consensus on precisely how QoL should be measured, there is evidence of some cross-national trade in instruments, such as the adaptation of the Schedule for the Evaluation of Individual Quality of Life (SEIQOL) for use in the Netherlands (Peeters *et al.*, 2005; see also Chapters 3, 6 and 8).

UNDERSTANDING QoL IN OLD AGE

In the light of the wide spectrum of disciplines involved in research on QoL in old age and their competing models, is it possible to draw any conclusions about how it is constituted? The answer is 'yes', but because of the lack of either a generally agreed definition or a way to measure it, such conclusions must be tentative. Firstly, although there is no agreement on these two vital issues, few would dissent from the idea that QoL should be regarded as a dynamic, multifaceted, and complex concept, which must reflect the interaction of objective, subjective, macro, micro, positive, and negative influences. Not surprisingly, therefore, when attempts have been made to measure it, QoL is usually operationalised pragmatically as a series of domains (Hughes, 1990; Grundy and Bowling, 1999).

Secondly, QoL in old age is the outcome of the interactive combination of life course factors and immediate situational ones. For example, prior employment status and midlife caring roles affect access to resources and health in later life (Evandrou and Glaser, 2004). Fernández-Ballesteros et al. (2001) combined both sets of factors in a theoretical model of life satisfaction. Recent research suggests that the influence of current factors such as network relationships may be greater than the life course influences, although, of course, the two are interrelated (Wiggins et al., 2004). What is missing, even from the interactive approaches, is a political economy dimension. QoL in old age is not only a matter of individual life courses and psychological resources but must include some reference to the individual's scope for action – the various constraints and opportunities that are available in different societies and to different groups, for example, by reference to factors such as socio-economic security, social cohesion, social inclusion, and social empowerment (Walker and van der Maesen, 2004). Hence, a consideration of the overarching and framing macro conditions, which is a matter of course in general QoL research and is the case in most of the contributions to this volume, should also become accepted practice in research on QoL in old age (see, e.g. Heyl et al., 2005).

Thirdly, some of the factors that determine QoL for older people are similar to those for other age groups, particularly with regard to comparisons between midlife and the third age. However, when it comes to comparisons between young people and older people, health and functional capacity achieve a much higher rating among the latter (Hughes, 1990; Lawton, 1991). This emphasises the significance of mobility as a prerequisite for an active and autonomous old age (Banister and Bowling, 2004; Mollenkopf et al., 2005), as well as the role of environmental stimuli and demands, and the potential mediating role of technology, in determining the possibilities for a life of quality (Mollenkopf and Fozard, 2004; Wahl et al., 1999; see also Chapter 7). In practice, with the main exception of specific scales covering physical functioning, QoL in old age is often measured using scales developed for use with younger adults. This is clearly inappropriate when the heterogeneity of the older population is taken into account, especially so with investigations among very frail or institutionalised older people. Older people's perspectives and implicit theories are often excluded by the common recourse to predetermined measurement scales in QoL research. This is reinforced by the tendency to seek the views of third parties when assessing QoL among very frail and cognitively impaired people (Bond, 1999). Communication is an essential starting point to involving older people and understanding their views, and recent research shows that this can be achieved successfully among even very frail older people with cognitive impairments (Tester et al., 2004).

Fourthly, the sources of QoL in old age often differ between groups of older people. The most common empirical associations with QoL and well-being in old age are good health and functional ability, a sense of personal adequacy or usefulness, social participation, intergenerational family relationships, availability of friends and social support, and socio-economic status (including income, wealth, and housing) (Lehr and Thomae, 1987; Mayer and Baltes, 1996; Knipscheer *et al.*, 1995; Bengtson *et al.*, 1996; Tesch-Römer *et al.*, 2001; Gabriel and Bowling, 2004a, b; see also Chapter 2). Still, different social groups have different priorities. For example, Nazroo *et al.* (2004) found that black and ethnic minority elders valued features of their local environment more than their white counterparts (see also Chapter 9). Differences of priority have been noted in Spain between older people living in the community and those in institutional care, with the former valuing social integration and the latter, the quality of the environment (Fernández-Ballesteros, 1998b). Other significant priorities for older people in institutional environments are control over their lives, structure of the day, a sense of self, activities, and relationship with staff and other residents (Tester *et al.*, 2004). This emphasises the importance of the point made earlier about the need to communicate with frail older people in order to understand their perceptions of QoL: although some recent research has begun to address this (Gerritsen *et al.*, 2004), the QoL of the very old is still a relatively neglected area of gerontology (see Chapters 3 and 13). Comparative European research also points to different priority orders among older people in different countries: e.g. the greater emphasis on the family in the South compared to the North (Walker, 1993; Polverini and Lamura, 2005). Another example of variations within Europe is the greater impact of objective living conditions on subjective QoL in former socialist countries like East Germany and Hungary compared to the more developed and affluent countries of most of the northern, western and southern parts of Europe (Mollenkopf *et al.*, 2004).

Fifthly, while there are common associations with QoL and well-being, it is clear that subjective self-assessments of psychological well-being and health are more powerful than objective economic or sociodemographic factors in explaining variations in QoL ratings (Bowling and Windsor, 2001; Brown *et al.*, 2004). Two sets of interrelated factors are critical here: on the one hand, it is not the circumstances *per se* that are crucial but the degree of choice or control exercised in them by an older person; on the other hand, whether or not the person's psychological resources, including personality and emotional stability, enable him or her to find compensatory strategies – a process that is labelled 'selective optimisation with compensation' (Baltes and Baltes, 1990). There is some evidence that the ability to operationalise such strategies, e.g. in response to ill health, disability, or bereavement, is associated with higher levels of life satisfaction and QoL (Freund and Baltes, 1998). Feelings of independence, control and autonomy are essential for well-being in old age (see Chapter 3). Moreover, analyses of the Basle Interdisciplinary Study of Aging show that psychological well-being is more strongly associated with a feeling of control over one's life than with physical health and capacity among the very elderly than among the young-old (Perrig-Chiello, 1999).

With this contextual background in mind, we hand the baton over to the authors of the subsequent chapters who deal with the various components of QoL in old age. Our concluding chapter highlights the main knowledge gaps and the next steps for theory and methodology in this field.

NOTES

1. The following sections include parts of an article published previously in the *European Journal of Ageing* (Walker, 2005a).

REFERENCES

Andersson, L. (2005) 'Sweden: Quality of Life in Old Age I' in A. Walker (ed.) *Growing Older in Europe*, Maidenhead, UK, Open University Press, pp.105–127.

Andrews, F.M. (ed.) (1986) *Research on the Quality of Life*, Michigan, University of Michigan.

Andrews, F.M. and Withey, S.B. (1976) 'Developing Measures of Perceived Life Quality: Results from Several National Surveys', *Social Indicators Research*, 1, 1–26.

Baltes, P.B. and Baltes, M.M. (eds) (1990) *Successful Ageing: Perspectives from the Behavioural Sciences*, New York, Cambridge University Press.

Banister, D. and Bowling, A. (2004) 'Quality of Life for the Elderly: The Transport Dimension', *Transport Policy*, 11, 105–115.

Beaumont, J.G. and Kenealy, P.M. (2004) 'Quality of Life: Perceptions and Comparisons in Healthy Old Age', *Ageing and Society*, 24, 755–770.

Bengtson, V., Rosenthal, C., and Burton, L. (1996) 'Paradoxes of Family and Aging' in R. Binstock and L. George (eds) *Handbook of Aging and the Social Sciences*, San Diego, Academic Press, pp.263–287.

Bigelow, D.A., McFarlane, B.H., and Olson, M.M. (1991) 'Quality of Life of Community Mental Health Programme Clients: Validating a Measure', *Community Mental Health Journal*, 27, 43–55.

Bond, J. (1999) 'Quality of life for People with Dementia: Approaches to the Challenge of Measurement', *Ageing and Society*, 19, 561–579.

Bowling, A. (1994) 'Social Networks and Social Support Among Older People and Implications for Emotional Well-being and Psychiatric Morbidity', *International Review of Psychiatry*, 9, 447–459.

Bowling, A. (1997) *Measuring Health*, Buckingham, Open University Press.

Bowling, A. (2004) 'A Taxonomy and Overview of Quality of Life' in J. Brown, A. Bowling, and T. Flynn (eds) *Models of Quality of Life: A Taxonomy and Systematic Review of the Literature*, Sheffield, University of Sheffield, FORUM Project.

Bowling, A. and Windsor, J. (2001) 'Towards the Good Life', *Journal of Happiness Studies*, 2, 55–81.

Bradburn, N.M. (1969) *The Structure of Psychological Well-being*, Chicago, Aldine Press.

Brandtstädter, J. and Renner, G. (1990) 'Tenacious Goal Pursuit and Flexible Goal Adjustment: Explication of Age-related Analysis of Assimilative and Accommodative Strategies of Coping', *Psychology and Ageing*, 5, 58–67.

Brown, J., Bowling, A., and Flynn, T. (2004) *Models of Quality of Life: A Taxonomy and Systematic Review of the Literature*, Sheffield, University of Sheffield, FORUM Project (http://www.shef.ac.uk/ageingresearch).

Calman, K.C. (1984) 'Quality of Life in Cancer Patients: A Hypothesis', *Journal of Medical Ethics*, 10, 124–127.

Clarke, P.J., Marshall, V.W., Ryff, C.D., and Rosenthal, C.J. (2000) 'Well-being in Canadian Seniors: Findings from the Canadian Study of Health and Aging', *Canadian Journal on Aging*, 19, 139–159.

Cooper, K., Arber, S., Fee, L., and Ginn, J. (1999) *The Influence of Social Support and Social Capital in Health*, London, Health Education Authority.

Deeg, D.J.H., Bosscher, R.J., and Broese van Groenou, M.I. (2000) *Ouder Warden in Nederland*, Amsterdam, Thela Thesis.

Diener, E. and Suh, E. (1997) 'Age and Subjective Well-being: An International Analysis', *Annual Review of Gerontology and Geriatrics*, 17, 304–324.

European Commission (2003) *Draft Joint Inclusion Report, Statistical Annex*, COM (2003) 773 final, Brussels, European Commission.

Evandrou, M. and Glaser, K. (2004) 'Family, Work and Quality of Life: Changing Economic and Social Roles Through the Lifecourse', *Ageing and Society*, 24, 771–792.

Farquhar, M. (1995) 'Elderly People's Definitions of Quality of Life', *Social Science and Medicine*, 41, 1439–1446.

Fernández-Ballesteros, R. (1998a) 'Quality of Life: Concept and Assessment' in J.G. Adair, D. Belanger, and K.L. Dion (eds) *Advances in Psychological Science*, East Sussex, UK, Psychology Press, pp.387–406.

Fernández-Ballesteros, R. (1998b) 'Quality of Life: the Differential Conditions', *Psychology in Spain*, 2, 57–65.

Fernández-Ballesteros, R., Zamarrón, M.D., and Marciá, A. (1996) *Calidad de Vida en la Vejez en Distintos Contextos*, Madrid, IMERSO.

Fernández-Ballesteros, R., Zamarrón, M.D., and Ruiz, M.A. (2001) 'The Contribution of Socio-demographic and Psychosocial Factors to Life Satisfaction', *Ageing and Society*, 21, 25–43.

Filipp, S.H. and Ferring, D. (1998) 'Regulation of Subjective Well-being in Old Age by Temporal and Social Comparison Processes?', *Zeitschrift für Klinische Psychologie – Forschung und Praxis*, 27, 93–97.

Freund, A.M. and Baltes, P.B. (1998) 'Selection, Optimization and Compensation as Strategies of Life Management: Correlations with Subjective Indicators of Successful Ageing', *Psychology and Aging*, 13, 531–543. Erratum (1999) 14, 700–702.

Gabriel, Z. and Bowling, A. (2004a) 'Quality of Life in Old Age From the Perspectives of Older People' in A. Walker and C. Hagan Hennessy (eds) *Growing Older: Quality of Life in Old Age*, Maidenhead, UK, Open University Press, pp.14–34.

Gabriel, Z. and Bowling, A. (2004b) 'Quality of Life from the Perspectives of Older People', *Ageing and Society*, 24, 675–692.

Gerritsen, D., Steverink, N., Ooms, M., and Ribbe, M. (2004) 'Finding a Useful Conceptual Basis for Enhancing the Quality of Life of Nursing Home Residents', *Quality of Life Research*, 13, 611–624.

Grundy, E. and Bowling, A. (1999) 'Enhancing the Quality of Extended Life Years', *Ageing and Mental Health*, 3, 199–212.

Heyl, V., Wahl, H.-W., and Mollenkopf, H. (2005) 'Visual Capacity, Out-of-Home Activities and Emotional Well-being in Old Age: Basic Relations and Contextual Variation', *Social Indicators Research*, 74(1), 159–189.

Hornquist, J. (1982) 'The Concept of Quality of Life', *Scandinavian Journal of Social Medicine*, 10, 57–61.

Hughes, B. (1990) 'Quality of Life' in S. Peace (ed.) *Researching Social Gerontology*, London, Sage, pp.46–58.

Kellaher, L., Peace, S.M., and Holland, C. (2004) 'Environment, Identity and Old Age: Quality of Life or a Life of Quality?' in A. Walker and C. Hagan Hennessy (eds) *Growing Older: Quality of Life in Old Age*, Maidenhead, UK, Open University Press, pp.60–80.

Knipscheer, C.P.M., de Jong Gierveld, J., van Tilburg, T.G., and Dykstra, P.A. (eds) (1995) *Living Arrangements and Social Networks of Older Adults*, Amsterdam, VU University Press.

Larson, R. (1978) 'Thirty Years of Research on the Subjective Well-being of Older Americans', *Journal of Gerontology*, 33, 109–125.

Lawton, M.P. (1983) 'Environment and Other Determinants of Well-being in Older People', *The Gerontologist*, 23, 349–357.

Lawton, M.P. (1991) 'Background: A Multidimensional View of Quality of Life in Frail Elders' in J.E. Birren, J. Lubben, J. Rowe, and D. Deutchman (eds) *The Concept and Measurement of Quality of Life in the Frail Elderly*, San Diego, Academic Press.

Lehr, U. and Thomae, H. (eds) (1987) *Formen Seelsichen Alterns*, Stuttgart, Enke.

Maslow, A. (1954) *Motivation and Personality*, New York, Harper.

Mayer, K.U. and Baltes, P. (eds) (1996) *Die Berliner Altersstudie*, Berlin, Akademie-Verlag.

Mayring, P. (1987) 'Subjektives Wohlbefinden im Alter', *Zeitschrift für Gerontologie*, 20, 367–376.

Michalos, A.C. (1986) 'An Application of Multi Discrepancies Theory (MDT) to Seniors', *Social Indicators Research*, 18, 349–373.

Michalos, A.C., Hubley, A.M., Zumbo, B.D., and Hemingway, D. (2001) 'Health and Other Aspects of the Quality of Life of Older People', *Social Indicators Research*, 54, 239–274.

Mollenkopf, H. and Fozard, J.L. (2004) 'Technology and the Good Life: Challenges for Current and Future Generations of Aging People' in H.-W. Wahl, R. Scheidt, and P. Windley (eds) *Aging in Context: Socio-physical Environments (Annual Review of Gerontology and Geriatrics*, 23, (2003), 250–279), New York, Springer.

Mollenkopf, H., Kaspar, R., Marcellini, F., Ruoppila, I., Széman, Z., Tacken, M., and Wahl, H.W. (2004) 'Quality of Life in Urban and Rural Areas of Five European Countries: Similarities and Differences', *Hallym International Journal of Aging*, 6(1), 1–36.

Mollenkopf, H. and Kaspar, R. (2005) 'Elderly People's Use and Acceptance of Information and Communication Technologies' in B. Jaeger (ed.) *Young Technologies in Old Hands: An International View on Senior Citizens' Utilization of ICT*, Copenhagen, DJOF Publishing, pp.41–58.

Mollenkopf, H., Marcellini, F., Ruoppila, I., Széman, Z., and Tacken, M. (eds) (2005) *Enhancing Mobility in Later Life – Personal Coping, Environmental Resources, and Technical Support: The Out-of-Home Mobility of Older Adults in Urban and Rural Regions of Five European Countries*, Amsterdam, IOS Press.

Nazroo, J., Bajekal, M., Blane, D., and Grewal, I. (2004) 'Ethnic Inequalities' in A. Walker and C. Hagan Hennessy (eds) *Growing Older: Quality of Life in Old Age*, Maidenhead, UK, Open University Press, pp.35–59.

Noll, H.H. (2002) 'Towards a European System of Social Indicators: Theoretical Framework and System Architecture', *Social Indicators Research*, 58, 47–87.

O'Boyle, C.A. (1997) 'Measuring the Quality of Later Life', *Philosophy Transactions of the Royal Society of London*, 352, 1871–1879.

Peeters, A., Bouwman, B., and Knipscheer, K. (2005) 'The Netherlands: Quality of Life in Old Age I' in A. Walker (ed.) *Growing Older in Europe*, Maidenhead, UK, Open University Press, pp.83–104.

Perrig-Chiello, P. (1999) 'Resources of Well-being in Elderly: Differences Between Young and Old and Old Old' in C. Hummel (ed.) *Les Science Sociales Face au défi de la Grande Vieillesse*, Geneva, Questions d'Age, pp.45–47.

Polverini, F. and Lamura, G. (2005) 'Italy: Quality of Life in Old Age I' in A. Walker (ed.) *Growing Older in Europe*, Maidenhead, UK, Open University Press, pp.55–82.

Roos, N.P. and Havens, B. (1991) 'Predictors of Successful Aging: A Twelve Year Study of Manitoba Elderly', *American Journal of Public Health*, 81, 63–68.

Scharf, T., Phillipson, C., and Smith, A.E. (2004) 'Poverty and Social Exclusion: Growing Older in Deprived Urban Neighbourhoods' in A. Walker and C. Hagan Hennessy (eds) *Growing Older: Quality of Life in Old Age*, Maidenhead, UK, Open University Press, pp.81–106.

Schilling, O. (2006) 'Development of Life Satisfaction in Old Age: Another View on the "Paradox"', *Social Indicators Research*, 75, 241–271.

Staudinger, U.M. and Freund, A. (1998) 'Krank und "arm" im Hohen Alter und trotzdem Guten Mutes?', *Zeitschrift für Klinische Psychologie*, 27, 78–85.

Suzman, R.H., Willis, D.P., and Manton, K.G. (eds) (1992) *The Oldest Old*, Oxford, Oxford University Press.

Tesch-Römer, C., von Kondratowitz, H.J., and Motel-Klingebiel, A. (2001) 'Quality of Life in the Context of Intergenerational Solidarity' in S.O. Daatland and K. Herlofson (eds) *Ageing, Intergenerational Relations, Care Systems and Quality of Life*, Oslo, Nova, pp.63–73.

Tester, S., Hubbard, G., Downs, M., MacDonald, C., and Murphy, J. (2004) 'Frailty and Institutional Life' in A. Walker and C. Hagan Hennessy (eds) *Growing Older: Quality of Life in Old Age*, Maidenhead, UK, Open University Press, pp.209–224.

Veenhoven, R. (1999) 'Quality-of-Life in Individualistic Society', *Social Indicators Research*, 48, 157–86.

Verbrugge, L.M. (1995) 'New Thinking and Science on Disability in Mid- and Late Life', *European Journal of Public Health*, 1, 20–28.

Wahl, H.W. and Mollenkopf, H. (2003) 'Impact of Everyday Technology in the Home Environment on Older Adults' Quality of Life' in K.W. Schaie and N. Charness (eds) *Impact of Technology on Successful Aging*, New York, Springer.

Wahl, H.-W., Mollenkopf, H., and Oswald, F. (1999) *Alte Menschen in ihrer Umwelt*, Wiesbaden, Westdeutscher Verlag.

Wahl, H.-W., Scheidt, R., and Windley, P. (eds) (2004) (*Annual Review of Gerontology and Geriatrics*, 23), *Aging in Context: Socio-physical Environments*, New York, Springer.

Walker, A. (1980) 'The Social Creation of Poverty and Dependency in Old Age', *Journal of Social Policy*, 9, 75–91.

Walker, A. (1993) *Age and Attitudes*, Brussels, European Commission.

Walker, A. (2002) 'A Strategy for Active Ageing', *International Journal of Social Security*, 55(1), 121–139.

Walker, A. (2005a) 'A European Perspective on Quality of Life in Old Age', *European Journal of Ageing*, 2(1), 2–12.

Walker, A. (2005b) *Growing Older in Europe*, Maidenhead, UK, Open University Press.

Walker, A. and van der Maesen, L. (2004) 'Social Quality and Quality of Life' in W. Glatzer, S. von Below, and M. Stoffregen (eds) *Challenges for Quality of Life in the Contemporary World*, The Hague, Kluwer.

Weidekamp-Maicher, M. and Reichert, M. (2005) 'Germany: Quality of Life in Old Age I' in A. Walker (ed.) *Growing Older in Europe*, Maidenhead, UK, Open University Press, pp.33–54.

Wenger, G.C. (1989) 'Support Networks in Old Age: Constructing a Typology' in M. Jeffreys (ed.) *Growing Old in the Twentieth Century*, London, Routledge, pp.166–185.

Wenger, G.C. (1996) 'Social Networks and Gerontology', *Reviews in Clinical Gerontology*, 6, 285–293.

WHOQoL Group (1993) *Measuring Quality of Life*, Geneva, World Health Organization.

Wiggins, R.D., Higgs, P., Hyge, M., and Blane, D.B. (2004) 'Quality of Life in the Third Age: Key Predictors of the CASP-19 Measure', *Ageing and Society*, 24, 693–708.

World Bank (1994) *Averting the Old Age Crisis*, Washington, World Bank.

2. QUALITY OF LIFE IN OLDER AGE

What older people say

INTRODUCTION

The increasing number of older people, with higher expectations of 'a good life' within society and with their high demands for health and social care, has led to international interest in the enhancement, and measurement, of quality of life (QoL) in older age. UK Government policy is also concerned with enabling older people to maintain their independence and active contribution to society and, in effect, to add quality to years of life. QoL has thus become commonly used as an endpoint in the evaluation of public policy (e.g. in the assessment of outcomes of health and social care). This indicates that a multifaceted perspective of QoL is required, with a shift away from single-domain approaches that focus only on single areas of life (e.g. physical health and/or functioning, mental health, social support, life satisfaction, and well-being) towards one that also reflects the views of the population concerned.

A measure of QoL requires a definition of the concept. QoL theoretically encompasses the individual's physical health, psycho-social well-being and functioning, independence, control over life, material circumstances, and external environment. It is a concept that is dependent on the perceptions of individuals, and is likely to be mediated by cognitive factors (Bowling, 2005a,b).

It reflects macro, societal, as well as micro, individual, influences, and it is a collection of objective and subjective dimensions which interact (Lawton, 1991). Lawton (1982, 1983a, b) developed a popular model and proposed that well-being in older people may be represented by behavioural and social competence (e.g. measured by indicators of health, cognition, time use, and social behaviour), perceived QoL (measured by the individual's subjective evaluation of each domain of life), psychological well-being (measured by indicators of mental health, cognitive judgements of life satisfaction, positive-negative emotions) and the external, objective (physical) environment (housing and economic indicators). He thus developed a quadripartite concept of the 'good life' for older people (Lawton, 1983a), which he later changed to 'quality of life' as the preferred overall term, accounting for all of life. However, there is no consensus within or between disciplines about conceptual definitions or measurement of QoL.

Most investigators have based their concepts and measures on experts' opinions, rather than those of lay people (Rogerson *et al.*, 1989; Bowling, 2001). Consequently, there is little empirical data on the extent to which the items included in most measurement scales have any relevance to people and their everyday lives. In addition, a pragmatic approach prevails in the literature, clarification of the

15

H. Mollenkopf and A. Walker (eds.), Quality of Life in Old Age, 15–30.
© 2007 *Springer.*

concept of QoL is typically bypassed and justified with reference to its abstract nature, and the selection of measurement scales often appears *ad hoc* (Carver *et al.*, 1999). This has resulted in many investigators adopting a narrow, or discipline-bound, perspective of QoL, and selecting single-domain measures such as scales of physical functioning, mental health, broader health status, life satisfaction, and so on. Some have used broader measures that aim to assess a person's needs for resources or services, which also overlap with QoL domains. Others have selected combinations of single-domain measures, in an attempt to adopt a broader view of QoL, although this often results in lengthy schedules with high respondent and researcher burden (see Bowling, 2005a and Haywood *et al.*, 2004 for reviews of measures).

Moreover, the models of QoL that have been developed are not consistent. Some have incorporated a needs-based satisfaction model, according to Maslow's (1954, 1968) hierarchy of human needs for maintenance and existence (physiological, safety and security, social and belonging, ego, status and self-esteem, and self-actualisation). Higgs *et al.* (2003), for example, based their model of QoL in older age on self-actualisation and self-esteem. In contrast, traditional US social science models of QoL have been based primarily on the overlapping, positive, concepts of 'the good life', 'life satisfaction', 'social well-being', 'morale', 'social tempera-ture', or 'happiness' (Andrews, 1986; Andrews and Withey, 1976; Lawton, 1996). The focus among psychologists is on psychological resources (Baltes and Baltes, 1990). The World Health Organization quality of life (WHOQoL) Group adopted a multifaceted approach, while emphasising subjective perceptions, values and cultural context, from which was developed their WHOQoL (100 items and brief version) (WHOQoL Group, 1993; Skevington, 1999; Skevington *et al.*, 2004). Stenner *et al.* (2003), on the basis of 'importance items', agreed with the WHOQoL Group during the development and testing of the WHOQoL-100. The analyses showed that the common factors that emerged included areas relating to family relationships, independence, a 'can do' and positive approach to life, health and reli-gious trust. The version for use with older people – the WHOQoL-OLD – is being field tested (http://www.euro.who.int/ageing/quality). Brown *et al.* (2004) reviewed the literature on QoL and developed a taxonomy of QoL comprising the following components: objective indicators; subjective indictors; satisfaction of human needs; psychological characteristics and resources; health and functioning; social health; social cohesion and social capital; environmental context; ideographic approaches.

What about the public views? Brown *et al.*,'s (2004) review concluded that older people's views of QoL overlapped with theoretical models. However, theoretical models need to be multidimensional in order to encompass people's values and perceptions. At present, QoL measures are based on the disciplines and perspectives of investigators (see earlier). Bowling's (1995a,b) earlier research on the most important things in people's lives, based on a random sample of adults in Britain, found that people prioritised their finances, standard of living, and housing; relationships with relatives and friends; their own health; the health of a close other;

and social and leisure activities. But there were variations by age and gender. For example, social relationships and work have been reported to be prioritised more by younger than by older adults, and health and family, more by people aged 65 and over (Brown *et al.*, 1994; Farquhar, 1995; Bowling, 1995a,b, 1996; Bowling and Windsor, 2001). Also, people who have social care needs, particularly those living in nursing and residential homes, might prioritise the ability to control their lives and the way they structure their days as most important (Qureshi *et al.*, 1994). Fry's (2000) research, based on a combination of survey data and in-depth interviews with older people living in Vancouver, reported that people valued most their personal control, autonomy and self-sufficiency, their right to pursue a chosen lifestyle, and their right to privacy.

It is increasingly important to develop a multidimensional model of QoL, for use in both descriptive and evaluative research (e.g. in health and social policy), which reflects the views of the population concerned. Such a model also needs to be developed on the basis of results from longitudinal and repeated cross-sectional research in order to measure any dynamic features, response shift, cohort, and ageing effects. Longitudinal research is also required to ascertain whether attempting to engineer gains in subjective QoL is a realistic policy goal.

AIMS AND METHODS

The primary aim of this chapter is to explore the constituents of perceived QoL in older age in order to deconstruct the concept of QoL. The second aim is to make comparisons between older people's perceptions of QoL and the content of existing QoL measures.

The Sample

Data were derived from a national interview survey of QoL in older age in Britain. Age 65 and over was taken to denote older age. The sample of people aged 65 and over was derived from four quarterly Office for National Statistics (ONS) Omnibus Surveys. Of the sample of 1,299 eligible respondents sifted by ONS from the Omnibus Survey, the overall response rate for the four QoL Surveys was 77% (999) (range over the four surveys: 69–83%); 19% refused to participate; and 4% could not be contacted during the interview period. The interviews were conducted in respondents' own homes. The sociodemographic characteristics of the sample were similar to those from mid-year population estimates for Great Britain (estimated from the last census). Full details of the method and sample have been published elsewhere (Bowling *et al.*, 2002, 2003; Bowling and Gabriel, 2004; Gabriel and Bowling, 2004).

Further in-depth interviews about QoL were carried out 12–18 months later with a subsample of 80 of the 999 participants in the QoL Survey. They were purposively sampled, using a quota matrix based on respondents' sociodemographic characteristics and health status, QoL ratings, and region of residence. The aim was to interview a broad cross section of respondents in order to obtain a better understanding of people's interpretations of QoL.

Open-Ended Survey Questions on QoL and Analysis

A series of open-ended questions were asked at the beginning of the interview in order to elicit people's views of the things that gave their lives quality, the things that took quality away from their lives, their priorities, and how QoL could be improved for themselves as well as for other people their age. Also included was a self-rating of QoL overall on a 7-point Likert scale, ranging from 'as good as can be' to 'as bad as can be'. The open-ended questions were used as the opening survey questions in order to prevent respondent bias towards the other, more specific questions and scales included in the questionnaire (see Bowling *et al.*, 2002 for details of the structured items and scales).

The detailed coding frames for the open-ended survey responses were developed after AB and two coders independently read all of the scripts of responses, and were refined as coding took place. The coding was carried out by two coders and checked by an independent researcher. Main themes and detailed subthemes were coded in order to capture the essence of people's definitions and exactly what made the QoL good and bad, and how life could be improved. These were entered onto SPSS[10] and merged with the main quantitative data-set. The main themes only are presented here. Detailed subcategories for each theme were also coded and, along with verbatim statements from respondents as examples, have been reported elsewhere (Bowling *et al.*, 2003).

In-Depth Interviews and Analyses

The subsequent in-depth interviews with a subsample of respondents were based on a semi-biographical interview technique, whereby the interviewer asked respondents first to describe key events in their histories, including marriage, work, and/or parenthood where relevant. This aimed to facilitate people talking about QoL in the context of their overall life, and to enhance the researchers' understanding of people's perspectives on life. Then the interviewer used a checklist and asked respondents what they thought of when they heard the words 'quality of life', to describe their QoL, what gave their life quality and what took quality away from it, how it could be improved, what would make it worse, and about any changes since the survey interview. The interviews were audio-recorded, transcribed, categorised by one researcher, and checked by an independent researcher. They were analysed using NU*DIST (version 5).

RESULTS

The Survey Data

About half, 48% (480), of the QoL Survey respondents were female and 52% (519) were male. The comparable figures for respondents aged 65 and over to the British General Household Survey (GHS) in 2000 were 51% female and 49% male (Walker *et al.*, 2001). Of the QoL Survey respondents, 62% (624) were aged 65–74 and 38% (375) were aged 75 and over; the comparable figures for the 2000 GHS were 58%

and 42% respectively. A total of 98% (983) of QoL respondents were white, as would be expected from national statistics.

Similar proportions of men and women, within each age group, rated their overall QoL as good or less than good (see Figure 1). Just 4% (21) of males and 6% (32) of females rated their QoL as 'so good it could not be better'; 78% (404) of males and 70% (362) of females labelled it as 'very good' or 'good'; and the remainder rated it as 'alright', 'bad', 'very bad', or 'so bad it could not be worse'. Associations with QoL ratings have been reported elsewhere (Bowling *et al.*, 2002).

The main themes categorised from responses to the open-ended survey question on the constituents of the 'good things' that gave quality to life were, in order of magnitude: social relationships (81%), social roles and activities (60%), solo activities (48%), health (44%), psychological outlook and well-being (38%), home and neighbourhood (37%), financial circumstances (33%), and independence (27%). Smaller numbers mentioned a wide range of other things. Poor health was most often mentioned as the thing that took 'quality away' from their lives (by 50%). Other commonly mentioned things that took quality away from life were home and neighbourhood (30%), financial circumstances (23%), and psychological outlook (17%). Good health, followed by better finances (i.e. having enough or more money), were the two most frequently mentioned things that respondents said would improve their QoL. While a common core of main constituents of QoL clearly emerged from the data, there was more variance in the subthemes underlying these, reflecting individual circumstances. These are displayed, under each main theme, in Appendix 1.

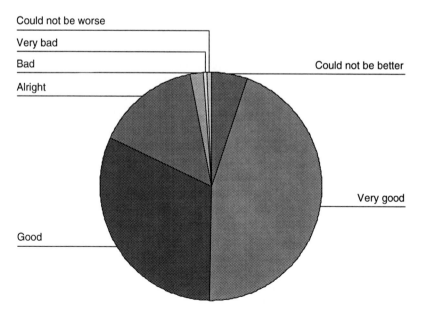

Figure 1. Quality of life ratings

In-Depth Follow-Up Interviews

Of the 80 survey respondents who were re-interviewed in-depth 12–18 months later, 40 were male and 40 female; 26 were aged 65 < 70, 20 were 70 < 75, 20 were 75 < 80, and 14 were 80 and over. As many as 37 were married and the rest were single: never married (5), married but separated (1), divorced (6), or widowed (31). They were purposively selected using a grid to ensure representation of people across social classes, with different levels of abilities in activities of daily living, and in QoL ratings, with a range of income, and with a wide geographical spread.

The main themes categorised from the in-depth interviews with the subsample of respondents on the 'good things' that gave quality to life were, in order of magnitude: social relationships (96%), home and neighbourhood (96%), psychological outlook and well-being (96%), solo activities (93%), health (85%), social roles and activities (80%), financial circumstances (73%), and independence (69%). Poor home and neighbourhood, poor health, and poor social relationships were the themes most often mentioned as those that took quality away from their lives (by 84%, 83%, and 80% respectively). These were followed by psychological outlook and well-being (63%), not having enough money (53%), and losing independence (46%). A range of other areas were mentioned by smaller numbers. The most commonly mentioned things that respondents said would improve their QoL were having a better home and neighbourhood, enough money, and better health.

Comparison of Lay Views and Regression Modelling

In sum then, the most frequent QoL themes raised by older people, both in the main survey and in the in-depth follow-up were: social relationships, social roles and activities, other leisure pursuits and activities enjoyed alone, health, psychological outlook and well-being, home and neighbourhood, financial circumstances, and independence. Modelling, using multiple regression, of the main independent predictors of QoL ratings (using the theoretically important structured scales and items administered in the survey) confirmed the importance of most of these themes (see Bowling et al., 2002 for details of approach and results). The main independent indicators of self-rated good QoL in the regression model, which explained most of the variance in QoL ratings, were: (downward) social comparisons between oneself and others, positive social expectations of life, optimism, better reported health and functional status, more social activities and social support, less reported loneliness, better ratings of the quality of local facilities (in their area of residence) and perceived safety of the area. None of the sociodemographic and socio-economic indicators were statistically significant in the model. Although self-efficacy (and sense of control over life), level of income, and other indicators of socio-economic status were not significant in the modelling, it is important to include these areas in the concept and measurement of QoL given that older people themselves raised financial circumstances and independence in, and control over, their daily lives, as important to their QoL.

COMPARISON OF MAIN AREAS OF QoL MENTIONED WITH QoL SCALES
USED WITH OLDER PEOPLE

Comparisons with the elicited main QoL themes were made with generic QoL scales developed for use with older people. The Short Form 36 (SF-36, version 2) was also included in this comparison. Although the latter was developed as a scale of broader health status, not QoL, it is often used as a proxy for health-related QoL (HRQoL) and even broader QoL; Ware *et al.* (1993) accepted this usage as the instrument was believed to contain key areas of HRQoL. There are numerous other scales of broader health status, and of needs for a wide range of services, as well as a number of utility measures which aim to assess QoL, but which mainly tap health and functioning (see Bowling, 2005a and Haywood *et al.*, 2004 for reviews). These were not included in the comparisons due to their main focus on health and functioning, and lack of reference to QoL in their conceptual development. Single-domain scales were also not included in the comparisons, by definition (e.g. where they measure only physical health and/or functioning, psychological health, well-being, or resources, social support, etc.). Inevitably this leaves a small number of instruments which are generic in their approach to QoL, and which were developed specifically to measure QoL in older populations (or, in the case of the SF-36, as a *proxy* measure of QoL and HRQoL).

The comparisons are shown in Table 1 and indicate that key areas of QoL, emphasised by older people themselves, are neglected in most measures aiming to tap QoL, as well as in the most well-used proxy measure (SF-36). Neglected areas include leisure and social activities enjoyed alone, home and neighbourhood (social capital), finances, and independence. The WHOQoL-OLD is the most comprehensive scale. Moreover, most of the detailed subthemes which emerged from the QoL survey and follow-up are conspicuously absent in these measures (see summary list of these in Appendix 1).

CONCLUSION

The survey and in-depth follow-up study reported here indicated that, overall, QoL is built on a series of interrelated drivers (main themes), while individuals emphasise varying constituent parts (subthemes). The core components, and the central planks, of QoL in older age, which were consistently emphasised by these methods, were psychological variables (e.g. social expectations and comparisons, optimism–pessimism), health and functional status, personal social support and social activity, and neighbourhood social capital. The lay models also emphasised the importance of financial circumstances and independence, which need to be incorporated into a definition of broader QoL.

Comparisons of these main themes with generic QoL scales used with older people show that several areas were neglected including: leisure and social activities enjoyed alone, home and neighbourhood (social capital), finances, and independence. The omission of key areas of life emphasised by lay people is consistent with Bowling's (1995a, b) comparisons of public perceptions of the important things in life with commonly used scales of broader health status. Some instruments also included

TABLE 1. Comparison of main themes derived from the QoL survey with QoL scales developed for use with older people, and the SF-36 (commonly used proxy measure for QoL) †

Lay themes: main QoL survey themes (good or bad):	Measurement scales:							
	Geriatric QoL Questionnaire (Guyatt et al. 1993)	QoL cards (Rai et al. 1995)	QoL Profile - Seniors version (Raphael et al. 1995a b)	SF-36, version 2 (Ware et al. 1997)+	LEIPAD (de Leo et al. 1998a; 1998b)	WHOQoL-OLD (Skevington 1999;±) ++	QoL - well-being, meaning and value (Sarvimäki and Stenbock-Hult 2000)	CASP-19 (Higgs et al. 2003; Hyde et al. 2003)*
Social relationships and support		√	√‡‡	√	√	√	√	
Social/leisure roles and activities/social participation/exclusion			√‡‡	√		√‡‡	√	
Solo leisure pursuits and activities		√						
Physical health, including physical functioning/ independence in functioning/self care	√		√‡‡	√	√	√	√	
Psychological a) psychological health, emotional/spiritual wellbeing/pleasure/ satisfaction with life/ areas of life	√	√	√‡‡	√	√	√	√	
b) psychological outlook/ growth/resources			√‡‡			√	√	√
Home/living area and neighbourhood/ environment/community			√‡‡			√	√‡‡	√
Finances/economic						√‡‡	√‡‡	

Independence and/or
  autonomy/control
  (not ADL)

Overall QoL rating

Other main themes not
  elicited as a main theme
  in QoL survey:

Cognitive functioning

Work capacity/limitations

Pain

Energy

Sexual activity/functioning

Activities over time (past,
  present, future)

Use of time

Sensory ability

Death and dying

Personality

Social desirability

Religion/spirituality

'Personal life'

Life meaning

Individualised items

† by definition, this excludes individualised measures and the many unidimensional measurement scales of health, physical functioning, psychological morbidity, social health, emotional and subjective well-being, including life satisfaction (see Bowling 2004 for details of these)

+ the SF-36 is included here because it is the most commonly used proxy measure of (health related) quality of life, although it was designed as a broader health status instrument

± and see http://www.euro.who.int/ageing/quality

++ The WHOQoL-OLD comprises the WHOQoL-100 with seven additional items

†† These are sub-domains of broader themes (main domains)

* The CASP-19 aims to tap control, autonomy, self-realisation and pleasure. Thus while items ask about family, health, finances they are asked in relation to their influence on these core domains only, and are thus indexed just under these themes

themes or items on areas not raised by respondents to the QoL survey reported here, notably sexual functioning and specific health symptoms. It is possible that older people felt inhibited during the interviews to mention sexual activity, and this highlights the need to be flexible towards questionnaire design and aim to include areas of practical and theoretical importance, as well as those derived from lay views.

The WHOQoL-OLD is the most comprehensive instrument considered, but it is still undergoing field testing, is very long, and its Likert scale format throughout appears tiresome, although there is no evidence that it has adversely affected response to date. The conclusion from this research is that a generic, truly multidimensional QoL measure for use with older populations, with minimal respondent burden, needs to be developed for use in the evaluation of health and social care outcomes in this group. The next step is to develop a multidimensional measure of QoL which includes the most pertinent and sensitive items listed in Appendix 1.

To conclude, it is important to develop a concept and measure of QoL for use in both descriptive and evaluative research (e.g. in health and social policy) which also reflects the views of the population concerned. Longitudinal and repeated cross-sectional research on QoL is also needed, with designs that can take response shift, effects of cohort and of ageing, as well as personality, into account. The latter is essential in order to address the issue of whether attempting to engineer gains in subjective QoL is a realistic policy goal, in view of the literature referred to concerning personality effects on perceived well-being (Lykken and Tellegen, 1996). The research reported here also indicates that there is a need to move beyond the common emphasis on health and functioning, which is prevalent in much of the QoL literature. A broader perspective of QoL is important for multisector policy evaluation, and a better understanding of the quality of later life is essential if we are aiming to enhance it.

Acknowledgements I thank the ONS Omnibus Survey staff and ONS Qualitative Research Unit for help with designing the questionnaire, for conducting focus groups with older people to inform the questionnaire design, for sampling and overseeing the Quality of Life Interview and processing the data. I am also grateful to Zahava Gabriel for conducting and analysing the in-depth interviews, Matthew Bond for advice with statistical analysis, Anne Fleissig and Lee Marriott-Dowding for assistance with the open coding and the development of the coding frames, ONS interviewers and the respondents themselves. Those who carried out the original analysis and collection of the data hold no responsibility for their further analysis and interpretation. Material from the ONS Omnibus Survey, made available through ONS, has been used with the permission of the Controller of The Stationery Office. The data-set is held on the Data Archive at the University of Essex. The research was funded by the Economic and Social Research Council (Growing Older Programme award no.L480254003 [Quality of Life]). The Quality of Life Questionnaire was also part-funded by grants, held collaboratively, by Professor Christina Victor and Professor John Bond (L480254042; Loneliness and Social Isolation, also part of the ESRC Growing Older Research Programme), and by Professor Shah Ebrahim (Medical Research Council Health Services Research Collaboration [Health and Disability]), and I gratefully acknowledge their support.

APPENDIX 1. OLDER PEOPLE'S VIEWS OF GOOD AND BAD AREAS
OF QoL: THEMES AND SUBTHEMES FROM THE QoL SURVEY
OPEN-ENDED QUESTIONS AND THE IN-DEPTH INTERVIEWS

Social relationships (good and bad):
Does not want to bother/worry others by asking for help
Enjoys seeing family happy/achieve/progress
Family too busy to spend time with him/her
Feels secure knowing has family help when needed
Friends/neighbours/others too busy to spend time with him/her
Gives help/support to friends/family
Has acquaintances in the neighbourhood
Has (no) company (for mixing, conversation/to be nice to one)
Has (no) contact with family
Has (no) contact with (great) grandchildren/other people's children
Has (no) contact with family by phone/post
Has (no) friends to do things with
Has (no) family to do things with
Has (no) good/close friends
Has good reciprocal relationship with family
Has good reciprocal/bad relationship with friends/neighbours
Has good (close/loving/caring/supportive/someone to love) relationships vs has poor
relationships (including arguments)
Has (no) partner
Has pet for company/to love/to depend on them/misses pet that has passed away
Has (no) practical help/regular help from family (e.g. with shopping, housework,
bathing, gardening)
Has (no) practical help from friends
Has (no) telephone contact with friends
Sees (does not) friends
Misses friends/family members who have died
Misses spouse who has died
Not as able to see others so much socially because losing health/growing older
and frailer
Worries about/responsibility for children

Social roles and social (group) activities (good only):
Attendance at cultural events (e.g. theatre/concerts/cinema)
Attendance at local events/meetings
Attendance at place of worship
Committee member of local group
Gambling (e.g. horses, bingo)

Holidays/weekends away
Involvement in voluntary work
Meals/drinks out
Mental pursuits to keep mind alert (evening classes, quizzes, bridge)
Outings/day trips
Performance in local groups (drama, art, choir, music)
Regular role helping friends, family, neighbours (child care, collecting things, e.g. from shops)
Shopping with someone else
Sports/physical activities/keeping fit/exercising/dancing; having fun; meeting people through social activities; watching sports matches
Walking dog to meet others/caring for pet

Other activities enjoyed alone (good only):
Collecting (e.g. stamp, coin, book collecting/other)
Cooking/eating new foods/diet
Crafts (e.g. woodwork, embroidery, restoring antiques, sewing, knitting, crochet, painting, flower arranging)
Cultural interests/art/theatre/architecture enjoyed alone
Do-it-yourself jobs
Drinking at home
Gardening or allotment/indoor plants
Hobbies (other)
Listening to cassettes/radio/music
Mental activities/pursuits done alone (e.g. doing crosswords, jigsaws, competitions/ writing)
Photography/videoing
Playing an instrument alone (e.g. piano, organ)
Reading books, poetry
Reading newspapers/keeping up to date with current affairs
Sports/physical activities/exercise/keeping fit/walking alone (e.g. going for a jog/ walking the dog for exercise)
Watching sport on TV
Watching TV/videos
Wildlife (e.g. feeding and watching birds, badgers, squirrels, foxes, butterflies)

Health (good and bad):
Close other's (ill) health
(Dis)comfort due to restricted functioning
Fit enough (not) to do what one wants
Fit enough (not) to get out and about
Good coping ability/coping ability when ill
(Un)able to drive; sleep
Has (not) access to health care
Has energy

Has health relative to others vs losing health
(In)ability to look after self and home
(In)ability to communicate
In pain
Restrictions on diet
Tiredness/lack of energy

Psychological well-being (good and bad):
Achieving (not) goals/able to do what one wants
Being (not) busy, occupied
Confident/loss of confidence due to having good/poor family/health/neighbourhood
Copes with the bad things in his/her life (not health)
Enjoys life (does not); depressed due to bereavement
Enjoys role/unhappy due to lack of role
Enthusiasm (not) for future/fear of the future; mentally alert/memory deteriorating
Feels (un)lucky
Lonely (not)
Positive/negative disposition (e.g. happy, content or depressed/unhappy/discontent)
Positive/negative memories of past (e.g. job/achievements/family)
Strength/lack of from beliefs – spiritual – to help face future
Stress-free/stressed

Home and neighbourhood (good and bad):
Adequacy of local facilities (e.g. shops/post-office/market); proximity to family
Age mix within the neighbourhood
Enjoyment/pleasure (lack of) from home (e.g. comfortable/spacious/nice)
Adequacy of local library/mobile library service
Adequacy of local services (e.g. council, police, repairs, street lights, refuse)
Adequacy of public transport
Adequacy of social services support
Friendliness of area/community feel
Feeling of being (un)safe/(in)secure
Home delivery services
Home/domestic help (private/social services)
Level of satisfaction with sheltered housing
Likes (dislikes) living alone/with others
Noise, youth vandalism (none) in the neighbourhood
Pleasantness of landscape/surroundings
Relationships with neighbours
Security from local help line number for use in emergencies
Stress of home (e.g. repairs needed/stairs/household chores difficult)

Financial circumstances (good and bad):
Able (not) to afford hobbies/pastimes/pets, travel/holidays, luxuries, health care;
debt/solvency

Able (not) to pay for basics, to afford essentials (e.g. heating, balanced diet, clothes, telephone), to run/maintain car/petrol, to pay for help/upkeep or repairs to house)
Adequacy of pension, income, standard of living
Adequacy relative to others/parent/self in past/expectations
Low financial expectations

Independence (good and bad):
(In)ability to do things *for oneself* (e.g. looking after home, garden – independence stressed not health)
(In)ability to please oneself
Pleased/depressed due to ability/inability to do things for self
Freedom from lack of restrictions on time/flexibility of time when retired to do things one wants/too many time pressures
Independent (not independent) due to (not) being able to drive/afford to drive a car; good/poor public transport

Other (good and bad):
Global society and politics: world news; crime rates in general; government policies (e.g. pensions); values/standards of behaviour in society in general/ageist attitudes; cruelty to animals
Other bad (e.g. weather, too much traffic, pollution, poor TV, does not use e-mail
Other good (e.g. good preparation for retirement, cigarettes give relief and pleasure, supermarkets deliver food, new computer, uses e-mail, new knowledge, stocks and shares)

REFERENCES

Andrews, F.M. (ed.) (1986) *Research on the QoL,* Michigan, University of Michigan Institute for Social Research.
Andrews, F.M. and Withey, S.B. (1976) *Social Indicators of Well-being, American's Perceptions of Life Quality,* New York, Plenum Press.
Baltes, P.B. and Baltes, M.M. (eds) (1990) *Successful Aging: Perspectives from the Behavioral Sciences,* New York, Cambridge University Press.
Bowling, A. (1995a) 'What Things Are Important in People's Lives? A Survey of the Public's Judgements to Inform Scales of Health-related Quality of Life', *Social Science and Medicine,* 10, 1447–1462.
Bowling, A. (1995b) 'The Most Important Things in Life: Comparisons Between Older and Younger Population Age Groups by Gender', *International Journal of Health Sciences,* 6, 169–175.
Bowling, A. (1996) 'The Effects of Illness on Quality of Life', *Journal of Epidemiology and Community Health,* 50, 149–155.
Bowling, A. (2001) *Measuring Disease: A Review of Disease Specific Quality of Life Measurement Scales,* 2nd edn, Buckingham, Open University Press.
Bowling, A. and Windsor, J. (2001). Towards the good life. A population survey of dimensions of quality of life. *Journal of Happiness Studies,* 2, 55–81.
Bowling, A. (2005a) *Measuring Health: A Review of QoL Measurement Scales,* 3rd edn, Maidenhead, UK, Open University Press.
Bowling, A. (2005b) *Ageing Well: Quality of Life in Old Age,* Maidenhead, UK, Open University Press.
Bowling, A., Bannister, D., Sutton, S., *et al.* (2002) A Multidimensional Model of QoL in Older Age, *Ageing and Mental Health,* 6, 355–371.

Bowling, A. and Gabriel, Z. (2004) 'An Integrational Model of Quality of Life in Older Age: A Comparison of Analytic and Lay Models of Quality of Life, *Social Indicators Research*, 69, 1–36.

Bowling, A., Gabriel, Z., Dykes, J., *et al.* (2003) 'Let's Ask Them, a National Survey of Definitions of QoL and Its Enhancement among People Aged 65 and Over', *International Journal of Aging and Human Development*, 56, 269–306.

Brown, J., Bowling, A., and Flyn, T. (2004) *Models of Quality of Life: A Taxonomy, Overview and Systematic Review of Quality of Life*, Sheffield, Department of Sociological Studies, European Forum on Population Ageing Research.

Brown, J.P., O'Boyle, C.A., McGee, H.M., *et al.* (1994) 'Individual Quality of Life in the Healthy Elderly', *Quality of Life Research*, 3, 235–244.

Carver, D.J, Chapman, C.A, Salazar, T., *et al.* (1999) 'Validity and Reliability of the Medical Outcomes Study Short Form-20 Questionnaire as a Measure of Quality of Life in Elderly People Living at Home', *Age and Ageing*, 28, 169–174.

de Leo, D., Diekstra, R.F.W., Lonnqvist, J., *et al.* (1998a) LEIPAD, an Internationally Applicable Instrument to Assess Quality of Life in the Elderly, *Behavioural Medicine*, 24, 17–27.

de Leo, D., Diekstra, R.F.W., Lonnqvist, J., *et al.* (1998b) LEIPAD *Questionnaire: Compendium of Quality of Life Instruments*, Chichester, West Sussex, Wiley.

Farquhar, M. (1995) 'Elderly People's Definitions of Quality of Life', *Social Science and Medicine*, 41, 1439–1446.

Fry, P.S. (2000) 'Whose Quality of Life Is It Anyway? Why Not Ask Seniors to Tell Us about It?', *International Journal of Aging and Human Development*, 50, 361–383.

Gabriel, Z. and Bowling, A. (2004) 'Perspectives on Quality of Life in Older Age, Older People Talking', *Ageing and Society* (Special Issue ESRC GO Programme commission), 24, 675–691.

Guyatt, G.H., Eagle, D.J., Sackett, B., *et al.* (1993) 'Measuring Quality of Life in the Frail Elderly', *Journal of Clinical Epidemiology*, 46, 1433–1444.

Haywood, K.L., Garratt, A.M., Mackintosh, A.E., and Fitzpatrick, R. (2004) *Health Status and Quality of Life in Older People: A Structured Review of Patient-assessed Health Instruments*, Report from the Patient-assessed Health Instruments Group to the Department of Health.

Higgs, P., Hyde, M., Wiggins, R., and Blane, D. (2003) 'Researching Quality of Life in Early Old Age, the Importance of the Sociological Dimension', *Social Policy and Administration*, 37, 239–252.

Hyde, M., Wiggins, R.D., Higgs, P., and Blane, D. (2003) 'A Measure of Quality of Life in Early Old Age, the Theory, Development and Properties of a Needs Satisfaction Model (CASP-19)', *Ageing and Mental Health*, 7, 186–194.

Lawton, M.P. (1982) 'Competence, Environmental Press and Adaptation of Older People' in M.P. Lawton, P.G. Windley, and T.O. Byerts (eds) *Aging and Environment: Theoretical Approaches,* New York, Springer.

Lawton, M.P. (1983a) 'Environment and Other Determinants of Well-being in Older People', *Gerontologist*, 23, 349–357.

Lawton, M.P. (1983b) 'The Varieties of Well-being', *Experimental Aging Research*, 9, 65–72.

Lawton, M.P. (1991) 'Background: A Multidimensional View of Quality of Life in Frail Elders' in J.E. Birren, J. Lubben, J. Rowe, and D. Deutchman (eds) *The Concept and Measurement of Quality of Life in the Frail Elderly,* San Diego, CA, Academic Press.

Lawton, M.P. (1996) 'Quality of Life and Affect in Later Life' in C. Magai, S.H. McFadden, *et al.* (eds) *Handbook of Emotion, Human Development, and Aging*, San Diego, CA, Academic Press.

Lykken, D. and Tellegen, A. (1996) 'Happiness Is a Stochastic Phenomenon', *Psychological Science*, 7, 186–189.

Maslow, A. (1954) *Motivation and Personality*, New York, Harper.

Maslow, A.H. (1968) *Toward a Psychology of Being*, 2nd edn, Princeton, NJ, Van Nostrand.

Qureshi, H., Nocon, A., Thomson, C. (1994). *Measuring outcomes of community care for users and carers: A review*. York: Social Policy Research Unit, University of York.

Rai, G.S., Kelland P, Rai, S.G.S., and Wientjes, HJ.F.M. (1995) 'Quality of Life Cards: A Novel Way to Measure Quality of Life in the Elderly', *Archives of Gerontology and Geriatrics*, 21, 285–289.

Raphael, D., Brown, I., Renwick, R., *et al.* (1995a) 'The Quality of Life of Seniors Living in the Community: A Conceptualization with Implications for Public Health Practice', *Canadian Journal of Public Health*, 86, 228–233.

Raphael, D., Smith, T.F., Brown, I., and Renwick, R. (1995b) 'Development and Properties of the Short and Brief Versions of the Quality of Life Profile: Seniors Version', *International Journal of Health Sciences*, 6, 161–168.

Rogerson, R.J., Findlay, A.M., Coombes, M.G., and Morris, A. (1989) 'Indicators of Quality of Life', *Environment and Planning*, 21, 1655–1666.

Sarvimäki, A. and Stenbock-Hult, H.B. (2000) 'Quality of Life in Old Age Described as a Sense of Well-being, Meaning and Value', *Journal of Advanced Nursing*, 32, 1025–1033.

Skevington, S.M. (1999) 'Measuring Quality of Life in Britain, Introducing the WHOQoL-100', *Psychomatic Research*, 47, 449–459.

Skevington, S.M., Lotfy, M., and O'Connell, K.A. (2004) 'The World Health Organization's WHOQoL-BREF Quality of Life Assessment, Psychometric Properties and Results from International Field Trials: A Report from the WHOQoL Group', *Quality of Life Research*, 13, 299–310.

Stenner, P.H.D., Cooper, D., and Skevington, S.M. (2003) 'Putting the Q into Quality of Life: The Identification of Subjective Constructions of Health-related Quality of Life Using Q Methodology', *Social Science and Medicine*, 57, 2161–2172.

Walker, A., Maher, J., Coulthard, M., et al. (2001). Living in Britain. Results from the 2000 General Household Survey. London: The Stationary Office.

Ware, J.E., Snow, K.K., Kosinski, M., and Gandek, B. (1993) *SF-36 Health Survey, Manual and Interpretation Guide*, Boston, MA, The Health Institute, New England Medical Center.

Ware, J.E., Snow, K.K., Kosinski, M., and Gandek, B. (1997) *SF-36 Health Survey, Manual and Interpretation Guide*, Revised edn, Boston, MA, The Health Institute, New England Medical Center.

WHOQoL Group (1993) *Measuring Quality of Life, the Development of the World Health Organization Quality of Life Instrument (WHOQoL)*, Geneva, World Health Organization.

SECTION II

KEY ASPECTS OF QUALITY OF LIFE

3. WELL-BEING, CONTROL, AND AGEING

An empirical assessment

INTRODUCTION

Quality of life (QoL) is at the same time a very concrete and elusive concept. It is concrete in the sense that most people have quite clear conceptions about what is *not* a good life. There is more variation, and less agreement, on what are the positive aspects of life. Among the reasons for this is that we as human beings share some basic needs which disturb us and make us unhappy when they are not met, while we as individuals have personal tastes and preferences which tend to direct our dreams about the good life in different directions. This being said, the convergent images of the negative sides of life, and the more divergent images of the positive will, in reality, interact and be modified by social structures, cultural norms, and shared experiences such as ageing. They do, however, represent contrasting perspectives on what QoL is and how it could be studied: one based on an indirect approach, the other on a more direct one. The first will tend to focus on circumstances that may make a good life possible, while the other will study the good life more directly in the form of subjective well-being or happiness.

These two traditions have dominated research in the field. The first originated in economics and focused on living standards in terms of income and material goods (Motel-Klingebiel, 2004). Sociologists added social indicators that were assumed to be important like housing, health, education, and social support (Erikson and Uusitalo, 1987; Ringen, 1995). The common feature is a focus on (more or less) objective circumstances that may enable a good life. This approach is normatively grounded in a conception of well-being as freedom or capability – the ability and power to reach valuable goals (Sen, 1993; Ringen, 1995). It works on the assumption that people are rational and will try to realize what they consider to be good for themselves, which is a reasonable assumption for most people and in most situations, but not without exceptions.

The second tradition aims at QoL more directly by simply asking people how they feel (Diener *et al.*, 1999; see also Chapter 2). The subjective well-being approach is primarily an arena for psychologists, but includes also branches of sociology and the health sciences. Among the arguments for this perspective is to accept people's own perceptions, to avoid paternalism, and to allow a larger range of diversity in the conceptions of a good life.

A third line bridges the two by studying how well-being is influenced by living conditions and vice versa. Campbell *et al.* (1976) have argued that subjective well-being is a function of both direct and indirect effects of structural conditions, but typically, the effects of objective living conditions are filtered through subjective

H. Mollenkopf and A. Walker (eds.), Quality of Life in Old Age, 33–47.
© 2007 *Springer.*

evaluations of these conditions (Smith *et al.*, 1999). Living conditions have an impact on individual well-being mainly through personal perceptions, for example relative to expectations and comparative standards. This may explain the often moderate correlations found between 'objective' and 'subjective' living conditions. Relative deprivation is, so to speak, a better indicator for subjective well-being than actual deprivation.

The role of personal control beliefs may be another bridge between the approaches, at least in two ways. One is that a sense of control may be a motivating factor for goal-directed activity, and then increase the probability of reaching desirable goals and thereby feeling better. This would be a case of 'primary control' within the terminology of Heckhausen and Schultz (1995), but could equally well be seen as an internalised sense of what Sen (1993) calls capabilities, and which to Sen is a characteristic of the person–environment relationship. A second and contrasting possibility for the bridging role of control beliefs between living conditions and well-being is that control beliefs may function as perceptive filters for how living conditions are interpreted. Low control may induce a person to give in to external constraints one has little control over. A strategy of 'accommodative flexibility' (to the realities) would then be more adaptive for well-being than one of 'assimilative persistence', to phrase it in the terminology of Brandstädter and Renner (1990).

RESEARCH QUESTIONS

We shall explore the possibly changing role of personal control for well-being with increasing age. Personal control is found to be an important predictor of subjective well-being in adulthood (DeNeve and Cooper, 1998; Smith *et al.*, 2000). Greater personal control – be it measured as internal control, self-efficacy, primary control, mastery, or similar constructs – is associated with both cognitive and emotional well-being (Skinner, 1996), although perhaps differently so for different types of control beliefs (Kunzmann *et al.*, 2000; Krause and Shaw, 2003). Less is known about the nature of the relationship, and even though the relationship between personal control and well-being has been found to be significant, the correlation is far from perfect, often varying in the area of 0.20 to 0.40 (Peterson, 1999). The challenge is to explore when and why the association is stronger or weaker, which is the approach taken here, and then specifically how this association varies with age (Slagsvold *et al.*, 2003).

A control-enhancing experiment by Langer and Rodin (1976) indicates how an increase in personal control can have a positive effect on well-being even among frail elders. This finding may seem contradictory to an adaptation theory like that of Brandstädter and Renner (1990), in which 'giving in' to external constraints (accommodation) is assumed to be more adaptive than goal-directed coping when resources decline. It indicates at least that instrumental control can be beneficial also when competencies are indeed low. Schultz and Heckhausen (1997) argue that there is a shift in balance from primary (instrumental, direct) to secondary (psychological, indirect) control strategies in later life, but primary control has still a motivational primacy over secondary control. Even the very frail may therefore benefit from an enhancement of control, as in the Langer and Rodin experiment among nursing home patients. Schultz and Heckhausen contrast their 'continuity position' to theories that advocate a fundamental change in motivational systems in late life, like Cumming and Henry's (1961) disengagement theory, Erikson's (1968) theory of ego development, and Carstensen's (1991) socio-emotional selectivity theory. The role of personal (primary) control for well-being in old age, or more generally the role of personal control when resources decline, should be stronger – and more direct – under the Schultz and Heckhausen position than as seen by these other theories in which goals like disengagement, ego integrity, or emotional security are expected to serve the person better.

Do we find support for one or the other of the two perspectives? Does the impact of personal control on well-being decline with age, or continue to be important even when remaining resources are low as indicated by the Langer and Rodin experiment? Or more generally – does personal control become less, or possibly even more, important for subjective well-being with increasing age in later life? These questions are important because they have consequences in terms of different practice recommendations for well-being in old age. They may also help us understand the 'stability despite loss' paradox (Schaie and Lawton, 1998; Diener et al., 1999; Smith et al., 1999; Kunzmann et al., 2000). Subjective well-being is often found to remain stable in late life despite losses in external and internal resources, or even to increase when the age-related losses are controlled for (adjusted life satisfaction). A sense of primary control is found to decline with age (Mirowsky, 1997), and if this is an adaptive response for well-being, it might contribute to explain why age-related losses are seemingly not a threat to well-being. Zapf (1984) sees satisfaction with poor living conditions as a prototypical response in older years, while younger people – with higher expectations – are more inclined to be dissatisfied even when living conditions are fair (Daatland, 2005).

Maintaining high subjective well-being in older years despite a loss of resources may be taken as evidence of resilience and adaptive capacity among elders, and as a strategy to maintain self-esteem when autonomy is threatened (Baltes and Baltes, 1990). Psychoanalysts might see this as a self-defensive strategy (rationalization, resignation). Sociologists would more likely seek the explanation in relative deprivation and reference group theory. George et al. (1985) belong to this tradition and have suggested that the level of life satisfaction is based on an assessment of actual to expected conditions. Deviations from the expected norm represent a relative deprivation that threatens well-being. This mechanism is elaborated by Michalos (1985) in the form of multiple discrepancies theory, in which it is the gap between aspirations and achievements that explains the subjective experiences, not the actual level of living itself.

Analytical Approach

Analyses will be carried out in several steps, first to establish whether or not the basic presuppositions are valid. The relationship between control and well-being may vary across different dimensions of well-being; hence the need to establish how personal control and the different dimensions of well-being are interrelated. Secondly, we must clarify if personal control is indeed inversely related to age (in later life) as it should according to the logic of the argument. A third step is to explore the association between age and well-being. Is the level of well-being rather stable across age as indicated by the 'stability despite loss' paradox, and if so, is it stable for all aspects (dimensions) of well-being? The fourth step is to separate the effects of age and personal control on well-being from possible confounding factors. The fifth and final step is to test the main research question, namely whether the importance of personal control for well-being is declining in higher ages.

Study and Data

The questions are explored with data from the Norwegian Life Course, Ageing and Generation Study (NorLAG), which was carried out at Norwegian Social Research (NOVA), and financed by the Norwegian Research Council. Random samples of persons aged 40–79, living in their own homes in 30 urban and rural communities all over Norway, were interviewed over telephone (computer-assisted telephone interviews, $n = 5,589$, response rate 67.3%) in 2002–2003. The respondents were also invited to answer a postal questionnaire on a broad range of issues, including personal control beliefs and several measures of subjective well-being ($n = 4,169$, response rate 74.6%). Data collection was carried out by Statistics Norway, which also added information about household and family structure from public data registries. The data reported here refer mainly to the sample that responded to both the interview and the questionnaire ($n = 4,169$, response rate 50.2%). This sample is, according to Statistics Norway, a fair representation of the targeted populations, but somewhat biased by age and gender in that men and the oldest people (70–79) have slightly lower response rates (Holmøy, 2004).

Measurements

Subjective well-being. This involves both cognitive and emotional aspects; some would also add the existential aspect. The emotional aspects are often split into positive (inspired, interested, etc.) and negative (worried, nervous, etc.) affects, which are found to be separate dimensions that are only weakly correlated, if at all (Watson *et al.*, 1988). What makes one happy is not simply the lack of what makes one miserable, and vice versa. The cognitive, or judgemental, component of well-being refers to how satisfied one is with life as a whole (global life satisfaction) or with different aspects of life, such as health, family life, and oneself for that matter (domain-specific life satisfaction).

In this study life satisfaction is measured using the Satisfaction with Life Scale (Pavot *et al.*, 1991). The scale has five items that tap global life satisfaction, and a five-item response scale ranging from 'strongly agree' to 'strongly disagree'. Internal consistency is reasonably high (Cronbach's alpha = 0.76) in the present study. The emotional aspects are measured using the Positive and Negative Affect Scale (PANAS). The original scale has 20 items (Watson *et al.*, 1988), but this study employs a 10-item variant adapted by Kercher (1992). The shorter version of PANAS is found to have an appropriate factor structure and high discriminant validity for both positive and negative affects (Hillerås *et al.*, 1998; Mackinnon *et al.*, 1998).

Many studies concentrate on one (cognitive) or the other (emotional) aspect of well-being; some include both for comparative reasons. The NorLAG study added several other dimensions such as depression, anxiety, and self-esteem. Depression and anxiety are not included here, as they are both closely related to negative affect ($r = 0.60$ in this study), and therefore do not add substance to the analyses. Self-esteem is added as a distinct well-being dimension, and then as scores on the Rosenberg (1965) 10-item self-esteem scale. How the person feels about himself or herself may be a rather stable characteristic, and as such an aspect of his or her personality. One may, however, also experience higher or lower self-esteem in response to living conditions or situational constraints. Self-esteem then appears as a more or less variable state (Næss, 2001), and it is in this capacity that self-esteem may be seen as an outcome indicator for well-being and not simply as an enabling factor. This also illustrates the sometimes vague line between input and outcome variables in analyses of subjective well-being. We shall then analyse the relationships for four dimensions and measurements of well-being: life satisfaction, positive affect, negative affect, and self-esteem.

Personal control beliefs. These and similar constructs have been defined and measured in numerous ways (Skinner, 1996). For this chapter we employ a global belief in *personal mastery*, as this is measured by the Mastery Scale of Pearlin and Schooler (1978). The scale has seven items, and a five-item response scale ranging from 'strongly agree' to 'strongly disagree' (Cronbach's alpha = 0.72). Mastery is an indicator for 'personal control over desirable outcomes', which is found to be an important contributor to well-being even in very old age, and possibly more so than a 'control over undesirable outcomes' (Kunzmann *et al.*, 2002).

Covariates. Subjective well-being responds to the objective living conditions, but is far from an actual reflection of them. These 'external conditions', the so-called bottom-up influences on well-being, explain only a smaller part of the variance in subjective well-being, according to Diener *et al.* (1999). The top-down influences from internal factors like personality traits and cognitions, e.g. control beliefs, explain far more. Several studies have found personality characteristics to be among the major predictors of emotional well-being. Positive affect tends to be positively related to extraversion, and negative affect to neuroticism (Diener and Lucas, 1999).

Greater personal control – be it measured as internal control, self-efficacy, primary control, or mastery as here – is associated with both cognitive and emotional well-being according to Peterson (1999). We therefore include both external and internal factors in the analytical model as this allows us to separate top-down and bottom-up influences on well-being, and to separate the impact of control beliefs from (other) personality characteristics. Personality is assessed by the so-called 'Big-Five' Factor Model of Personality (McCrae and Costa, 1999). Only four of the five dimensions (personality traits) are included here: extroversion, neuroticism, openness to experience, and agreeableness. The fifth factor, conscientiousness, loads rather evenly on all four indicators of well-being and adds little to the variation. The external factors cover demographic variables like gender, age, civil status (single or partnered), number of children, and selected indicators of living conditions: education, (physical) health, and income. Health is indicated by scores on the SF-12 scale, developed by Gandek et al. (1998) from the larger SF-36 version of Ware and Sherbourne (1992). SF-12 also gives a score for mental health, which is not included here, as mental health is closely related to negative affect. Income is represented as personal income after tax, which is a better approximation of actual monetary resources for men than for women, as quite a few women in these cohorts are (economically) supported by their husbands. We have still found it appropriate to use these more objective indicators of living conditions instead of subjective indicators (perceived health and economy), as the latter, by definition, would be higher correlated with (subjective) well-being. This was also found to be the case when we ran the analyses with these (subjective) indicators (not presented here).

FINDINGS

Dimensions of Well-Being

How the four indicators of well-being are interrelated is illustrated in Table 1. The pattern corresponds reasonably well with earlier findings in that life satisfaction is positively related to positive affect ($r = 0.28$) and negatively to negative affect ($r = -0.34$). The latter two are not correlated ($r = -0.03$), which also corresponds to theory and earlier findings.

Self-esteem is positively related to both life satisfaction and positive affect, and not higher ($r \sim 0.40$), to suggest that they tap different dimensions of well-being. The association between self-esteem and negative affect is of the same size, but

TABLE 1. Pearson Correlations Between Age, Mastery, and Indicators of Well-Being

	1	2	3	4	5	6	7
1. Age							
2. Gender (1 = female)	−0.01						
3. Mastery	−0.28***	−0.06***					
4. Life satisfaction	0.03*	0.02	0.41***				
5. Positive affect	−0.22***	0.00	0.40***	0.28***			
6. Negative affect	−0.10***	0.11***	−0.35***	−0.34***	−0.03		
7. Self-esteem	−0.12***	−0.10***	0.57***	0.43***	0.41***	−0.36***	

negatively so. When the correlations accommodate so well to earlier findings, they also add credibility to the present study.

Age and Mastery

Table 1 also shows that age and mastery (control) are indeed negatively correlated in this age bracket ($r = -0.28$). Figure 1 illustrates that the relationship is more or less linear. Mean scores show that mastery is already declining from age 40 onwards, but slightly more so from age 55–60 and over, and would probably decline even more steeply after age 80, which was the upper limit included in the study. Mastery is, however, only weakly (but significantly) related to gender, with women having the (slightly) lower scores, as indicated by the negative correlation with gender ($r = -0.06$) in Table 1.

Age and Well-Being

The oft-reported stability in the level of well-being in the second half of life is partly confirmed, but more so for life satisfaction than for the emotional and self-appraisal sides of well-being. Life satisfaction levels are nearly stable, even when they increase slightly with age ($r = 0.03$, Table 1). The large sample size makes the modest correlation statistically significant. The affective dimensions are, however, more variable with age, with a negative trend for positive affect and a positive trend for negative affect as far as well-being is concerned. Positive affect is lower in the

Figure 1. Mean mastery scores by age (n)

higher age groups, and so also is negative affect, which is a good thing in terms of well-being. The two affective dimensions taken together indicate that the main trend is one of lower emotionality in older ages, both for good and bad. The stronger association with age for positive affect relative to negative affect is in line with earlier studies, and is often attributed to positive affect having more of a mood (and situational) character (Diener and Suh, 1997). Negative affect owes its greater stability to a more dispositional nature. Self-esteem seems to be more or less as robust to age changes as negative affect, but with a negative slope, implying that self-esteem has a modest decline ($r = -0.12$) with age within this age bracket.

It would have been more appropriate to talk about *age differences* rather than *age changes* (increase or decline) as these are cross-sectional data. The observed differences may be cohort contrasts and need not be reactions to ageing. We need longitudinal data in order to separate the two effects. We can, however, separate age effects (based on cohort or ageing) from other age-related differences. Whether or not the observed differences in well-being are unique age effects and remain after control for age-related differences in health, living conditions, and internal factors like mastery is explored in the multiple regression analyses below.

Mastery and Well-Being

We need, first, to add a brief comment about the relationship between mastery and well-being (Table 1). Earlier studies have found personal control (mastery) to be a major predictor of both the cognitive and emotional sides of well-being, often reported in the range of $r = 0.20$–0.40 (Peterson, 1999; Kunzmann et al., 2002). These relationships in our study are somewhat stronger, namely between $r = 0.40$ for life satisfaction, positive, and negative affect, and $r = 0.57$ for self-esteem.

The high value in the latter case illustrates that mastery feeds self-esteem and vice versa over time. Each of them can be seen as an enabling factor or as an outcome variable depending on what mechanisms are under scrutiny. Self-esteem is, in this case, selected as an aspect of well-being, but is closely related to a sense of mastery, to the extent that they cover much of the same ground.

AGEING, MASTERY, AND WELL-BEING

The possibly changing role of mastery for well-being with increasing age is explored in a series of multiple regressions of the four dimensions of well-being on age, mastery, and the external (living conditions) and internal (personality) covariates. The analyses allow us first to test the association between age and well-being after control for the age-related variation in health, living conditions, and personality characteristics. Is this 'adjusted well-being' equally high, or even higher, among the older than among the younger as suggested by some earlier studies, and if so – is this the case on all dimensions of well-being?

Secondly, we briefly consider the impact of external (bottom-up) and internal (top-down) influences on well-being. Indeed do personal dispositions, be it in the form of personality traits or control beliefs, explain more of the variation in well-being than living conditions do, as suggested by Diener et al. (1999) and others?

The third step is the test of the main research question – whether or not control beliefs (mastery) become less important for well-being with increasing age. If so, this might help us understand the so-called 'stability despite loss' paradox: that well-being remains high in late life despite a loss of resources and control. For this purpose we include the interaction between age and mastery as a separate factor in the analyses. A negative effect of this factor on well-being indicates that mastery has less impact on well-being with increasing age, a positive effect that the importance of mastery is growing with age. The results of the analyses are given in Table 2 in the form of two regression models for each indicator of well-being. Model 1 includes only age, mastery, and the interaction item, while model 2 includes all variables.

In response to the first question, age does, in fact, come out as a positive and unique factor for well-being on three of the four dimensions, and most strongly so for life satisfaction ($\beta = 0.20$). Self-esteem is positively related to age as well, but only slightly so ($\beta = 0.04$), while negative affect seems to become less of a problem with age ($\beta = -0.16$), which adds to the positive well-being trend with increasing age. The only negative slope is found for positive affect, which tends to decline with age even after control for age-related risks such as poorer health and lower mastery ($\beta = -0.07$). A comparison of the coefficients in Tables 1 and 2 indicates that risks such as poorer health, single status, and a loss of mastery tend to lower the level of well-being among older people. Given the same level of mastery, older people in this age bracket tend to report *higher* well-being than the younger, particularly in terms

TABLE 2. OLS Regression of Well-Being on Age, Mastery, with Controls for External and Internal Factors. Standardised Regression Coefficients (n)

	Life satisfaction		Positive affect		Negative affect		Self-esteem	
Age	0.16***	0.20***	-0.12***	-0.07***	-0.21***	-0.16***	0.04**	0.04**
Mastery	0.45***	0.33***	0.37***	0.24***	-0.41***	-0.23***	0.59***	0.42***
Age*mastery	-0.04**	-0.04**	-0.04*	-0.03*	0.02	0.01	-0.03**	-0.03
Gender (1 = female)		0.11***		0.03		0.01		-0.02
Health		0.16***		0.05**		-0.03		-0.01
Income		0.03a		0.02		0.02		0.03
Education		0.01		0.13***		0.04**		0.09***
Civil status (1 = partner)		0.23***		0.03		-0.05**		0.04**
Number of children		0.04*		0.00		0.01		0.00
Agreeableness		0.06**		0.09***		0.03*		0.04**
Neuroticism		-0.10***		-0.05**		0.43***		-0.24***
Openness		0.00		0.26***		0.03*		0.10***
Extraversion		0.04*		0.08***		0.03*		0.08***
R^2	0.19***	0.30***	0.17***	0.30***	0.16***	0.31***	0.33***	0.42***
(n)	(4050)	(3179)	(4037)	(3169)	(4054)	(3178)	(4082)	(3201)

[a] $p = 0.053$
* $p < 0.05$; ** $p < 0.01$; *** $p < 0.001$

of (higher) life satisfaction and (lower) negative affect. This is also illustrated in Figure 2, which gives the ordinary mean scores on well-being with age, and the adjusted means, when differences in mastery levels are adjusted for. The adjusted means illustrate the increase in well-being with age on both life satisfaction and negative affect. Trends are similar for self-esteem (not shown in Figure 2), while positive affect has a slight decline with age, although less so for adjusted compared to ordinary means.

The positive and unique contribution of (increasing) age to well-being is masked by the associated risks of ageing, implying that if one grows old 'successfully', and is not suffering these risks, one may be more content with life as a senior than in younger age. The findings indicate that if and when older people have low subjective well-being, this is for the most part explained by risk factors such as these. Again it needs be emphasized that these conclusions must be reserved for the younger-old, as the 80+ were not included in the study. We may also need to reserve the conclusions to the Scandinavian setting, where living standards are fair for older people, and access to services are comparatively generous.

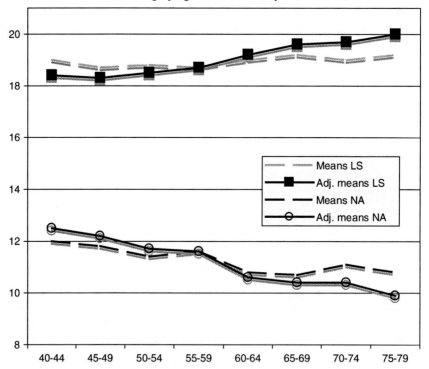

Figure 2. Means and adjusted means of life satisfaction (LS) and negative affect (NA) by age. Adjusted means are adjusted for differences in mastery levels (UNIANOVA analysis)

The external conditions in terms of health, income, and family resources seem to have far more impact on the judgemental aspects of well-being (life satisfaction) than on emotional well-being and self-esteem (Table 2). The latter (self-esteem) is mainly explained by mastery and neuroticism. When coefficients reach 0.40 and over, as is the case for neuroticism on negative affect (0.43) and for mastery on self-esteem (0.42), they are close to reporting from two sides of the same coin (construct). The other regression coefficients are more moderate and reasonable.

Mastery comes out as a powerful and unique predictor for all four aspects of well-being, but more so for life satisfaction ($\beta = 0.33$) and self-esteem ($\beta = 0.42$) than for the two affective dimensions ($\beta \sim 0.24$). There is, however, little support for the idea that mastery grows less important for well-being with age. To be sure, the interaction factor of age and mastery is negatively related to two of the well-being dimensions (life satisfaction and positive affect), but only weakly so ($\beta = -0.04$). Hence, mastery may grow slightly less important for some aspects of well-being with age, but the effect is neither consistent nor considerable, and is far from able to explain the high levels of well-being in older years. Separate regression analyses (not shown here) for each 5-year age group illustrate this. Mastery comes out as the most important factor for well-being in each age group, and more or less equally so at 45, 60, and 75. The trend may shift after 80, when resources decline even more, and instrumental mastery may therefore become less adaptive, but until then, the most reasonable conclusion seems to be that mastery is more or less equally important for well-being for all age groups from middle age (the forties) to early old age (the seventies).

The 'Paradox' of Later Life Satisfaction

Mastery is found to be a major predictor for all dimensions of well-being. As mastery declines with age, while well-being levels do not (or only slightly so), a likely explanation for this 'paradox' could be that mastery grows less important for well-being with increasing age. The hypothesis seems not only logical, but also in line with theories about the shift from primary to secondary control strategies in later life.

However, the hypothesis is not supported by the present findings. Mastery seems to be more or less equally important for well-being all through the age range (40–79) studied here. Older people, or more precisely older Norwegians, have nearly the same level of well-being as midlifers even when their sense of mastery is lower. Given the same level of mastery, older people, in fact, have higher well-being on all measures except positive affect. Why is this?

The findings are not actually in contrast to those of adaptation theories that argue for a shift from primary to secondary coping strategies in response to a loss of competence, but they argue for the persistent positive effect of instrumental control in line with Schultz and Heckhausen (1997). Secondary coping may be important for well-being in old age, but so also are more instrumental strategies. The findings correspond also with Kunzmann et al. (2002), who find that personal control (over desirable outcomes) is an important contributor to well-being even in very old age, when resources are low and environmental constraints are correspondingly high.

'Giving in' may be an adaptive response to environments you have no control over, but the essence of the findings is that a sense of actual control over the environment – however small – may add considerably to well-being even among people with low capabilities, which is what Langer and Rodin (1976) found in their experiment in a nursing home. Patients were given the opportunity to decide over 'trivial' things like flowers and decorations, and responded with higher well-being. This finding indicates that mastery may in some cases be even more important for autonomy and self-respect when competencies are low, and mastery is not taken for granted.

Primary and secondary control are not alternative capacities, they are both part of any person's repertoire, although differently balanced, and differentially beneficial according to competence and environmental constraints. When the young-old do so well here, it may be because they enjoy the benefits of both strategies. The younger may be too inclined to trust instrumental strategies, perhaps even beyond their actual capacity, while the very old may have lost too much in active mastery, which they are not able to compensate for.

The 'paradox' of being satisfied with life in older years, therefore, need not be so paradoxical, but in the presumed paradox lies also a potential source of benefit if many older people find that life is in fact better than they expected. Older people are subject to the same ageing stereotypes as any other age group (Levy, 2003), and their internalised low expectations may have added to their well-being if their worst fears are not met.

We are still left with an explanatory problem. Why are well-being levels so high among older people? The findings challenge our very conceptions of what a good life is and how these qualities change over the life course. We may simply take the findings on face value, and admit older people the 'right' to be satisfied with life, but maybe more in the form of worry-free contentment (life satisfaction) than as inspired excitement (positive affect). The relative absence of emotional excitement which would be a source of tragedy for the young is perhaps a relief in old age, if we believe philosophers like Plato and Cicero. These are perhaps rather trivial statements, but they point towards the necessity of further exploring the different 'qualities of life' (Veenhoven, 2000). We may then need to assess more sophisticatedly what the better balance is for a good life between activity and reflection, externality and internality, and continuity and change.

CONCLUSION

Well-being is hard to comprehend, and is not easily captured in the dimensions and measurements such as those we have used here. The role of age and ageing for well-being is equally complex. It may be better than expected to grow older, if not really old, but most people would probably find the unique and positive effect of age on well-being strange. Younger people would more likely see it as a matter of resignation. We need to confront values and attitudes behind such reactions, and to further develop our theories and measurements of well-being and QoL, and what indeed age and ageing have to do with it. The larger crossroads of life tend to help us separate the important from the trivial, and to extract what really counts from things of minor importance, but the final test of a qualitatively good life may not be clear to us until we reach the very end of our life, if at all.

We need to add some reservations and limitations of the study. First, the very old are not included. A larger age range would have made it possible to explore the role of mastery for well-being better. As it is, our conclusions must be reserved for those aged 40–79. Second, the sample may be slightly biased in the direction of representing the healthy and more satisfied seniors better than those with fewer resources. Third, the association between mastery and well-being may be sensitive to different aspects of these phenomena. What is missing, in particular, is an indicator of secondary control in addition to that of primary control (mastery) included here. Fourth, the findings may not generalize to countries in which living standards and welfare state arrangements are poorer. QoL and what is considered 'successful ageing' may indeed also vary across cultures (Torres, 2005), which adds to the doubts about the external validity of the findings. These questions need therefore be explored in other cultures and contexts.

REFERENCES

Baltes, P.B. and Baltes, M.M. (1990) 'Psychological Perspectives on Successful Aging: The Model of Selective Optimization with Compensation', in P.B. Baltes and M.M. Baltes (eds) *Successful Aging: Perspectives from the Behavioural Sciences*, Cambridge, Cambridge University Press.

Brandstädter, J. and Renner, G. (1990) 'Tenacious Goal Pursuit and Flexible Goal Adjustment: Explication and Age-related Analysis of Assimilative and Accommodative Strategies of Coping', *Psychology and Aging*, 5(1), 58–67.

Campbell, A., Converse, P.E., and Rodgers, W.L. (1976) *The Quality of American Life: Perceptions, Evaluations, and Satisfactions*, New York, Russel Sage Foundation.

Carstensen, L. (1991) 'Socio-emotional Selectivity Theory: Social Activity in Life-span Context', *Annual Review of Gerontology and Geriatrics*, 11, 195–217.

Cumming, E. and Henry, W. (1961) *Growing Old*, New York, Basic Books.

Daatland, S.O. (2005) 'Quality of Life and Ageing', in M.L. Johnson (ed.) *The Cambridge Handbook of Age and Ageing*, Cambridge, Cambridge University Press, pp.371–377.

DeNeve, K.M. and Cooper, C.H. (1998) 'The Happy Personality: A Metaanalysis of 137 Personality Traits and Subjective Well-being', *Psychological Bulletin*, 124, 197–229.

Diener, E. and Suh, M.E. (1997) 'Subjective Well-being and Age: An International Analysis', *Annual Review of Gerontology and Geriatrics*, 17, 304–324.

Diener, E. and Lucas, R.E. (1999) 'Personality and Subjective Well-being', in D. Kahneman, E. Diener, and N. Schwaraz (eds) *Well-being: The Foundations of Hedonic Psychology*, New York, Russel Sage foundation, pp.213–229.

Diener, E., Suh, E.M., Lucas, R.E., and Smith, H.L. (1999) 'Subjective Well-being: Three Decades of Progress', *Psychological Bulletin*, 125, 276–302.

Erikson, E.H. (1968) 'Generativity and Ego Integrity', in B.L. Neugarten (ed.) *Middle Age and Aging: A Reader in Social Psychology*, Chicago, University of Chicago Press.

Erikson, R. and Uusitalo, H. (1987) 'The Scandinavian Approach to Welfare Research', in R. Erikson, E.J. Hansen, S. Ringen, and H. Uusitalo (eds) *The Scandinavian Model: Welfare States and Welfare Research*, New York, M.E. Sharpe, pp.177–193.

Gandek, B., Ware, J.E., Aronsen, N.K., *et al.* (1998) 'Cross-validation of Item Selection and Scoring of the SF-12 Health Survey in Nine Countries: Results from the IQOLA Project', *Journal of Clinical Epidemiology*, 11, 1171–1178.

George, L.K., Okun, M.A., and Landerman, R. (1985) 'Age as a Moderator of the Determinants of Life Satisfaction', *Research on Aging*, 7, 209–233.
Heckhausen, J. and Schultz, R. (1995) 'A Life-span Theory of Control', *Psychological Review*, 102(2), 284–304.
Hillerås, P., Jorm, A., Herlitz, A., and Winblad, B. (1998) 'Negative and Positive Affect among the Very Old: A Survey on a Sample Aged 90 Years or Older', *Research on Aging*, 20, 593–610.
Holmøy, A. (2004) *Undersøkelse om livsløp, aldring og generasjon (LAG) Dokumentasjonsrapport*, Oslo, Statistisk sentralbyrå, Notater, 2004/24.
Kercher, K. (1992) 'Assessing Subjective Well-being in the Old-Old', *Research on Aging*, 14, 131–168.
Krause, N. and Shaw, B.A. (2003) 'Role-specific Control, Personal Meaning, and Health in Late Life', *Research on Aging*, 25(6), 559–586.
Kunzmann, U., Little, T.D., and Smith, J. (2000) 'Is Age-related Stability of Subjective Well-being a Paradox? Cross-sectional and Longitudinal Evidence from the Berlin Aging Study', *Psychology and Aging*, 15(3), 511–526.
Kunzmann, U., Little, T.D., and Smith, J. (2002) 'Perceiving Control: A Double-edged Sword in Old Age', *Journal of Gerontology, Psychological Sciences*, 57B(6), P484–P491.
Langer, E.J. and Rodin, J. (1976) 'The Effects of Choice and Enhanced Personal Responsibility for the Aged: A Field Experiment in an Institutional Setting', *Journal of Personality and Social Psychology*, 34, 191–198.
Levy, B.R. (2003) 'Mind Matters: Cognitive and Physical Effects of Aging Self-stereotypes', *Journal of Gerontology, Psychological Sciences*, 58B(5), P203–P211.
Mackinnon, A., Jorm, A., Christensen, H., Korten, A., Jacomb, P.A., and Rodgers, B. (1998) 'A Short Form of the Positive and Negative Affect Schedule: Evaluation of Factorial Validity and Invariance across Demographic Variables in a Community Sample', *Personality and Individual Differences*, 27, 405–416.
McCrae, R.R. and Costa, P.T. (1999) 'A Five Factor Theory of Personality', in L.A. Pervin and O.P. John (eds) *Handbook of Personality: Theory and Research*, New York, Guilford Press, pp.139–153.
Michalos, A.C. (1985) 'Mulitple Discrepancies Theory' (MDT), *Social Indicators Research*, 16(4), 347–414.
Mirowsky, J. (1997) 'Age, Subjective Life Expectancy, and Sense of Control: the Horizon Hypothesis', *Journal of Gerontology, Psychological Sciences*, 52(3), P25–P34.
Motel-Klingebiel, A. (2004) 'Quality of Life and Social Inequality in Old Age', in S.O. Daatland and S. Biggs (eds) *Ageing and Diversity: Multiple Pathways and Cultural Migrations*, Bristol, Policy Press, pp.189–205.
Næss, S. (2001) *Livskvalitet som psykisk velvære*, Oslo, NOVA, 3–2001.
Pavot, W., Diener, E., Colvin, C.R., and Sandvik, E. (1991) 'Further Validation of the Satisfaction with Life Scale: Evidence for the Cross-method Convergence of Well-being', *Journal of Personality Assessment*, 57, 149–161.
Pearlin, L.J. and Schooler, C. (1978) 'The Structure of Coping', *Journal of Health and Social Behavior*, 19, 2–21.
Peterson, C. (1999) 'Personal Control and Well-being', in D. Kahneman, E. Diener, and N. Schwaraz (eds) *Well-being: The Foundations of Hedonic Psychology*, New York, Russel Sage Foundation, pp.288–301.
Ringen, S. (1995) 'Well-being, Measurement and Preferences', *Acta Sociologica*, 38, 3–15.
Rosenberg, M. (1965) *Society and the Adolescent Self-image*, Princeton, NJ, Princeton University Press.
Schaie, K.W. and Lawton, M.P. (1998) 'Focus on Emotion and Adult Development', *Annual Review of Gerontology and Geriatrics*, 17, New York, Springer.
Schultz, R. and Heckhausen, J. (1997) 'Emotion and Control: A Life-span Perspective', *Annual Review of Gerontology and Geriatrics*, 17, 185–205.
Sen, A. (1993) 'Capability and Well-being', in M. Nussbaum and A. Sen (eds) *The Quality of Life*, Oxford, Clarendon Press, pp. 30–53.
Skinner, E.A. (1996) 'A Guide to Constructs of Control', *Journal of Personality and Social Psychology*, 71, 549–570.

Slagsvold, B., Daatland, S.O., Clausen, S.E., and Hansen, T. (2003) 'Personal Control and Subjective Well-being: The Impact of Ageing', Barcelona, Paper at the 5th European Congress of Gerontology.

Smith, J., Fleeson, W., Geiselmann, B., Settersten, R.A. Jr., and Kunzmann, U. (1999) 'Sources of Well-being in Very Old Age', in P.B. Baltes and K.U. Mayer (eds) *The Berlin Ageing Study*, Cambridge, Cambridge University Press, pp. 450–471.

Smith, G.C., Kohn, S.J., Savage-Stevens, S.E., Finch, J.J., Ingete, R., and Lim, Y-O. (2000) 'The Effects of Interpersonal and Personal Agency on Perceived Control and Psychological Well-being in Adulthood', *The Gerontologist*, 40(4), 458–468.

Torres, S. (2005) 'Making Sense of the Construct of Successful Ageing: The Migrant Experience', in S.O. Daatland and S. Biggs (eds) *Ageing and Diversity: Multiple Pathways and Cultural Migrations*, Bristol, Policy Press, pp.125–140.

Veenhoven, R. (2000) 'The Four Qualities of Life: Ordering Concepts and Measures of the Good Life', *Journal of Happiness Studies*, 1, 1–39.

Ware, J.E. and Sherbourne, C.D. (1992) 'The MOS 36-item Short-form Health Survey (SF-36)', *Medical Care*, 30(6), 473–484.

Watson, D., Clark, L.A., and Tellegen, A. (1988) 'Development and Validation of Brief Measures of Positive and Negative Affect: The PANAS Scales', *Journal of Personality and Social Psychology*, 54, 1063–1070.

Zapf, W. (1984) 'Individuelle Wohlfahrt: Lebensbedingungen und wahrgenommene Lebensqualität', in W. Glatzer and W. Zapf (eds) *Lebensqualität in der Bundesrepublik Deutschland. Objekive Lebensbedingungen und subjektives Wohlbefinden*, Frankfurt/Main, Campus.

4. SOCIAL RESOURCES

INTRODUCTION

Assessing and studying the quality of life (QoL) in adulthood has attracted the attention of many researchers concerned with life course and ageing issues. In fact, the history of gerontology is in many ways an attempt to understand issues of life quality as people grow old (George, 2006). Measures developed to indicate QoL are multiple, including subjective and objective indictors, as well as complex approaches that draw on several dimensions (Andrews and Withey, 1976; Campbell *et al.*, 1976; Fernandez-Ballesteros, 2002; George, 2006; Kahn and Juster, 2002; Skevington *et al.*, 2004). Subjective ratings, however, are those that seem most sensitive to social resources.

In this chapter we consider the role of social resources, defined as social relations, in QoL. It is said that 'no man is an island'. Research on interpersonal relationships and social relations has certainly demonstrated this to be the case. People with few social resources tend to be marginalized, stigmatized, and have lower life quality, while people who are interconnected, tied to others, and generally engaged in interactive activities tend to be more positively adjusted, and have better mental and physical health. In addition, just as 'no man is an island', researchers have also recognized truth in the statement 'the child is father to the man'. This is the proverbial, if somewhat inappropriately gender-specific, version of lifespan developmental theory. We propose that this perspective is essential to understanding the role and nature of social relations in the QoL of older people. To begin, we highlight the importance of a lifespan developmental perspective to the Convoy Model of Social Relations. This model provides a multidimensional perspective on those characteristics that influence the development and effect of social relations across the lifespan, both of which are critical to understanding the social resources available in late life and how these contribute to life quality. We then consider the link social relations have to subjective ratings of life satisfaction, happiness, depression, and health, constructs that both scholars and policymakers often find to be useful indicators of life quality (George, 2006; Kahn and Juster, 2002; Skevington *et al.*, 2004).

CONVOY MODEL OF SOCIAL RELATIONS

As researchers began to study social relations and recognize their importance, it became clear that existing theoretical perspectives were insufficient to grasp their complex nature. The Convoy Model was developed to permit examination of personal and situational characteristics that influence social relations across the lifespan

49

H. Mollenkopf and A. Walker (eds.), Quality of Life in Old Age, 49–64.
© 2007 *Springer.*

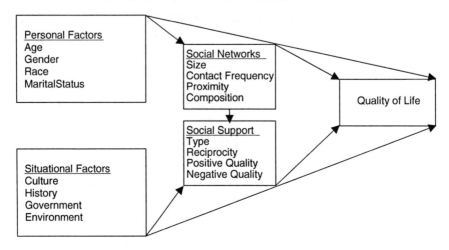

Figure 1. Convoy Model of Social Relations

(Figure 1). The term 'convoy' is derived from the early anthropological work of Plath (1980), who studied groups of boys and girls in Japanese villages. He observed that these children grew up together and came to look to each other for comfort, protection, and guidance. He recognised that each cohort formed a special bond among themselves, which became critical to the socialization and developmental experiences of its members. He coined the term 'convoy', reminiscent of the military term used to portray the safe passage or escort offered to vulnerable soldiers, e.g. trucks or ships. Plath's convoys represented the same type of safe passage or protection that conjoint members of a cohort offered each other as they confronted each developmental milestone, aged, and matured. Later Kahn and Antonucci (Kahn, 1979; Kahn and Antonucci, 1980; Antonucci, 1986) further developed the concept to incorporate both the growth and maturity that Plath recognized and to extend this perspective across the lifespan, incorporating both a developmental and organizational framework. Although early developmental research focused on mother–infant attachment (Ainsworth, 1989; Bowlby, 1988), this work was later extended to include both the older child and other attachment figures including adult friendship and romantic partnership. The Convoy Model weds this extended attachment perspective with the adult literature focusing on interpersonal relationships and social support (Kahn and Antonucci, 1980) arguing for a lifespan integration of both perspectives. The contribution of lifespan theory is essential to understanding how social resources and social relations influence QoL in old age.

Lifespan theory postulates that development over the life course involves adaptation to both normative and non-normative events experienced during one's lifetime (Baltes et al., 1998). Our focus on late life highlights the need to attend to the effect of cumulative life experiences, thus making a life course perspective particularly appropriate. As Figure 1 illustrates, the Convoy Model proposes that both individual and situational characteristics shape the social network and support, that each individual experiences.

We highlight personal (age, gender, race, marital status) and situational (culture, history, government, environment) characteristics, shown in the gerontological literature to be particularly critical to health and well-being. We especially incorporate the role of family in late life, both because it is the longest relationship and organizational membership to which most older people belong and because it fundamentally influences the history, past and current experiences, of each individual in late life.

SOCIAL RESOURCES

In this chapter social relations are identified as a key resource over the life course, a form of social capital that potentially influences QoL. We elaborate on the term 'social relations', an umbrella term that represents two dimensions of social resources: structure of social network, as well as type and quality of social support (positive and negative), defining and providing illustrative examples of each aspect.

Network Structure

The term 'social network' describes a structure of individuals with designated relationships to the focal person. Social networks vary in terms of the number of network members, frequency of contact, geographic proximity, and composition. Networks represent an available resource, a source of help in times of trouble, comfort in times of pain, and information in times of need. At the same time, as Granovetter (1973) noted, sometimes there is strength in weak ties. Thus, having a multifaceted, diffuse, large network may be considerably more helpful in solving a problem than one which is dense, family-based, and small.

A great deal of evidence suggests that social networks may provide resources in times of need; yet social networks can become a source of conflict, stress, and sometimes contribute to poor life quality. One important finding is that larger social networks do not guarantee happiness; nor do they necessarily create a buffer from stress. For example, larger social networks are associated with greater stress among women, especially due to role strain. Women often experience role overload, a situation in which demands are placed on individuals that exceed their capacity to cope with them. Informal caregiving to seriously impaired relatives along with the responsibilities of work and children may result in a significant amount of stress. Having more social contacts 'not only provides more potential resources but also may create additional demands on time and increase the probability of interpersonal conflicts' (Cohen and Syme, 1985, p.12).

Network characteristics affect the probability of support (Pugliesi and Shook, 1998; Taylor et al., 1996) as well as the quality and type of support (Antonucci and Akiyama, 1987; Seeman and Berkman, 1988). They also indicate the extent to which individuals participate in a number of social activities (Morgan, 1989).

Support Type

Types of support may include forms that are both tangible and psychological. Tangible support includes resources that facilitate instrumental support needs such as aid with financial challenges and those due to health or functional limitations (e.g. transportation, house cleaning, bathing). Yet high levels of tangible social support may also contribute to poor QoL. For example, research with older people demonstrates that such high levels of support from caregivers may cause stress and thus be harmful to the older person's QoL. Too much support may foster dependency, causing the receiver to lose autonomy and develop low self-esteem.

Psychological resources include emotional types of support. Pearlin (1985) argued that supportive others increase feelings of personal achievement which not only aid in coping with specific events but also increase a general sense of mastery. Similarly, Antonucci and Jackson (1987) suggested that there is a lifespan developmental benefit to the constant repetition of supportive interactions. Receiving support over time leads to viewing oneself as worthy, cared for, and loved. Under optimal conditions, this exchange provides the individual with a sense of personal efficacy which is critical as the individual faces the challenges or crises of life.

Reciprocity refers to the exchanges that occur between and among people, which are often lifelong and dynamic within a family context. Reciprocity may be defined as 'equal or comparable exchanges of tangible aid, affection, advice or information between individuals or groups' (Akiyama et al., 1997, p.164). The norms that guide such exchanges are an important aspect of social relations, particularly when considering QoL in old age. Antonucci (1990) coined the term 'Support Bank' to emphasize long-term reciprocity, suggesting that deposits are made over the life course, in the form of support provided. In later life, withdrawals may be made when needed, and the resulting nature of exchanges is evaluated according to the lifetime experience of providing and receiving support. While norms of reciprocity may vary by culture, research continues to demonstrate its centrality to QoL in old age (e.g. Lewinter, 2003; Litwin, 2004; Silverstein et al., 2002).

Relationship Quality

It is well established that social relationships influence individuals' psychological well-being by providing love, intimacy, reassurance of worth, tangible assistance, and guidance. Relationship quality refers to both positive and negative aspects of relationships. Across the lifespan, lacking high-quality relationships is associated with negative physical and psychological consequences such as anxiety, depression, loneliness, and poor health. The presence of a social relationship does not necessarily indicate that the relationship is supportive. The overall quality of the relationship is more important than its mere existence, although the former clearly depends on the latter.

Morgan (1989) conducted focus group discussions with recently widowed women and men to explore whether social networks actually ease adjustment to widowhood. The discussions revealed that a significant portion of negative interactions occurred in family relations. Support from family often produces stress when there are discrepancies between expectations and actual performance. Thus, in considering whether relationships can act as buffers against stress, it is important to consider their negative as well as positive potential.

In an attempt to uncover the processes by which declines in mobility and functional health influence well-being, researchers have explicitly examined what role, if any, social relations may play. Hellström et al. (2004) address the influence of social relations by comparing those who receive help with daily living to those who do not receive such help in Sweden. As expected, those who do not receive help describe a higher QoL than those who receive help. Additionally, while no gender differences emerged with regard to total QoL among those receiving help, more women reported lower QoL than men within the group not receiving help. The authors conclude that living at home and receiving help does not necessarily protect older people from having a poor QoL (measured by an index of questions such as: how do you feel about your life today; do you often think life could be less monotonous; do you often feel depressed because every day is the same; I am very satisfied with my life at present). Results point to the necessity of evaluating QoL differently depending on whether or not the older person is receiving help: 'It is likely that low QoL is the result of a combination of emerging need for help from others and suffering caused by various symptoms' (Hellström et al., 2004, p.591). But in some cultures, like Japan, it is said that elders expect to receive help and do not 'suffer' because of it (Akiyama et al., 1997).

Social relations have also been examined as a potential mediator between challenging situations and QoL. Newsom and Schulz (1996) examine support structure and perceived support as a mediator to the association between physical decline and life satisfaction as well as depression. Findings illustrate that subjective evaluation of support (i.e. perceived) 'is the key factor in mitigating depressive symptoms and lower life satisfaction' (Newsom and Schulz, 1996, p.41). Decreased functional health was associated with lower levels of perceived and actual support; physical impairment was linked with fewer family and friendship contacts, less belonging, less tangible (instrumental) support, and a tendency to provide less material assistance to others. Tangible support, however, was the most predictive of QoL. Additionally, contact frequency with family and friends was no longer significantly associated with QoL when perceived support variables were included in the statistical model. Some aspects of social relations represent a resource and pathway to QoL among those who are less physically functional. Specifically, perceived support may help older persons who are less functionally healthy to maintain a sense of security and efficacy over seemingly uncontrollable circumstances (Antonucci and Jackson, 1987).

The Convoy Model highlights the significance and influence of social relations on QoL; yet personal and situational characteristics influence the type of social relations people develop. Research evidence clearly indicates that these characteristics are associated with social resources more broadly. We turn next to a consideration of select personal and situational characteristics in late life, and their influence on social relations and QoL.

In this section we review age, gender, race, and marital status, as illustrative of how personal characteristics can affect both social relations and QoL. While all of these characteristics are either lifelong or long-term in nature it is clear that their influence is different at various points in the life cycle. We focus here on late life.

Age

It is commonly agreed that at all ages, and especially in late life, the most important social relations are familial. In late life, and with increased longevity, older people are likely to have the advantage of living in multiple-generation families. This potentially increases the number of social relations with which they might be engaged. On the other hand, age is also accompanied by an increased likelihood that mates, peers, and parents would have died. This point is clearly demonstrated in Martin *et al.*'s (1996) comparison of social and psychological resources among the oldest old. In this unique study, people in their sixties, eighties, and centenarians were interviewed. They found that with age people reported many fewer resources, specifically fewer telephone contacts, less likelihood of having a confidant, more loneliness, and not as much visiting of friends and relatives as they wished. Interestingly, and in keeping with the higher probability of functional limitations, with age people reported receiving more help from others. As one might predict, age was also associated with less education and poorer health. Living arrangements were not linear in that people in their eighties were more likely to live alone than people in their sixties, but centenarians were less likely to live alone than 80-year-olds. No doubt this finding reflects the fact that while 60-year-olds are most likely to still be married, 80-year-olds are not, although they may still be relatively healthy and thus able to live alone. By the age of 100, the ability to live alone is much more limited and people are most likely to live with their children. The association of age with QoL is interesting in that as early as 1976, Campbell, Converse, and Rodgers noted that with age people became more satisfied with life, although they concomitantly reported being less happy. At the same time, it has been consistently reported that with age, people report less negative affect and less negative interaction, except with their spouse, with whom no age changes are reported (Akiyama *et al.*, 2003). Campbell *et al.* (1976) have argued, among others, that these differences indicate that cognitive appraisal of life improves with age, while affective or emotional appraisal tends to plateau or dampen. As these examples illustrate, age is also important because of its association with other variables such as the reduction of peer relationships, changes in living arrangements, and/or changes in functional and physical health.

The nature and function of social relations are considered highly critical over the life course. In old age, social relations constitute an especially important resource due to the higher likelihood that older people lose physical resources and strength, which in turn yield more need for the attention of others. While the loss of physical strength undoubtedly occurs at every life stage, there emerge higher proportions of individuals in old age reporting illness, chronic disease, and fair to poor health. As a result, relations with others are considered both an indicator of, and a potential influence on, well-being. We turn next to a consideration of gender.

Gender

While biological differences are most directly associated with sex differences, lifetime socialization, experiences, and context are responsible for gender differences. With age, some differences become more pronounced and others disappear. Women live longer than men; therefore, they are more likely to experience at least part of their late life as widows. Unfortunately, widowhood is often associated with resource deficits. Antonucci *et al.* (2002) examined differences between men and women in four nations and found that women were more likely to experience widowhood, illness, and financial strain. Given that women live longer than men, these negative consequences of age for women may seem counterintuitive. Some have suggested by way of explanation that women suffer longer with chronic, non-fatal illnesses while men are more likely to experience acute illnesses, which are often, relatively quickly, fatal.

Women are also more likely to have never been married, although this is a relatively small group of either men or women (Bureau of the United States Census, 2001). On the positive side, women are more likely to have children with whom they have active, if sometimes ambivalent, relationships. Children are clearly an important resource for health care, assistance with daily living, and, with increasing age, for living arrangements. All of these resources are often associated with higher QoL and well-being.

On the other hand, women, at least in most Western industrialized countries, experience higher levels of depression than do men. They also report lower QoL and this difference appears to increase with age (Pinquart and Sorensen, 2000). It is not clear to what degree these differences are related to the unique and distinct place of gender in society. Women who are currently in late life are less educated, and less likely to have been employed outside the home. As this experience changes for women, gender differences can be expected to change as well. Given that some circumstances reported by women – e.g. closeness to, and number of, children – could be considered advantages, and other circumstances – e.g. less education and lower income – could be considered disadvantages, it is not clear how comparable gender differences will be to future gender patterns as women and men reach old age with vastly different lifetime experiences. At the same time, it should be noted that there are some areas where, even across nations, there do not appear to be gender differences. For example, network size, the number of people to whom you feel close, and the people identified as close were found to be generally the same for men and women in France, Japan, and the USA. Gender differences and similarities are clearly a complex phenomenon in which many causal dynamics are not clearly understood.

Race

In the USA and many other countries ethnic and racial minorities face a lifetime of discrimination and disadvantage. Nevertheless, the association between race and QoL is not simple. In the USA, where one might expect African Americans to report significantly lower levels of QoL, due to a history of prejudice and discrimination, this is not always the case. African Americans often indicate that they are as satisfied, if not more, with life as the white majority (Jackson *et al.*, 1986). Nevertheless, it is clear that minority status is usually associated with fewer resources. Most minorities have lower life expectancies, lower levels of education, and lower socio-economic status. Research examining the social relations of African Americans compared with those of whites in the Detroit metropolitan area (Ajrouch *et al.*, 2001) found that African Americans had fewer close social ties, a finding clearly associated with the lower likelihood of African Americans being married, both over their lifetime and in late life. On the other hand, much research on African Americans in the USA, while recognizing a predominance of resource deficits, also acknowledges the important role of the Church (Taylor *et al.*, 2004). Many African Americans feel that the Church is an important source of friendship and support that provides a critical sense of belongingness and connection to a higher being (Chatters and Taylor, 1989).

Recent research contrasting African American and Afro-Caribbean Americans suggests that the unique histories of these two groups result in differences in QoL as they age. Jackson and Antonucci (2005) have shown that immigrants from the Caribbean have better mental and physical health than non-immigrants, despite their similar ancestry. However, age moderates this relationship, with older immigrants who have lived in the USA longer having better QoL than recent older immigrants. Hence, race, too, paints a complicated picture. It should be noted that many researchers have documented the likelihood that apparent race differences disappear once education, occupation, or socio-economic status are considered (Antonucci, 2001).

Marital Status

Whether one is married, never married, divorced, or widowed can be a life-defining status. Certainly each marital category is associated with the availability of different resources (Waite, 1995). Of course, being married is the most common status, but with age the probability of being married changes. Marital status can also be associated with cohort or country-specific history. For example, fewer people married during the Great Depression or during the two World Wars.

Being married in late life is usually associated with greater stability, income, and education. In addition to these 'objective' characteristics, a spouse is likely to have been a lifelong partner with whom one is especially familiar and in whom one has considerable trust. In times of ill health, and psychological, physical or emotional vulnerability, a spouse is a resource who, under ideal circumstances, shares with the individual a lifetime of memories and beliefs, which not only communicates that someone else thinks highly of you but also manages to instill in you a sense of competence, self-worth, and ability. This common effect of being married should not be underestimated. As intangible resources such as love, commitment, and support are increasingly shown to affect health and well-being, their contribution to QoL should not be ignored. In old age, the availability of such resources, often but not always found among spouses, offers an important reserve to rely upon when faced with the problems and challenges of old age.

SITUATIONAL CHARACTERISTICS

In this section we review four types of situational characteristics that can also affect QoL both directly and indirectly through their influence on the experience and expectation of social relations. We consider culture, history, government, and environment as illustrative examples.

Culture

Assessing QoL in various cultural contexts is important. Subjective ratings of QoL provide critical information about the extent to which societies meet the needs of their members (Kahn and Juster, 2002). Yet outside the USA empirical research utilizing established QoL measures remains scarce due to data limitations (George, 2006). Numerous studies examine the state of social relations among older populations in various cultural contexts (e.g. Ajrouch *et al.*, 2005; Antonucci *et al.*, 2002; Fernandez-Ballesteros, 2002; Melchior *et al.*, 2003). In considering the influence of culture, we examine aspects of social relations and their influence on QoL within both cross-national and immigrant contexts.

A recent study in Spain (Fernandez-Ballesteros, 2002) examined multiple aspects of social relations compared to other European countries (Denmark, the Netherlands, the UK, Italy, and Greece). Spain reports few older people living alone, and many living with children (either in their home or their child's). Among those aged 65+, Spanish older people report high contact frequency with family, friends, and neighbours, more than older people in northern European countries. While family are the number one source of support in times of sickness or need, neighbours are those with whom older people in Spain report highest frequency of contact, even more than with children who live in the same area. Children living in the same area and friends are seen the second most often, with about the same frequency. This finding suggests that social resources are quite diverse and that neighbours and friends play an important role in the lives of older people in Spain.

In the same study, QoL indicators were examined. According to Fernandez-Ballesteros (2002), the most important indicator of QoL in old age among Spaniards is health, followed by independence and income, with social relations figuring last. This ranking was the same for men and women regardless of age among those 65 and over. Finally, the association between social relations and QoL including various health measures (illness, self-rated health, pain) was examined. Results reveal a weak association between social relations and health among older adults in Spain, contrary to the growing body of work suggesting that social relations are important influences on the health and well-being of individuals. Some have hypothesized that the role of family is so central in Spain that people, somewhat counterintuitively, tend to underestimate its importance.

QoL examinations often assume that the identified dimensions are equally significant across time, space, and culture. As Skevington *et al.* (2004) have noted, however, whether a particular dimension is evaluated as 'important' signifies a critical, yet underacknowledged, aspect of QoL assessments. Importance ratings of QoL assessments allow for a full estimation of its various forms. For instance, family presence in times of sickness may be more important for some cultural groups than others. Anthropological and sociological work suggests that the family unit is central to Arab American daily life. Identity and responsibility are determined by family affiliation (Aswad, 1997), suggesting that family must be recognized as the context within which health care decisions are made. Arab Americans often look to family in times of sickness or crisis, and often become distressed if family members are not present (Meleis, 1981). The centrality of family to QoL is evident in studies focusing on other ethnic and immigrant groups within the USA as well (e.g. Cantor *et al.*, 1994).

One specific example of cultural differences can be recognized in family size and family relations. Different family structures affect all family members, though perhaps older people more. Large families suggest greater availability of others to provide help, as does less geographic mobility. Some cultures have a tradition of the eldest son staying in the family home to care for his parents, even after he marries. In these cultures, for example, in Japan, the daughter-in-law is the typical caregiver. This stands in startling contrast to the USA where the daughter is the usual caregiver. Of course, times are changing, leading to changing customs. In Japan, as in the USA, more older people are living alone, thus facilitating the provision of care by daughters to their parents.

Another example of culture influencing social relations can be seen in cultural norms concerning reciprocity. In the USA there are strong norms of reciprocity, although it has now been clearly documented that people actually prefer to provide somewhat more than they receive, especially within close, long-term, and ongoing relationships. While the idea of a Support Bank – building up a savings account of support upon which to draw in times of need – explains relations in Western societies, e.g. USA, France, and Germany, the rules concerning reciprocity, especially within families, are significantly different elsewhere. In Japan, it is always assumed that children are indebted to their parents, i.e. their relationship is non-reciprocal. It is a debt children are expected to spend their lifetime repaying (Akiyama *et al.*, 1997).

With world events including war, famine, and ethnic/racial persecution in modern times, culture is influencing our lives in many ways. Relevant to late life is the fact that these circumstances (along with more accessible transportation) are increasing immigration around the world as more and more people flee difficult circumstances. For example, the USA, historically known as a country of immigrants, currently hosts more immigrants than at any other time in its history. In addition, unlike in previous times when migration flowed predominantly from one part of the world, e.g. Northern or Southern Europe, the USA is currently receiving large numbers of immigrants from multiple regions including Asia, South and Central America, Africa, and the Middle East. Each of these streams of immigration brings with them distinct cultural traditions, which are then either blended with, or superimposed upon, American culture and values.

Clearly, the broader cultural context within which older people live fundamentally influences their social networks, the support that is exchanged, and the evaluation or judgement of their support relations. We turn next to consider a related characteristic, history, or historical context.

History

The historical context within which an individual lives provides another important situational context for understanding social relations. As noted above, during the Great Depression people had fewer children. Elder (1998) demonstrated that they experienced unique family dynamics due to the toll of their difficult life circumstances. There are other historical periods that are known to have had an effect on social relations. Distinctly different marital patterns are evident during the World Wars. The Berlin Ageing Study (Baltes and Mayer, 1999) documented an unusual number of never married men and women, as well as a greater number of people who never had children. The people in the Berlin Ageing Study came of age when men were off fighting and dying in a war that was eventually lost, resulting not only in the loss of many lives but also in an extended period of difficult post-war recovery. The pattern of their lifetime social relations were fundamentally influenced by these events over their entire life course.

Other historical periods are also noteworthy. The Korean and Vietnam Wars in the USA were, in contrast to the World Wars, distinctly unpopular. They led to a period of discontent in the USA, a period of marked breakdown in normative social relations. Interestingly, and often unrecognized, is the fact that there was much less intrafamily generational conflict than often presumed. So-called rebellious young people tended to have supportive parents who essentially shared their views, though often to a lesser degree (e.g. Kenniston, 1968). It was an unusual time of discontent, and with the end of these unpopular wars, a feeling of empowerment grew as this same group of people came to believe that they could, and had, shaped the course of history. Those very people are now becoming old. The young people have experienced the fall of the Berlin wall, the end of the Soviet Union, the terrorist attacks of 9/11, and two wars in the Middle East. These historical events will similarly influence their familial and non-familial social relations.

Government

Recent research has shown that government and public policy can fundamentally influence support relations, among families and older people. A recent comparative study across five European nations and the USA is illustrative (Brown *et al.*, 2006;

Lowenstein and Ogg, 2004). People of all ages in the USA, Germany, UK, Israel, Norway, and Spain were asked who they thought should provide financial, personal, and instrumental care for older people. Interestingly, people in Norway and Israel were most likely to report that these needs should be met by the government, whereas people in the USA and Germany were least likely to say so. The reverse was found when people were asked about filial obligations, i.e. to what extent should children be responsible for their parents. While people of all ages endorsed the idea of filial obligation, interestingly, young people were more likely to endorse this concept than older people. At the same time, people from the USA were more likely to support notions of filial obligation than people from any other country. Given the well-known family orientation of countries like Spain and Israel, this may, at first, seem counterintuitive. However, upon careful inspection it becomes clear that when governments provide resources, such as health care, institutional care, social and medical services, people feel quite comfortable accepting them. At the same time, it is noteworthy that regardless of government intervention, most families continue to have individually based affective relations with their elders (Lowenstein and Ogg, 2004). Countries where the government provides clear resources have citizens who reconstruct their social convoys to accommodate these provisions. It is not that the relationship is lost because of government intervention, but rather that appropriate adaptations are made.

Environment

We consider environment in its broadest context to include the general pleasantness or unpleasantness of the neighbourhood, and the affluence of the people and businesses in the community. It is an unfortunate reality that more pleasant environments, e.g. those that are safer, cleaner, and with greater affluence, afford more open, accepting, and interactive social relations. The influence of socio-economic status on the individual is well recognized. One can apply a similar assessment of the socio-economic status of a neighbourhood. When older people live in a poor or deteriorating community, it is likely that available services will be few, the number of people available or willing to provide support will be limited, and the security that a person will feel walking to the store or standing on a corner waiting for a bus will be minimal (Balfour and Kaplan, 2002).

The meaning of QoL may depend on the contextual and environmental challenges of a given group. For example, an exploratory study was conducted in rural Bangladesh (Nilsson *et al.*, 2005). Qualitative interviews and analysis led to the identification of two overarching themes: to have a role and to have a function. Most telling was the finding that QoL for older people in rural Bangladesh is 'embedded in social relations' (Nilsson *et al.*, 2005, p.371). To have a role included matters of reciprocity within the family unit, providing meaning and security. Older people feared that the inability to provide threatened the likelihood of receiving instrumental support from their children due to limited resources. A functional role incorporated both physical and social elements, but was dependent on strength and ability to work, which in turn relies on good health, obtained from good food. As a result, nutrition represented a basic determinant of QoL among older people living in rural Bangladesh. An active role and physical functionality were thought to depend on good health, social networks and social support, as well as a secure financial situation, each of which represents dimensions in other cultural contexts, but within the Bangladeshi context are contingent on basic needs such as ample nutrition.

This situational characteristic allows us to consider the enormous differences in people who are born to affluence compared to those born with less socio-economic capital. When old age is associated with poverty, these environmental constraints are sorely felt. Consider a poor neighbourhood that reduces or eliminates bus services. If older people lost bus transportation, they would be unable to visit any area to which they could not walk. Their social relations and QoL would be significantly limited. They would be likely to feel compelled not to leave the house at night, since many poor areas are often unsafe. At the same time, unlike in wealthy areas, it is unlikely that a park would be available in which to pass a pleasant afternoon or enjoy a peaceful walk in a lovely surrounding. Numerous characteristics can be incorporated under this environmental label. While personal safety might be one aspect of the environment, unregulated industry can violate the safety of a neighbourhood by exposing its members, for example, to chemical toxicity or pathogens. Less environmentally safe areas are likely to lead to constrained social exchanges, which become limited as a result of a hostile and unsafe neighbourhood (Krause, 1993).

CONCLUSION

Social relations appear as one element of QoL in many indices (e.g. WHOQoL-BREF), or are otherwise acknowledged as a key contextual aspect that influences other QoL indicators (Hellström *et al.*, 2004; Nilsson *et al.*, 2005). Whether considered a characteristic that indicates QoL or a major influence on QoL, social relations undoubtedly represent a critical part of the human experience.

Using the Convoy Model of Social Relations as a guiding framework, we described the ways in which social resources may be influenced by both personal characteristics and larger contextual or situational forces. The association between social resources and QoL was then reviewed in various national contexts, in order to provide a fuller picture of the ways in which social resources are significant to QoL in later life.

Resources available from relationships forged between and among persons represent a uniquely human characteristic, often providing support during times of need and ultimately influencing life experiences in a myriad of ways. Social resources are marked by both continuity and change across the lifespan, and usually accumulate over time. Ranging from broader structural characteristics such as network size, contact frequency, geographic proximity, and network composition, to the more intimate emotional and instrumental aspects of support provided and received, as well as expressions of both positive and negative relations, the multidimensionality of social resources has been illustrated as a means of addressing the complex ways in which they influence QoL. The role of such resources is undoubtedly critical to developing a more nuanced understanding of the ways that social relations influence QoL in late life.

REFERENCES

Ainsworth, M.D.S. (1989) 'Attachments beyond Infancy', *American Psychologist*, 44, 709–716.

Ajrouch, K.J., Antonucci, T.C., and Janevic, M.R. (2001) 'Social Networks among Blacks and Whites: The Interaction between Race and Age', *Journal of Gerontology: Social Sciences*, 56B(2), S112–S118.

Ajrouch, K.J., Blandon, A., and Antonucci, T.C. (2005) 'Social Networks among Men and Women: The Effects of Age and Socioeconomic Status', *Journal of Gerontology: Social Sciences*, 60, 311–317.

Akiyama, H., Antonucci, T.C., and Campbell, R. (1997) 'Exchange and Reciprocity among Two Generations of Japanese and American Women', in J. Sokolovsky (ed.) *The Cultural Context of Aging: Worldwide Perspectives*, West Port, CT, Bergin & Garvey, pp.162–178.

Akiyama, H., Antonucci, T.C., Takahashi, K., and Langfahl, E.S. (2003) 'Negative Interactions in Close Relationships across the Lifespan', *Journal of Gerontology: Psychological Sciences*, 58B, 70–79.

Andrews, F.M. and Withey, S.B. (1976) *Social Indicators of Well-being: America's Perception of Life Quality*, New York, Plenum Press.

Antonucci, T.C. (1986) 'Social Supports Networks: A Hierarchical Mapping Technique', *Generations*, 4, 10–12.

Antonucci, T.C. (1990) 'Social Support and Social Relationships', in R.H Binstock and E. Shanas (eds) *Handbook of Aging and Social Sciences*, New York, Van Nostrand Reinhold.

Antonucci, T.C. (2001) 'Social Relations: An Examination of Social Networks, Social Support and Sense of Control', in J.E. Birren and K.W. Schaie (eds) *Handbook of the Psychology of Aging*, 5th edn, New York, Academic Press, pp.427–453.

Antonucci, T.C. and Akiyama, H. (1987) 'Social Networks in Adult Life and a Preliminary Examination of the Convoy Model', *Journal of Gerontology*, 42, 519–527.

Antonucci, T.C. and Jackson, J.S. (1987) 'Social Support, Interpersonal Efficacy, and Health: A Life Course Perspective', in L.L. Carstensen and B.A. Edelstein (eds) *Handbook of Clinical Gerontology*, New York, Pergamon Press, pp.291–311.

Antonucci, T.C., Lansford, J.E., Akiyama, H., Smith, J., Baltes, M.M., Takahashi, K., Fuhrer, R., and Dartigues, J.F. (2002) 'Differences between Men and Women in Social Relations, Resource Deficits, and Depressive Symptomatology during Later Life in Four Nations', *Journal of Social Issues*, 58, 767–783.

Aswad, B.C. (1997) 'Arab American Families', in M.K. DeGenova (ed.) *Families in Cultural Context*, Mayfield, CA, Mayfield Publishing.

Balfour, J.L. and Kaplan, G.A. (2002) 'Neighborhood Environment and Loss of Physical Functioning in Older Adults: Evidence from the Alameda County Study', *American Journal of Epidemiology*, 155(6), 507–515.

Baltes, P.B. and Mayer, K.U. (1999) *The Berlin Ageing Study: Ageing from 70 to 100*, Cambridge, Cambridge University Press.

Baltes, P.B., Lindenberger, U., and Staudinger, U.M. (1998) 'Life-span Theory in Developmental Psychology', in R.M. Lerner (ed.) *Handbook of Child Psychology*, Vol. 1: *Theoretical Models of Human Development*, New York, Wiley, pp.1029–1143.

Bowlby, J. (1988) *A Secure Base: Parent–Child Attachment and Healthy Human Development*, New York, Basic.

Brown, E., Jackson, J.S., and Faison, N. (2006) 'Work and Retirement Experiences of Aging Black Americans', in P. Wink and J.B. James (eds) *The Crown of Life: Dynamics of the Early Post Retirement Stage*, New York, Springer, pp.39–60.

Bureau of the United States Census (2001) *Current Population Reports*, US Department of Commerce (www.census.gov).

Campbell, A., Converse, P.E., and Rodgers, W.L. (1976) *The Quality of American Life*, New York, Russell Sage Foundation.

Cantor, M.H., Brennan, M., and Sainz, A. (1994) 'The Importance of Ethnicity in the Social Support Systems of Older New Yorkers: A Longitudinal Perspective (1970–1990)', *Journal of Gerontological Social Work*, 22, 95–128.

Chatters, L.M. and Taylor, R.J. (1989) 'Age Differences in Religious Participation among Black Adults', *Journal of Gerontology: Social Sciences*, 44, S183–S189.

Cohen, S. and Syme, S.L. (1985) *Social Support and Health*, Orlando, FL, Academic Press.

Elder, G. (1998) 'The Life Course and Human Development', in R.M. Lerner (ed.) *Handbook of Child Psychology*, Vol. 1: *Theoretical Models of Human Development*, New York, Wiley, pp.939–991.

Fernandez-Ballesteros, R. (2002) 'Social Support and Quality of Life among Older People in Spain', *Journal of Social Issues*, 58, 645–659.

George, L.K. (2006) 'Perceived Quality of Life', in R.H. Binstock and L.K. George (eds) *Handbook of Aging and the Social Sciences*, 6th edn, Amsterdam, Academic Press, pp.321–338.

Granovetter, M.S. (1973) 'The Strength of Weak Ties', *American Journal of Sociology*, 78, 1360–1380.

Hellström, Y., Persson, G., and Hallberg, I.R. (2004) 'Quality of Life and Symptoms among Older People Living at Home', *Journal of Advanced Nursing*, 48, 584–593.

Jackson, J.S. and Antonucci, T.C. (2005) 'Physical and Mental Health Consequences of Aging in Place and Aging out of Place among Black Caribbean Immigrants', *Research in Human Development*, 2, 229–244.

Jackson, J.S., Chatters, L.M., and Neighbors, H.W. (1986) 'The Subjective Life Quality of Black Americans', in F.A. Andrews (ed.) *Research on the Quality of Life*, Ann Arbor, MI, University of Michigan.

Kahn, R.L. (1979) 'Aging and Social Support', in M.W. Riley (ed.), *Aging from Birth to Death: Interdisciplinary Perspectives*, Boulder, CO, Westview Press.

Kahn, R.L. and Antonucci, T.C. (1980) 'Convoys over the Life Course: Attachment, Roles, and Social Support', in P.B. Baltes and O. Brim (eds) *Life-Span Development and Behavior*, Vol. 3, New York, Academic Press, Reprinted (1989) in J. Munnichs and G. Uildris (eds) *Psychogerontologie*, Deventer, Van Loghum Slaterus, pp.81–102.

Kahn, R.L. and Juster, T. (2002) 'Well-being: Concepts and Measures', *Journal of Social Issues*, 58, 627–644.

Kenniston, K. (1968) *Young Radicals*, New York, Harcourt, Brace & World.

Krause, N. (1993) 'Neighborhood Deterioration and Social Isolation in Later Life', *International Journal of Aging and Human Development*, 36(1), 9–38.

Lewinter, M. (2003) 'Reciprocities in Caregiving Relationships in Danish Elder Care', *Journal of Aging Studies*, 17, 357–377.

Litwin, H. (2004) 'Intergenerational Exchange Patterns and Their Correlates in an Aging Israeli Cohort', *Research on Aging*, 26, 202–223.

Lowenstein, A. and Ogg, J. (eds) (2004) *Old Age and Autonomy: The Role of Service Systems and Intergenerational Family Solidarity*, Final Report, Israel, Center for Research and the Study of Aging, University of Haifa.

Martin, P., Poon, L.W., Kim, E., and Johnson, M.A. (1996) 'Social and Psychological Resources of the Oldest Old', *Experimental Aging Research*, 22, 121–139.

Melchior, M., Berkman, L.F., Niedhammer, I., Chea, M., and Goldberg, M. (2003) 'Social Relations and Self-reported Health: A Prospective Analysis of the French Gazel Cohort', *Social Science and Medicine*, 56, 1817–1830.

Meleis, A.I. (1981) 'The Arab-American in the Health Care System', *American Journal of Nursing*, 81, 1108.

Morgan, D.L. (1989) 'Adjusting to Widowhood: Do Social Networks Really Make It Easier?', *The Gerontologist*, 29, 101–107.

Newsom, J.T. and Schulz, R. (1996) 'Social Support as a Mediator in the Relations between Functional Status and Quality of Life in Older Adults', *Psychology and Aging*, 11, 34–44.

Nilsson, J., Grafström, M., Zaman, S., and Kabir, Z.N. (2005) 'Role and Function: Aspects of Quality of Life of Older People in Rural Bangaldesh', *Journal of Aging Studies*, 19, 363–374.

Pearlin, L.I. (1985) 'Social Structure and Processes of Social Support', S. Cohen and S.L. Syme (eds) *Social Support and Health*, New York, Academic Press, pp.43–60.

Pinquart, M. and Sorensen, S. (2000) 'Influences of Socioeconomic Status, Social Network, and Competence on Subjective Well-being in Later Life: A Meta-analysis', *Psychology and Aging*, 15, 187–224.

Plath, D. (1980) *Long Engagements: Maturity in Modern Japan*, Stanford, Stanford University Press.

Pugliesi, K. and Shook, S.L. (1998) 'Gender, Ethnicity, and Network Characteristics: Variation in Social Support Resources', *Sex Roles*, 38, 215–238.

Seeman, T.E. and Berkman, L.F. (1988) 'Structural Characteristics of Social Networks and Their Relationship with Social Support in the Elderly: Who Provides Support?' *Social Science Medicine*, 26, 737–749.

Silverstein, M., Conroy, S.J., Wang, H., Giarrusso, R., and Bengtson, V.L. (2002) 'Reciprocity in Parent–Child Relations over the Adult Life Course', *The Journals of Gerontology: Series B: Psychological Sciences and Social Sciences*, 57B, S3–S13.

Skevington, S.M., O'Connell, K.A., and the WHOQOL Group (2004) 'Can We Identify the Poorest Quality of Life? Assessing the Importance of Quality of Life Using the WHOQoL-100', *Quality of Life Research*, 13, 23–34.

Taylor, R.J., Chatters, L.M., and Levin, J. (2004) *Religion in the Lives of African Americans: Social, Psychological, and Health Perspectives*, Thousand Oaks, CA, Sage Publications.

Taylor, R.J., Hardison, C.B., and Chatters, L.M. (1996) 'Kin and Nonkin as Sources of Informal Assistance', in H.W. Neighbors and J.S. Jackson (eds) *Mental Health in Black America*, Thousand Oaks, CA, Sage, pp.130–145.

Waite, L.J. (1995) 'Does Marriage Matter?' *Demography*, 32, 483–508.

5. ECONOMIC RESOURCES AND SUBJECTIVE
WELL-BEING IN OLD AGE

THE IMPORTANCE OF ECONOMIC RESOURCES FOR SUBJECTIVE
QUALITY OF LIFE IN OLD AGE

In the light of demographic change and its accompanying implications for important societal functions, the economic and material circumstances of older people in Germany has been attracting increasing attention both in social gerontology and in the social and economic sciences as a whole (Naegele, 1991, 2003). The debate of recent years about the economically defined justice and solidarity between the generations, the public discussion about the demographically induced burdens on the social security systems as well as the growing importance of the economic power of older people for economic development[1] of society attests to this. However, it is noticeable that questions about the importance of economic resources for the life satisfaction and subjective well-being of older people go largely unheeded both in political and in scientific discussions. This paradoxical 'ambivalence' partly also characterises social gerontology in Germany: although economic resources are by common accord regarded as one of the most important dimensions of a good life in old age, priority is not currently given to the analysis of the relationship between material wealth and individual well-being in old age.

The implicit disregard of this subject matter has many reasons, the most prominent of which seems to be the relative affluence of older people in general compared to the past economic status of different age groups at different times. It is striking that, in comparison to younger age groups, the financial status of those aged over 60 has improved constantly in recent decades. Thus, for instance, the just-published Fifth Report on the Situation of the Elderly in Germany points to the higher economic resources of those aged over 60 and, in doing so, also emphasises their importance for the development of future consumer markets (BMFSFJ – Federal Ministry for Family Affairs, Seniors, Women and Youth, 2006). Organization for Economic Cooperation and Development (OECD) data corroborates this trend for other industrialised countries. According to this data, retired people have a disposable income that, on average, corresponds to 70–80% of the income of comparable groups of people in the later stages of their professional lives (OECD, 2001, p.22). Although income typically decreases when people withdraw from working life, the average disposable income of older people often only deteriorates slightly due to declining occupational expenditures, decreasing education and training costs for their children or mortgage loan costs (Casey and Yamada, 2002). There is a broad agreement in Germany today that poverty is at present (no longer) an explicit old-age risk (BMGS – Federal Ministry of Health and Social Security, 2005).

H. Mollenkopf and A. Walker (eds.), Quality of Life in Old Age, 65–84.
© 2007 *Springer.*

If one considers earlier debates on questions of material circumstances in old age, it appears that these were mainly conducted from the perspective of poverty – particularly with regard to very old age. Given this comparatively good financial situation of older people, the prevailing opinion seems to be that a (further) improvement of the quality of life (QoL) in old age can be achieved less by additional investments into economic resources than by investments into other resources, such as social or health care. Furthermore, echoing Chapter 1, geronto-logically oriented QoL research lacks a tradition, and perhaps also a publicly derived acceptance, for examining the specific importance of economic resources. In the past at any rate, the financial well-being of the older generation was only of political and/or scientific interest if a loss of QoL due to insufficient funds was ascertained. In contrast, the issue of further gains in QoL in old age thanks to income growth met with little public and scientific response.

A third explanation for the minor research interest might be the observation, stemming from QoL research, that there is an age-specific shift in the subjective importance of different life domains. Thus, while economic independence and material success rank among the most important life goals in early and middle adult-hood, non-material dimensions become more important with increasing age. This is especially true for the commodity health, the growing significance of which is based on the experience of its increasing vulnerability and finiteness. From late adult age onwards, health becomes one of the most central life themes and, in very old age, it becomes one of the most significant predictors of subjective well-being (Dittmann-Kohli et al., 2001; Smith et al., 1996; Michalos et al., 2001).

Consequently other life domains inevitably become less important. This particu-larly applies to paid employment and the financial motives it involves. This shift in the subjective value system is furthermore reinforced by the trend towards an increasingly earlier Entberuflichung (de-occupation) of old age that is observable in almost all modern industrial societies. For this reason earned income also inevitably becomes subjectively less important whereas alternative sources of income, which can, however, only be influenced to a limited extent, come to the fore. In the light of the growing number of older people and the increasing financial strain on social secu-rity systems – in particular the old-age security system – the issue of social transfers has become more pressing not only from the societal but also from the individual perspective of older people. In the face of empty pension funds and the like, questions regarding the security of pensions and one's own financial future in old age have contributed to sensitising a growing number of older people in Germany to financial issues of retirement. We can assume that the research about the importance of the financial dimensions for quality of life in old age will increase too (Naegele and Walker, 2006).

For many reasons, which we can only briefly outline here, the topic of material well-being in old age for a long time has not attracted much scientific interest. In order to represent the former and current state of knowledge in this respect, in the following section we will try to fill this gap and, among other things, present the state of research so far. In the process, we will first present selected data on the objective material life circumstances of older people in Germany. Second, we will go into the domain-specific satisfaction of older people with their economic situation on the basis of international studies. Furthermore, three kinds of – also statistically verifiable – relations will be described: (1) the relationship between economic resources and satisfaction with the economic situation in old age, (2) the role of these resources for the global subjective well-being of older people, and (3) the importance of the specific satisfaction with the individual financial situation for a general assessment of the subjective QoL in old age.

THE OBJECTIVE ECONOMIC SITUATION OF OLDER PEOPLE IN GERMANY

The Definition of Economic Resources

To characterise the current economic situation of older people, the concept of 'economic resources' is generally used. It comprises all the sources of income available to an older person or to a household of older people. Economic resources include all income streams, savings, financial as well as tangible assets that can also be employed to gain income (letting, leasing, or sale) (Smeeding, 1990). Moreover, the tangible assets that do not directly serve the acquisition of income but could principally be used for this purpose, by means of sale, are also counted among these. In some studies, even self-inhabited property is regarded as part of the economic resources, as it can, for example after the mortgage loan has been paid off, considerably lower one's living costs.

As far as the pragmatic operationalisation of the term by research is concerned the complete material and financial situation of a person is only seldom referred to in practice. In most cases, income is taken as a proxy to describe the overall economic resources of a person. This simplification is due to the fact that income can be measured comparatively easily.[2] For pragmatic reasons, we too employ the term in this way for our research. Moreover, different indicators of the economic circumstances such as income and wealth, often correlate with each other. Nonetheless this limited assessment of the nature of economic resources was criticised in the early stages of QoL research (Moon, 1977). Although income can, without doubt, be considered as the best single indicator to depict the available liquid financial resources, its implicit equation with the totality of economic resources leads to bias in the relationship between the objective indicators of material life and the indicators of subjective QoL. Given the fact that a large number of older people have personal assets but that there is an unequal distribution of wealth in old age, it would be desirable to include these within the scope of future studies.

In the following section, we will begin by giving an up-to-date review of (statistically accessible) data on regular incomes from the existing old-age security systems in Germany, fully aware of the fact that only part of the economic resources of older people in Germany is recorded. Subsequently, representative data about the distribution of poverty and wealth among older people in Germany will be presented. There is no representative data on further possible indicators.

The Most Important Sources of Income of Pensioners in Germany[3]

The existing old-age security systems at a glance. Due to its historical background old-age insurance in Germany is structured and organised in a very complex way. It consists of a conglomerate of different systems, institutions and performance principles. This differentiation is accompanied by differences regarding the organisation, the categories of people included, the respective insurance goals, the benefit requirements and levels, as well as the funding modalities. Despite equal personal characteristics, the benefits differ from system to system. The systems can be assigned to different levels: the pillar or level model of the German old-age security system.

- The *first level* comprises the standard systems which consist of:
  – the statutory pension insurance scheme
  – the civil servants' pension scheme
  – pension schemes for certain groups of self-employed professionals.
- The *second level* is made up of occupational pension schemes, which can be divided into
  – occupational pension schemes for employees in the private sector and the
  – supplementary insurance for the workers and employees in the public sector.
- The *third level* is represented by private retirement schemes. Here one has to differentiate between age-related asset formation and life insurances and private insurance funds.
- Lastly, if one takes into account the basic income grant for older people or for people with a reduced earning capacity, which is based on the welfare principle, one has a *fourth level*.

These levels contribute to the old-age incomes of the different population groups to varying degrees. In summary, the employment and occupation-related old-age security systems that link old-age insurance to employment and occupation status and, at the same time, align the annuity rate with the former earned income and the duration of gainful employment, characterise old-age provision in Germany.

In Germany, the statutory old-age insurance is of paramount importance for the composition of old age income. The average pensioner household (in 2003) received 85% of its income from pension benefits, 5% from benefits from occupational pension plans, and 10% from payments from private retirement schemes. However, there is a shift taking place in the direction of individual provision for old age and occupational pension schemes. The introduction of 'Riester pension' also marked the beginning of the lowering of the provision level of old-age insurance; the arising coverage gap is supposed to be closed by the expansion of private pension provision and of occupational pension schemes which are promoted by means of state subsidies and tax relief. The second and third levels have thus become a part of the governmental social security policy (Naegele and Walker, 2006).

The Standard Pension Systems

A defined group of people have been incorporated into the standard pension systems via compulsory membership. These systems have different entry requirements (age limits, qualifying periods, etc.). The standard old-age security systems are outlined briefly below.

1. Statutory old-age insurance. The statutory old-age insurance is a public law mandatory insurance for all employees, for certain groups of self-employed professionals as well as for further non-active population groups. It is financed by contributions on a pay-as-you-go basis and by additional tax subsidies. In 2004, the statutory old age insurance paid out more than 24 million pensions to approximately 19.8 million pensioners (including survivorship annuities).

2. Civil servants' pension scheme. The old-age provision for all civil servants, judges, and professional soldiers who are employed under public law as well as for their surviving dependants is regulated by the Civil Service Benefits Act and the Act on Soldiers' Pensions. It is financed from tax revenue. The number of old age beneficiaries totalled a little more than 1.4 million in 2003.

3. Pension schemes for the self-employed and for freelancers. Self-employed people are excluded from the statutory old-age insurance in Germany. They must provide for their own old age. However, there are exceptions from this basic principle. For farmers and for some groups of self-employed professionals there are special systems. Other groups of self-employed people in turn are mandatory members in old-age insurance.

Occupational Pension Schemes

The occupational pension schemes comprise occupational pension plans for employees in the private sector and the supplementary insurance for employees in the public sector. While the occupational pension schemes in the private sector are predominantly based on the employers' pension commitments and thus only cover some employees, supplementary insurance in the public sector is based on tariff agreements and incorporates all employees in the public sector and equivalent fields. It is characteristic of both types of schemes, that their benefits complement and supplement other pension income, as a rule those from the statutory old-age insurance. For a long time, the occupational old-age provision in the private sector played a subordinate role in Germany. Due to the paradigm change in the pension and old-age security policy initiated in 2000, which is characterised by a reduction in the pension level in the statutory old-age insurance and, at the same time, by the public promotion of occupational and private pension schemes, the latter have become more important. In March 2003, 10.3 million of the employees subject to social insurance contributions in the private sector had acquired an entitlement to a company pension; this corresponds to about 43% of all employees. Only just over one-third of business establishments have a pension scheme.

Private provision for old age. For a small group of older people the prospective old age income is mainly based on private provision for old age. But for a growing number of people private provision for old age has a supplementary function. It is difficult to grasp empirically the importance that the numerous forms of asset utilisation and asset release have for old age. For by no means all forms of asset formation are direct retirement provisions. This applies only to those forms that are explicitly geared to securing one's maintenance in old age and are not prematurely used for other purposes. As far as the private provision for old age in Germany is concerned life insurance plays an outstanding role. According to indications from the insurance industry, there were around 91.5 million contracts in 2003. As for private provision for old age as a whole, life insurances are very unequally distributed in the population and the upper income and occupation groups are overrepresented.

Other sources of income in old age. In addition to the above-mentioned benefits there are several sources of income that mostly supplement the benefits from the standard pension scheme. They only act as the main source of income in old age in exceptional cases. These include

- Income from (marginal/sideline) employment. Around 440,000 people aged 65 and above were still economically active in 2003.
- Income from war victims' benefits.
- Cash benefits from the nursing care insurance (nursing allowance).
- Pensions from the statutory accident insurance, which are, however, partly offset against pensions from the statutory old-age insurance.
- Housing subsidies as well as benefits from the needs-weighted basic income grant in old age and in cases of reduced earning capacity. At the end of 2003, around 270,000 people aged 65 and over received benefits from the needs-oriented basic social security. This corresponds to 1.8% of all older people.

Income Situation and Income Distribution in Old Age

Accumulation of pensions and total income. The pension payments from the different systems of old-age security only give limited information about the actual income and provision situation of older people. To get a comprehensive idea, we have to take into account that one person can draw various benefits from one or several systems. The total income at a personal level is thus determined by the accumulation of several annuities from different sources. At the same time there is (still) a substantial discrepancy in Germany between the old and the new federal states as far as income situation of older people is concerned. Therefore, we will – where required – differentiate between West Germany (old federal states) and East Germany (new federal states).

Regarding the current distribution of old-age incomes in Germany, findings of a recent representative study on 'Old-Age Provision in Germany' are now available, which we refer to later (Bieber and Klebula, 2005). They show that in both the old and the new federal states alike, most older men (91% and 99%) and older women (82% and 99%) draw pensions from the statutory old-age insurance. It is characteristic of the situation in the old federal states that the pension from the statutory old-age insurance is often supplemented by other benefits (Table 1). Forty-four per cent of men are also entitled to occupational pensions or to benefits from supplementary insurance or from the civil servants' pension scheme. Thirty-nine per cent have to rely solely on their own pension from the statutory old-age insurance. In the case of women, the widow's annuity plays a decisive role, while supplementary occupational pensions and benefits from the supplementary insurance (altogether in 10% of the cases) are of subordinate importance. About a quarter of older women draw a survivorship annuity in addition to their own pension. Thirty-six per cent of women have to rely solely on their own benefits from the statutory old-age insurance.

The picture is fundamentally different for the new federal states. There the benefits from the statutory old-age insurance are the sole old-age income for 88% of men and 54% of women. Occupational pensions, benefits from the supplementary insurance of the public sector or from the civil servants' pension scheme, are irrelevant, in this respect inexistent. This is due to the fact that such systems were almost completely lacking in the former German Democratic Republic (GDR).

TABLE 1. Forms of Accumulation of Pension Benefit Claims, 2003 (0 = <0.5 but >0) (Bieber and Klebula, 2005)

Old age benefits	Total (%)	Men (%)	Women (%)
Old federal states			
Solely own pension from the statutory old-age insurance	37	39	36
Own and derived pension from the statutory old-age insurance	13	1	21
Own pension from the statutory old-age insurance and occupational pension	14	29	4
Own pensions from the statutory old-age insurance and from the supplementary insurance of the public sector	7	10	6
Solely own civil servants' pension	2	5	1
Own pension from the statutory old-age insurance and civil servants' pension	2	5	0
No old age benefits	4	2	6
Miscellaneous	20	9	27
Total	100	100	100
New federal states			
Solely own pension from the statutory old-age insurance	68	88	54
Own and derived pension from the statutory old-age insurance	29	7	43
Own pension from the statutory old-age insurance and occupational pension	1	1	0
Own pensions from the statutory old-age insurance and from the supplementary insurance of the public sector	1	2	1
Miscellaneous	1	0	2
Total	100	100	100

Total income in old age. The accumulation of pensions has direct repercussions on the total income in old age. We will consider the average net income and will differentiate between the old and new federal states as well as between different types of households (Table 2).

The data represented in Table 2 point to some inequalities in income of certain groups of older people.

• Due to the lack of additional benefits from other old-age security systems, the net income of older people in the new federal states is on average markedly lower than that in the old federal states. Only East German wives are better off because married women in East Germany were (and still are) more likely to be economically active.

• The income of married men exceeds that of married women. The prevalence rate of own pensions from the statutory old-age insurance is, in fact, rising from cohort to cohort among women and is approaching that of men (almost 100% of wives in the new and 70% of wives in the old federal states have their own old-age

TABLE 2. Net Income of the Population Aged 65 Years and Over, 2003 (TNS Infratest, 2005, p.91 et seq.)

Net income	Married couples (%)	Single men (%)	Single women Unmarried (%)	Divorced (%)	Widowed (%)
Old federal states					
Average in euro	2,211	1,515	1,189	1,051	1,195
Growth 1999–2003	10.7	8.9	6.2	10.2	4.5
<200	0	0	1	0	0
200 to <300	0	1	3	3	1
300 to <500	0	2	4	3	4
500 to <700	1	5	12	16	9
700 to <1,000	4	13	27	33	27
1,000 to <1,500	20	39	26	29	38
1,500 to <2,000	26	22	15	11	13
2,000 to <2,500	20	10	7	3	5
2,500 to <5,000	26	8	4	2	3
5,000 and more	2	0	–	–	0
New federal states					
Average in euro	1,938	1,284	953	827	1,207
Growth 1999–2003	8.7	9.0	8.9	10.1	15.0
<200	–	–	–	–	0
200 to <300	–	0	0	1	–
300 to <500	0	1	2	2	1
500 to <700	0	3	17	30	4
700 to <1,000	1	21	45	47	19
1,000 to <1,500	14	54	31	18	61
1,500 to <2,000	47	18	5	1	14
2,000 to <2,500	27	2	–	1	2
2,500 to <5,000	10	1	–	0	0
5,000 and more	0	1	–	–	–
% of the old federal states	88	85	80	79	101

insurance claims). But the pensions are still low despite an upward trend in recent years so that the income situation of older couples is still largely determined by the husband's income (75.6% in the old and 63.9% in the new federal states).

- In the case of *single women* total income is composed in very different ways. While the main source of income of widows is the widow's pension, paid in addition to an own pension, unmarried women are only entitled to pensions after above-average length of employment and insurance.

- Furthermore, we have to take into account that these average values conceal a broad distribution of incomes. Thus in 2003, the proportion of households in the old federal states that had less than € 700 per month at their disposal were 8% of single men, 20% of unmarried women, 22% of divorced women, and 14% of the widows. This concerns those who, due to their unfavourable employment and insurance histories, only draw low own and derived pensions from the statutory old-age insurance and at the same time cannot claim benefits from other security systems. On the other hand, 8% of single men and 28% of married couples declare that they have a monthly net income of more than € 2,500.

It is not known what proportion of older people can maintain their income levels after withdrawing from working life, i.e. which proportion by dint of the combination of standard pension benefits and private as well as occupational supplementary benefits has an income that secures their standard of living, and how high the proportion of people is who have to cope with noticeable income losses upon retirement.

Poverty and wealth in old age. The majority of older people in Germany live in (marital) partner relations and keep house with their joint income. In order to be able to compare their income situation with that of single people, the personal incomes of married pensioners have to be combined and to be shown as per capita income. Likewise, the cost advantage that arises from financing a common household must be taken into account with a needs weighting. For married couples, the average income available to older people, measured by net equivalent income, is above the societal average. The income situation of households in younger age groups or in special life situations (unemployment, single parents) compares unfavourably with that of older households in Germany. Today older people are those who, given their generally good income situation, make one-time or constant private money transfers to their children and grandchildren.

Research shows that the poverty risk of older people in Germany is comparatively low and has dropped markedly in the recent past. The poverty line benchmark is 60% of the needs-weighted median national income: according to this benchmark, old-age pensioners had an 11.8% poverty rate in 2003 – compared to 13.5% of the total population. With a rate of 13.5%, older women are more affected than older men with a rate of 9.8%. The proportion of older people who are on welfare (assistance towards living expenses according to the Social Security Code XII) likewise is comparatively low and has decreased in the past years. In 2002, 1.3% of those aged 65 and over received subsistence allowances, the proportion in the total population, however, amounted to 3.3%. At the end of 2003, around 270,000 people aged 65 and over drew benefits from the needs-weighted basic income grant. The rate calculated from this number is higher than the previous welfare rate. But in 1990 the welfare rate of older people was still 2.3%.[4] The situation of poor older people is contrasted with that of pensioners whose income situation is characterised by wealth. Where asset formation takes place in the course of their lives older people can fall back on and bequeath financial, property, and productive assets. In the mid-1990s, the old households containing one person over the age of 65 possessed about a quarter of the total financial assets in Germany although they only made up 17% of the total households. A further 40% of pensioner households simultaneously also owned property. But wealth too is very unequally distributed in old age, even more so than income: in the group of two-or-more-person households with people aged over 65, the lower 50% of these households only owned 13% of the financial and property assets in 1998, while 33.7% of the private means belonged to the top 10% of these households.

THE IMPORTANCE OF ECONOMIC RESOURCES FOR QoL IN OLD AGE

The Satisfaction of Older People with Economic Resources

George (1992, p.72) defines satisfaction with economic resources as 'the subjective evaluations of the degree to which one's financial resources are adequate versus inadequate or bring satisfaction versus dissatisfaction'. The definition refers to two possible dimensions: on the one hand, satisfaction with the economic circumstances can reflect the subjectively perceived adequacy of one's income, standard of living, assets, wealth, and so on; while, on the other hand, satisfaction with economic resources can reflect a direct evaluation of the specific satisfaction (with these resources) in itself.[5] Empirically data on satisfaction with material or economic life is often collected by means of single-item measures.[6] Besides income, the standard of living, savings, or even the person's or household's economic situation as a whole serve as the basis of evaluation. In other studies, data on different aspects of the financial situation are collected and then compiled in an index (Lamura et al., 2003). Due to this diversity of operationalisations assessments of material and economic satisfaction are quite contradictory (Herzog and Rodgers, 1981).[7] While satisfaction with the economic situation, given its summarising nature, represents the most comprehensive measure of material well-being, income satisfaction normally refers to the level of income stemming from different sources. Satisfaction with the standard of living in turn refers to the quantity and quality of the available goods and services as well as to the rights to use public facilities or services free of charge.

Findings of the first representative studies of social indicator research show that older respondents are, on average, more satisfied with their material situation than their younger counterparts (Campbell et al., 1976; Andrews and Withey, 1976,

Lawton, 1983). This is also true for satisfaction with the standard of living (Herzog and Rodgers, 1981; Diener and Suh, 1997; Noll and Schöb, 2002). On the basis of a meta-analysis of 20 studies that were mainly conducted in the USA between 1960 and 1990, George (1992) pointed to the exceptionally high satisfaction older people felt about their financial situation in comparison to other age groups.

Representative data from Germany[8] confirm a pattern familiar from US research. Both the Welfare Survey and the Socio-Economic Panel show that satisfaction with income as well as with standard of living rises with increasing age. In 2002, the mean value of income satisfaction in those aged 65 and over in West Germany was $x = 7.0$ (min. = 0, max. = 10) and was thus almost a whole scale point higher than in the group of the 17–59 age group ($x = 6.3$, min. = 0, max. = 10) (Noll and Weick, 2004). A similar tendency can be seen for the standard of living satisfaction, the evaluations being on the whole even more positive (Noll and Schöb, 2002).

The findings of the Age Survey[9] also point to a particularly high satisfaction of older people with their material life circumstances. More often than other groups, the oldest respondents – those aged 70–85 – rated their standard of living as 'good' or 'very good' (Motel-Klingebiel, 2000, p.239). As the Age Survey is a combination of a cross-section sample with a longitudinal sample, it allows researchers to compare equal age groups at different points in time (1996 and 2002) as well as to draw conclusions about changes in the course of time. Thus the data of the Age Survey analysed for West Germany shows that the 55- to 69-year-olds in 2002 rated their standard of living more positively than their counterparts in 1996, a finding that can be explained by the fact that this age group has also objectively experienced noticeable income gains in the period under review. In contrast, the younger respondents' (40- to 54-year-olds) satisfaction value dropped in the reference period. The positive rating of the oldest group (70- to 85-year-olds), however, remained at an unchanged high level (Motel-Klingebiel, 2006; Tesch-Römer and Wurm, 2006).

German gerontologists continue to take an interest in the differences and the changes in the transition phase from old age to very old age. The Berlin Aging Study (BASE),[10] which focused on people between the ages of 70 and 103, is a key source. According to this study, satisfaction of the respondents with their financial situation slightly increases as they approach a very high age and is in the positive scale range for the whole sample. Thus age proves to be a direct predictor of satisfaction with one's financial situation: the older the interviewees, the more satisfied they are with their financial situation (Smith et al., 1996, p.514). Moreover, the study shows that the differences in satisfaction with economic resources between the different groups of older people were indeed significant but not particularly pronounced. Older people seem to differ less between themselves regarding their satisfaction with their financial situation than younger people – at least according to findings from the USA; here the standard deviation in satisfaction values of young adults was generally higher than that of the older population (Campbell et al., 1976; Andrews and Withey, 1976; George, 1992).

The Importance of Economic Resources for Satisfaction
with Economic Resources

Not only the data of the BASE points – especially among the very old – to a high number of people who report a high level of satisfaction with their financial situation despite a comparatively low material standard of living and despite the existence of material needs (the economic satisfaction paradox) (Walker, 1986; Smith *et al.*, 1996). Representative population surveys in the USA confirm this finding (Campbell *et al.*, 1976). Thus, if we were only to use subjective indicators for the general public, we would be in danger of underrating the effective financial needs of older people (Naegele, 1997; Clemens and Naegele, 2005). The BASE supports this conclusion. Interestingly, the level of the equivalent income used in the BASE model has an effect on the financial satisfaction only and has no predictive power for other dimensions of domain-specific satisfaction (Smith *et al.*, 1996, p.515). Although income level was the strongest predictor of material well-being, also in very old age, there must therefore be other explanatory variables of financial satisfaction.

The findings of a Canadian survey by Michalos *et al.* (2001), in which satisfaction with the standard of living was examined as a dependent variable, also point in this direction. The existence (or the lack) of financial difficulties (most notably because of high accommodation costs), satisfaction with one's own financial security, the existence of social support as well as satisfaction with the access to age-relevant information rank among the most important predictors in this respect: the higher satisfaction with these dimensions, the higher satisfaction of the (older) people with their standard of living. Financial worries proved to be the strongest predictor in this connection. This leads to the conclusion that the effect (in this case impairing) of 'negative influence variables' from the material life-domain on the subjective QoL of older people is stronger than the effect of 'positive influence variables'.

However, material well-being is not only the direct product of income level, but also continues to be the result of different cognitive comparison processes (perceived gaps), in which (older) people compare their given financial situation with their own goals, the situation in the past or even with the economic situation of other people (Michalos, 2003). On this issue, we can cite own research findings (Weidekamp-Maicher, in press). The survey of 50- to 85-year-old men and women in a large West German city in the period 2004/2005 revealed that the rating of the degree of need satisfaction ($\beta = 0.21$) and of the extent of the fulfilment of their own material life goals ($\beta = 0.24$) both had a significant influence on satisfaction with the standard of living. Contrary to expectations, the comparison of the current standard of living with that of the past, with the economic expectations for the future as well as the comparison of their own standard of living with that of people of the same age proved to be insignificant. Besides the level of the equivalent income ($\beta = 0.24$), satisfaction with the standard of living also depends on whether the person's level of income had changed in the last 1 or 2 years ($\beta = 0.12$) and whether they were able to save regularly ($\beta = 0.16$). As expected, both the level of equivalent income and the changes in income in the recent years had a significant and even stronger influence on income satisfaction (equivalent income: $\beta = 0.31$, change in income: $\beta = 0.33$), even if subjective evaluations of the degree of need satisfaction and the attainment of one's own material goals were introduced into the model.

The Direct and Indirect Importance of Material Resources for the Subjective QoL in Old Age

The empirical findings on the importance of economic resources which have been brought forward so far point to the fact that objective indicators of material life are, including for older people, important predictors for the domain-specific satisfaction with economic resources. Nevertheless this only covers a part of the potential predictors for global subjective QoL in old age. An open question is what influence (objective) economic resources have on global subjective well-being understood here as a multidimensional construct that encompasses at least two dimensions: (1) a cognitively oriented dimension that is referred to as 'life satisfaction' and (2) an emotional dimension that is termed 'happiness' (Diener et al., 2003). A further open question is what influence satisfaction with the economic resources has on subjective well-being. Thus people's assessment of the importance of different life domains in old age can change. This can, for instance, result in the fact that the subjective well-being of older people is affected less by economic factors than by satisfaction with other life domains, which are, according to the findings of the BASE study in particular, health, mobility, or social relations (Smith et al., 1996, p.515). This could be a further explanation for the above-mentioned 'economic satisfaction paradox', especially with very old people.

In order to be able to comment on these connections in more detail, we will briefly expand on the so-called bottom-up model of subjective QoL. According to this model, good living conditions that are, inter alia, represented by economic resources, lead to subjective well-being in two ways. The income available to a person can have a direct influence on the level of subjective well-being. However, previous research results using multivariate analyses show that income level only explains a very small part of the total variance in the level of subjective well-being, i.e. the influence of economic resources is of a more indirect nature (Campbell et al., 1976). Therefore, objective indicators – such as the level of economic resources – first have an effect on domain-specific satisfaction (e.g. with income, standard of living, and savings) and, second, an effect on global subjective well-being. Some findings even lead to the conclusion that the direct influence of economic resources on life satisfaction and happiness diminishes with increasing age while the significance of the indirect influence grows or remains unchanged (George, 1992).

According to the classic studies of Andrews and Withey (1976) as well as of Campbell et al. (1976), income level has both a direct and an indirect influence on the level of subjective well-being. In their studies both research teams used different kinds of domain-specific satisfaction: while Andrews and Withey measured satisfaction with the economic resources, Campbell and his colleagues examined satisfaction with the standard of living and the level of savings. Comparison of the study results confirms the influence of economic resources on global subjective QoL on the one hand but, on the other hand, it shows that not only income satisfaction but also other dimensions of material well-being too have to be borne in mind.

Relations Between the Level of Economic Resources and Global Subjective Well-Being in Old Age

Bivariate correlations on the relation between objective indicators of the economic resources and the degree of subjective well-being, which were obtained in studies with different age groups, show that correlation coefficients often turn out to be lower for older respondents than for younger ones. In the case of the study conducted by George (1992, p.77) the coefficients are not only lower for older people but also spread less than the values of younger people. From young adult age to middle adult age, the correlation coefficients are on the increase, to decrease again in late adult age. In other words, according to this study, subjective QoL, particularly in old age, is determined less and less by the direct influence of income (George et al., 1985).

More recent studies from the USA and the UK, however, only partly confirm these relations identified by George in 1992 (Mookherjee, 1997; Mullis, 1992; Smith *et al.*, 2004). Our own research also arrives at similar but more sophisticated results (Weidekamp-Maicher, in press). Here the correlations depend on the age of the people surveyed, on the one hand, and, on the other hand, on the indicators used in each case for evaluating the subjective well-being as well as on the employed measures for economic resources. All of the measures of economic resources used (the level of the individual income, of the household income, and of the equivalent income) show a medium-strong, significant, positive relation to life satisfaction (r = between 0.29 and 0.61). The correlations tend to be stronger for the younger respondents (50–69 years) than for the older interviewees (70–85 years). Conversely, the household and equivalent income relate more strongly to life satisfaction for older interviewees than for younger interviewees. Moreover, the indicators used for measuring the subjective well-being also play an important role: while the bivariate correlations with life satisfaction relate significantly for all examined age groups as well as for all income indicators, they are either markedly weaker or not significant at all for all age groups as far as positive and negative emotions (happiness) are concerned.

Furthermore, the significance of so-called relative income measures for the subjective well-being also plays a role in QoL research. These income measures include the allocations to certain income groups (income terciles, quartiles, quintiles, etc.) but also different concepts of relative poverty or even different conceptions of social stratification. In the context of a study conducted in six European countries on the relationship between material well-being and subjective QoL in late adult life (The European Study of Adult Well-being – ESAW[11]), the affiliation to a nationally defined income tercile was consulted and it showed that the respective relative income position within a society also has a share in influencing the subjective well-being of older people. As a result of its international orientation, this study also calls attention to a further interesting phenomenon: the moderating function of the national affiliation of those interviewed and, therefore, to the specifics of ageing and being old in different societies. Thus, for instance, in an analysis of representative population data for adults from 19 countries, which was conducted in 2000 by Diener and Oishi (2000),[12] national affiliation played an important role for the ascertainment of the level of the correlation coefficient between the socio-economic status and the level of global subjective well-being, for both direction and strength of the correlation coefficient are moderated by the national affiliation.

Causal models of QoL among older people support the thesis of the direct and indirect predictive power of income on global subjective well-being. Thus the first two waves of the German Age Survey in the first instance confirm a direct and positive influence of the equivalent income on life satisfaction. But, in the second wave in 2002, this direct influence disappears completely or is noticeably diminished, if subjective assessments of different life domains, among other the subjective evaluation of the standard of living, are introduced into the model (Tesch-Römer and Wurm, 2006).

Relationships Between the Satisfaction With Economic Resources and Global Subjective Well-Being in Old Age

In the last section of this chapter, we will delve into the question how satisfaction with income, with standard of living, or with overall economic situation affects subjective well-being life satisfaction or happiness. If one examines the findings of previous studies to this end, it initially appears that the bivariate correlations between satisfaction with economic resources and global indicators of subjective QoL regularly turn out to be higher than the relationship between objective economic resources and global subjective QoL – indeed independently of age. However, on the basis of meta-analyses, George (1992) comes to the conclusion that age also constitutes an important moderating variable in this case. According to her study, the level of the economic resources themselves not only have a less strong influence on the subjective well-being of older people but satisfaction with these resources also contributes less to explaining global subjective QoL than in the case of younger people. Recent studies call attention to the greater complexity of this relationship. Thus age does not seem to be solely responsible for the strength of the correlations between domain-specific satisfaction with income and subjective well-being but also the relative income level (Mookherjee, 1992; Smith *et al.*, 2004, p.800).

In addition, international comparative studies also point to the high significance of subjective evaluations of a person's financial situation for global subjective QoL in old age. Within the scope of ESAW it also became apparent that the coefficients between different indicators of material and subjective well-being were comparatively strong (Lamura *et al.*, 2003). Although material well-being here is primarily operationalised by means of subjective evaluations of material security, the list of question also encompasses a range of further items, *inter alia*, indications of satisfaction with one's own economic possibilities of using resources. The highest coefficient occurs between a factor that is referred to as 'satisfaction with the economic situation' and life satisfaction.[13] The relationship between the following two items is likewise comparatively high: subjective evaluation (1) whether the available financial resources suffice to supply the prevailing wants and (2) whether the interviewees are now and then able to afford some luxury in everyday life.

Other studies examine the causal relationships between satisfaction with the economic resources and global subjective well-being in old age. According to these the predictive power of the respective domain-specific satisfaction for subjective well-being of older people depends on which aspect of material life is included. Thus it seems that it is not so much satisfaction with income level but, rather, satisfaction with other aspects of material circumstances, such as income security or the level of living that make an important contribution to subjective well-being in old age, in particular satisfaction with standard of living has to be emphasised (Campbell *et al.*, 1976; Herzog and Rodgers, 1981). As satisfaction with level of savings also has an independent and statistically significant influence on life satisfaction according to these studies, which (further) aspects of the material QoL that are factored into the analyses with older people prove to be significant.

In our study (Weidekamp-Maicher, in press) satisfaction with standard of living also has a significant influence on the level of life satisfaction ($\beta = 0.28$) even after 15 other kinds of domain-specific satisfaction were introduced into the model. Satisfaction with the standard of living has a similarly high explanatory power as regards the positive affect ($\beta = 0.27$). But income satisfaction too decisively contributes to an explanation of both life satisfaction ($\beta = 0.25$) and positive affect ($\beta = 0.32$). The two waves of the German Age Survey made similar findings. Here the subjective evaluation of one's own standard of living has a strong influence, in fact, the strongest of all the used specific evaluations of different life dimensions, on subjective life satisfaction (Tesch-Römer and Wurm, 2006). This relationship becomes weaker though if the very old are focused on, for in their case satisfaction with their state of health becomes more important. This is also confirmed by the BASE findings, according to which satisfaction with the current financial situation 'only' constitutes the second strongest predictor of subjective well-being.[14] It is clearly surpassed by the importance of subjective health. The Canadian study by Michalos *et al.* (2001) found similar results: the higher the interviewees' age (compared to the average age of an examined sample), the more significant the satisfaction with state of health becomes for subjective QoL.

CONCLUSION

The available research findings on material well-being in old age confirm that older people are more satisfied with their financial situation than younger ones. Moreover, multivariate analyses point to the fact that the predictive power of income level for satisfaction with one's financial situation as well as for the level of the subjective well-being decreases with advancing age. Indeed there are age and cohort-specific evaluation patterns inasmuch as criteria other than income prevail in old age as regards subjective evaluation of QoL. Given the background of age-related constraints which emerge most often in very old age, these are satisfaction with the state of health, with mobility, and with social networks.

Furthermore, the research findings presented show that the absolute level of income plays a less important role for subjective evaluation of financial resources than the living standard it enables, i.e. whether the available economic resources suffice to attain a satisfactory living standard, to enable participation in social life despite mobility losses, or to compensate for rising costs of health or nursing care. Furthermore, the findings show how important it is to factor more dimensions than just income level into the applied socio-gerontological satisfaction and QoL research. Thus, in particular, satisfaction with income security or, for example, with savings.

In this respect, the findings cannot be interpreted to the effect that financial resources are increasingly losing their importance for a good QoL in old age. Quite the contrary: due to decreasing health and social resources, their objective significance is actually on the rise, for the more vulnerable social support networks become and the more health impairments hamper people from carrying out everyday activities, the more important other supporting resources become, including, among other things, income especially sufficient earnings to secure the necessary old-age income. Older respondents often overlook this compensating role of income and consequently underestimate the importance of economic resources. This also helps to explain the income satisfaction paradox. For possible social policy recommendations this means that indications of subjective income satisfaction should not be used as the sole criterion for assessing objective financial needs and income situations in old age.

If there is evidence that contemporary cohorts of older people in Germany predominantly rate their current income situation as positive, then this (still) reflects the current objective economic situation that it is the comparatively high provision level that the different old-age income systems (still) guarantee. Against the background of a growing number of precarious jobs (jobs with or without insufficient social security) the already-decided different cuts in the pension level of statutory old-age insurance and the introduction of private insurance elements into statutory old-age insurance system of Germany, the objective income situation of a growing group of older people could soon deteriorate (Bäcker et al., 2006). If QoL research then still registers high satisfaction values, it will be necessary to call to mind these connections so as to avoid reaching the wrong policy conclusions.

NOTES

1. The Fifth Report of the Federal Government on the Situation of the Elderly that focuses, among other things, on the economically defined potentials of elders and societal ageing can be regarded as symbolic for this discussion in Germany (*BMFSFJ*, 2006).
2. It is much more difficult to ascertain individual wealth holdings, as detailed and reliable data on the individual distribution of wealth is lacking. Global data, however, points to the increasing importance of wealth in old age and to the unequal distribution of the personal assets (Bäcker et al., 2006).
3. Parts 2.2 and 2.3 constitute a condensed summary of the respective remarks in Bäcker et al. (2006). In the case of the following income data discussed by us, we have to concentrate on recurring revenues from the standard old-age security systems because of a comparatively poor data basis.

However, we are well aware of the fact that we thus neglect a not insignificant part of the financial resources of older people, which has furthermore strongly risen in the past.

4. It must be taken into account, however, that the registration of the category of poor older people by the social welfare statistics is imprecise, as especially very old people still often do not claim benefits.

5. These two definitions are not identical. Thus the perceived adequacy, in this respect adequateness of income refers to the assessment of income level, which is made according to mostly one individual standard. In comparison, the personal income satisfaction is a judged conclusion that can, *inter alia*, be based on the criterion of the perceived adequacy. While income adequacy is thus as a rule only measured on the basis of a single criterion, several criteria are often incorporated into the assessment of income satisfaction.

6. Thus, for instance, the question in the Socio-Economic Panel: 'How satisfied are you with the income of your household?' The answers were given according to an 11-step rating scale from 0 (completely dissatisfied) to 10 (completely satisfied).

7. However, as a rule the different indicators of the objective material life circumstances strongly correlate with each other. Thus, a high income often goes hand in hand with a high standard of living, with higher savings or with high financial and tangible assets. Nonetheless it is true that single indicators can reflect different aspects of the material life circumstances, which in turn imply different levels in the domain-specific satisfaction.

8. The Welfare Survey and the Socio-Economic Panel are representative surveys on the quality of life, well-being, and welfare in Germany. The Welfare Survey is a sampling that has all in all been carried out seven times since 1978. The sample consists of adults above the age of 18 (approximately 3,000 interviewees in 1998). The surveys in the context of the Socio-Economic Panel have each been conducted once a year since 1984. The sample consists of adults above the age of 16 (altogether more than 12,000 interviewees per year).

9. The Age Survey is a sampling of people in middle adulthood and in old age (between the ages of 40 and 85). So far, two waves were conducted, in 1996 and 2002.

10. The Berlin Aging Study (BASE) is an interdisciplinary study about the very old age (Baltes and Mayer, 2001).

11. The European Study of Adult Well-being (ESAW) represents a European subgroup of a larger global study. Material security was used here as one of five key components of ageing well. The sample comprised people between the ages 50 and 89 from Austria, Italy, Luxembourg, the Netherlands, Sweden, and the UK.

12. The findings relate to all age groups and not exclusively to the group of older people.

13. Life satisfaction was here measured on the basis of a modified version of the Life Satisfaction Index (LSI-A) (Wood *et al.*, 1969).

14. In comparison, the level of the equivalent income did not prove to be a significant predictor of subjective well-being.

REFERENCES

Andrews, F.M. and Withey, S.B. (1976) *Social Indicators of Well-being: Americans' Perceptions of Life Quality*, New York, Plenum Press.

Bäcker, G., Bispinck, R., Hofemann, K., and Naegele, G. (2006) *Sozialpolitik und Soziale Lage*, 2nd edn, Wiesbaden, Westeutscher Verlag.

Baltes, P.B. and Mayer, K.U. (2001) (eds) *The Berlin Ageing Study: Aging from 70 to 100*, Cambridge, Cambridge University Press.

Bieber, U. and Klebula, D. (2005) 'First Results of the Study "Old Age Security in Germany – 2003" ("Alterssicherung in Deutschland – ASID") – Results Regarding the Complexity of the Monetary Situation in Old Age', *Deutsche Rentenversicherung*, 6–7/2005.

Bundesministerium für Familie, Senioren, Frauen und Jugend (BMFSFJ) (2006) *Potentiale des Alters in Wirtschaft und Gesellschaft: der 5. Altenbericht der Bundesregierung*, Berlin, Eigenverlag.

Bundesministerium für Gesundheit und Soziale Sicherung (BMGS) (2005) *Lebenslagen in Deutschland: Zweite Armuts-und Reichtumsbericht der Bundesregierung*, Bonn, Eigenverlag.

Campbell, A., Converse, P.E., and Rogers, W.L. (1976) *The Quality of American Life*, Ann Arbor, MI, Institute for Social Research.

Casey, B. and Yamada, A. (2002) *Getting Older, Getting Poorer? A Study of the Earnings, Pensions Assets and Living Arrangements of Older People in Nine Countries*, Paris, OECD.

Clemens, W. and Naegele, G. (2005) 'Lebenslagen im Alter' in A. Kruse and M. Martin (eds), *Enzyklopädie der Gerontologie. Alternsprozesse in multidisziplinärer Sicht*, Bern, Huber.

Diener, E. and Oishi, S. (2000) 'Money and Happiness: Income and Subjective Well-being Across Nations' in E. Diener and E.M. Suh (eds), *Culture and Subjective Well-being*, Cambridge, MA, MIT Press.

Diener, E. and Suh, E.M. (1997) 'Subjective Well-being and Age: An International Analysis', *Annual Review of Gerontology and Geriatrics*, 17, 304–324.

Diener, E., Scollon, Ch.N., and Lucas, R.E. (2003) 'The Evolving Concept of Subjective Well-being: the Multifaceted Nature of Happiness', *Advances in Cell Aging and Gerontology*, 15, 187–219.

Dittmann-Kohli, F., Bode, Ch., and Westerhof, G.J. (2001) *Die zweite Lebenshälfte – psychologische Perspektiven. Ergebnisse des Alterssurvey*, Stuttgart, Kohlhammer.

George, L.K. (1992) 'Economic Status and Subjective Well-being: a Review of the Literature and an Agenda for Future Research' in N.E. Cutler, D.W. Gregg, and M.P. Lawton (eds), *Aging, Money, and Life Satisfaction: Aspects of Financial Gerontology*, New York, Springer.

George, L.K., Okun, M.A., and Landerman, R. (1985) 'Age as a Moderator of the Determinants of Life Satisfaction', *Research on Aging*, 7, 209–233.

Herzog, A.R. and Rodgers, W.L. (1981) 'Age and Satisfaction: Data from Several Large Surveys', *Research on Aging*, 3, 142–165.

Lamura, G., Balducci, C., Melchiorre, M.G., Quattrini, S., Spazzafumo, L., Burholt, V., Weber, G., Ferring, D., Fortuijn, J.D., and Tallberg, J. (2003) *Ageing Well and Material Security in Europe*, European Study of Adult Well-being (ESAW), Comparative Report (unpublished paper).

Lawton, M.P. (1983) 'Environment and other Determinants of Well-being in Older People', *The Gerontologist*, 23(4), 349–357.

Michalos, A.C. (2003) 'Multiple Discrepancies Theory (MDT)' in A.C. Michalos (ed.) *Essays on the Quality of Life*, Dordrecht, The Netherlands, Kluwer.

Michalos, A.C., Hubley, A.M., Zumbo, B.D., and Hemingway, D. (2001) 'Health and other Aspects of the QoL of Older People', *Social Indicators Research*, 54, 239–274.

Mookherjee, H.N. (1992) 'Perceptions of Well-being by Metropolitan and Nonmetropolitan Populations in the United States', *Journal of Social Psychology*, 132(4), 513–524.

Mookherjee, H.N. (1997) 'Perception of Well-being among Older People in Non-metropolitan America', *Perceptual and Motor Skills*, 85, 943–946.

Moon, M. (1977) 'The Economic Welfare of the Aged and Income Security Programs' in M. Moon and E. Smolensky (eds), *Improving Measures of Economic Well-being*, New York, Academic Press.

Motel-Klingebiel, A. (2000) *Alter und Generationenvertrag im Wandel des Sozialstaates: Alterssicherung und private Generationenbeziehungen in der zweiten Lebenshälfte*, Berlin, Weißensee-Verlag.

Motel-Klingebiel, A. (2006) 'Materielle Lagen älterer Menschen – Verteilungen und Dynamiken in der zweiten Lebenshälfte' in C. Tesch-Römer, H. Engstler, and S. Wurm (eds), *Altwedern in Deutschland. Sozialer Wandel und individuelle Entwicklung in der zweiten Lebenshälfte*, Wiesbaden, Verlag für Sozialwissenschaften.

Mullis, R.J. (1992) 'Measures of Economic Well-being as Predictors of Psychological Well-being', *Social Indicators Research*, 26(2), 119–135.

Naegele, G. (1991) 'Anmerkungen zur These vom "Strukturwandel des Alters" aus sozialpolitikwissenschaftlicher Sicht', *Sozialer Fortschritt*, 5/6, 162–172.

Naegele, G. (1997) 'Lebenslagen älterer Menschen' in A. Kruse (ed.), *Psychosoziale Gerontologie, Bd. I: Grundlagen. Jahrbuch der Medizinischen Psychologie*, Göttingen, Hogrefe.

Naegele, G. (2003) 'The Economic Productivity of an Ageing Society: The German Case', 56th Annual Scientific Meeting of the Gerontological Society of America, San Diego, CA (unpublished paper).

Naegele, G. and Walker, A. (2006) 'Social protection: incomes, poverty and the reform of pension systems' in J. Bond *et al.* (eds) *Ageing in Society*, 3rd edn (forthcoming).

Noll, H.-H. and Schöb, A. (2002) 'Lebensqualität im Alter' in Deutsches Zentrum für Altersfragen (ed.), Expertisen zum 4. Altenbericht der Bundesregierung. Band 1: Konzepte, Forschungsfelder, Lebensqualität, Hannover, Vincentz.

Noll, H.-H. and Weick, S. (2004) 'Verluste an Lebensqualität im Alter vor allem immaterieller Art', *Informationsdienst Soziale Indikatoren*, 21, 7–11.

OECD (2001) *Ageing and Income. Financial Resources and Retirement in Nine OECD-Countries*, Paris, OECD.

Smeeding, T.M. (1990) 'Economic Status of the Elderly' in R.H. Binstock and L.K. George (eds), *Handbook of Aging and the Social Sciences*, San Diego, CA, Academic Press.

Smith, A.E., Sim, J., Scharf, T., and Phillipson, C. (2004) 'Determinants of QoL amongst Older People in Deprived Neighbourhoods', *Ageing and Society*, 24, 793–814.

Smith, J., Fleeson, W., Geiselmann, B., Settersten, R., and Kunzmann, U. (1996) 'Wohlbefinden im hohen Alter: Vorhersagen aufgrund objektiver Lebensbedingungen und subjektiver Bewertung' in K.U. Mayer and P.B. Baltes (eds), *Die Berliner Altersstudie*, Berlin, Akademie-Verlag.

Tesch-Römer, C. and Wurm, S. (2006) 'Veränderung des subjektiven Wohlbefindens in der zweiten Lebenshälfte' in C. Tesch-Römer, H. Engstler, and S. Wurm (eds), *Altwedern in Deutschland. Sozialer Wandel und individuelle Entwicklung in der zweiten Lebenshälfte*, Wiesbaden, Verlag für Sozialwissenschaften.

TNS Infratest (2005) Alterssicherung in Deutschland 2003 (ASID, 03). Tabellenband 3, Deutschland, München, Infratest Sozialforschung.

Weidekamp-Maicher, M. (in press), *Materielles Wohlbefinden im späten Erwachsenenalter und Alter*, Dissertation, University of Dortmund.

Walker, A. (1986) 'The Politics of Ageing in Britain' in C. Phillipson, M. Bernard, and P. Strong (eds) *Dependency and Interdependency in Old Age: Theoretical Perspectives and Policy Alternatives*, London, Croom Helm, pp.30–35.

Wood, V., Wylie, M.L., and Sheafor, B. (1969) 'An Analysis of a Short Self-report Measure of Life Satisfaction: Correlation with Rater Judgments', *Journal of Gerontology*, 24, 4, 465–469.

6. QUALITY OF LIFE IN OLD AGE, INEQUALITY,

AND WELFARE STATE REFORM

A Comparison between Norway, Germany, and England

INTRODUCTION

The ageing of societies has strong implications for social security and for social gerontological analysis, while diversity, social inequality, and social justice are traditional core themes of sociology and social policy. Both perspectives are interrelated, as demographic transitions may, on the one hand, be influencing the social status attached to age and cohort membership. On the other hand, current reforms of welfare state systems – partly in reaction to demographic shifts, partly motivated by changing ideologies and ongoing processes of globalization – may have effects on the average welfare situations of people of different ages and cohorts and the distribution of these situations within age groups and cohorts. In the current debate, distributive relations between generations as age groups as well as birth cohorts are discussed as a crucial problem of inequality and justice in modern society. The role of generational equity in this debate is becoming increasingly dominant, both in Germany and in other European societies (Schmähl, 2004). However, the meaning of this term is still unclear and is used quite arbitrarily in current debates. Popular societal discourse in particular is characterised by a muddle of disparate ideas about social justice, distributive norms, and patterns (Tesch-Roemer and Motel-Klingebiel, 2004; Clasen and von Oorschot, 2002). Beyond doubt, there are problems with distribution and equity in most modern welfare states, as demonstrated by contemporary disputes; the assumption is often found that the young could be disadvantaged by the setting up of modern welfare regimes, while the old face enormous gains from the expansion of social security systems (Bommier *et al.*, 2004; Price and Ginn, 2003). In such a zero-sum game the welfare of one age group clashes with welfare of another (e.g. Bäcker and Koch, 2003 – and featured years ago in the dispute between Preston (1984) and Easterlin (1987)). Even if this 'only' reflects simplified subjective perceptions of social change and 'real' problems, a legitimation deficit could result for the extensive redistributive role of modern welfare states. In the absence of a satisfactory reaction in key societal discourses, including scientific debates, a potential for conflict could be the result. To define the perspectives of such reforms for improving generational equity, various welfare regimes, and their redistributive components are discussed as blueprints within the national debates.

From a social gerontology perspective, however, intergenerational equity – and thus intergenerational cohesion in social security – is just one important issue among many. Another core question is the effect of such diverse redistributive systems on inequality patterns in later life. Both dimensions may have important consequences

H. Mollenkopf and A. Walker (eds.), Quality of Life in Old Age, 85–100.

for societies as a whole, since demographic developments as well as intergenerational relations, wider social networks, and well-being on an individual micro level or the societal macro level can be influenced by individual activities such as childbearing, care for older relatives, and employment activity being recognised and rewarded by the welfare state. Decisions on whether or not to adopt such activities are themselves influenced by their inequality repercussions or the (dis)advantages associated with them. Hence, reforms leading to further privatisation of social security may not only exacerbate equity problems but also boost conflicts between production and reproduction.

It is primarily the institutionalisation and the extent of societal redistribution of resources and, consequently, life chances that allow social scientists and social policy analysts to differentiate between different types of welfare systems, as for instance defined by Titmuss (1963, 1987) and Esping-Andersen (1990, 1999). It is a common characteristic of all of these systems that they basically rearrange distributive patterns between social groups at a certain point in time – employed and unemployed, healthy and sick people, men and women. But conventionally, welfare states do not explicitly redistribute between generations as birth cohorts, although provision for old age can be discussed as redistribution over the individual life course and, hence, over time, and pension schemes can contribute to intergenerational risk-sharing and diversification. Intergenerational redistribution by welfare state systems seems to be more a by-product of intra-cohort allocation of resources. While economic growth and prosperity for some time prevented a clash of both perspectives, the relationship between them became crucial during the economic and demographic crises of modern societies. Intergenerational equity as a political goal was on the agenda and came under discussion as the continuous improvement of the economic conditions for following birth-cohorts became more and more questionable. This affected both the moral and the political economy of ageing within modern welfare states. It also has implications for current welfare state reforms and the core dimension of intragenerational distributive patterns as such. Reforms based on generational equity figures, for example as implemented in Germany, mainly focus on a substantial reduction of social security levels in the contribution-and/or tax-based pay-as-you-go-pillar in favour of a basic provision and a strengthening of insurance principles within the public system of old age provision, in earlier as well as in recently established private pillars. Mid- and long-term effects on inequality relations in later life are likely and urgently need to be researched as a substantial social gerontological input to ongoing social policy debates.

THEORY AND MEASUREMENT OF QUALITY OF LIFE AND SOCIAL INEQUALITY

In order to discuss the link between welfare systems and inequality in later life empirically, it is useful to define a sphere which is essential for social inequality. Quality of life (QoL) emerges as a valuable category in this perspective, since it can be defined as a significant outcome of unequally distributed living conditions and opportunities in social environments. QoL is a multidimensional construct and includes material and non-material, objective and subjective, individual and collective aspects (Motel-Klingebiel *et al.*, 2003a). In relevant literature, two main traditions for the conceptualization and measurement of QoL can typically be observed (Noll, 1999): one is the 'level of living-approach' which comes from the Scandinavian research tradition and is based on a resource concept (Erikson, 1974). The second approach, widely employed in Anglo-American research on well-being, focuses on the subjects' interpretations and their evaluation of living situations.

Within social and behavioural ageing research the concepts 'QoL' and 'subjective well-being' are used (quite often interchangeably) to denote merely the subjective dimension of a good life as opposed to the objective living conditions or resources available to a person (Allardt, 1975; Bulmahn, 2000; Smith *et al.*, 1996, 1999). Subjective QoL can be considered as a reaction to experiences or living conditions with which a person is confronted over his or her life course (Diener, 2000; Diener and Suh, 1998; Diener *et al.*, 1999; Filipp, 2002; Noll and Schöb, 2002). Satisfaction also includes comparisons with expectations and goals, and it has been shown empirically that overall life satisfaction refers holistically to the entirety of a person's life situation, while specific evaluations refer to genuine differences in life domains (Campbell *et al.*, 1976). The distribution of QoL may reflect key patterns of social inequality and is structurally independent of possible shifts in the relevance of objective indicators. Social inequality is defined as given if access to available and desirable social assets and/or social positions is constrained – with limited and/or favoured opportunities for individuals, groups, or societies evolving from that (Kreckel, 1992). Such a definition goes beyond pure variability, as it includes normative aspects (e.g. positive or negative evaluations of differences and an agreement on a classification system) and is directly connected to classical theoretical concepts of social inequity (Atkinson, 1983; Bolte *et al.*, 1975; Sen, 1973). As mentioned above, QoL can basically be seen as a core outcome of access and restrictions (Fahey and Smyth, 2004; Motel-Klingebiel, 2004). As these may vary over the individual life course with its different institutionalized stages (Kohli, 1985), this definition is sufficiently sensitive to analyse the interaction between age and unequal distribution of assets and resources.

CONCEPTUAL FOCUS AND RESEARCH QUESTIONS

While the political economy of ageing and old age traditionally focuses on the structural characteristics that determine the patterns of allocation of social resources to older people and the potential marginalisation of the old (e.g. Estes *et al.*, 1996), the approach of this chapter is on the interaction between the intra- and inter-cohort or age group perspective. It is necessary to discuss which effects on intra-cohort distributions may derive from a stronger focus on inter-cohort differences as implemented by the 'intergenerational equity' debate. Basically, to test this empirically would call for a longitudinal perspective that analyses the results of long-term field studies. But that would require a historical analysis – for which insufficient data is available – or an experimental welfare state reform approach – which is obviously not a sound methodological manoeuvre. Instead an international comparative approach was chosen in this chapter to test the effects of different distributive arrangements on the within-cohort inequality in later life.

This chapter analyses the effects of different types of old age provision in the context of social security reform in modern welfare states – taking England, Norway, and Germany as key examples, with the German debate serving as a starting point. To discuss the effects of different types of social security systems on social inequality in later life, the distribution of QoL as a core outcome of welfare state intervention will be examined in this chapter.

Based on key assumptions on anticipated changes in paradigms of security and (re)distribution and of the anticipated relevance for inequality – where the German debate served as a conceptual basis – this chapter will examine the association between types of social security arrangements on the one hand and relative levels, variation, and social inequality in later life. The empirical analyses will therefore include the following queries: How do levels of QoL in later life depend on welfare state arrangements? Is the variation of objective and subjective QoL related to welfare state arrangements? What is the relevance of social structure indicators for this variation and how is it related to old age security systems? What can be learned for the perspectives of current debates on equity and social security reforms?

HYPOTHESES

As mentioned above, in the context of the effects of changing redistributive patterns a politico-economic approach involves a number of prospective assumptions. But a test based on longitudinal data would be hard to achieve. Instead, cross-sectional hypotheses are formulated in a welfare state comparative perspective. A comparison between Norway, England, and Germany – well-known and well-established examples of main welfare regimes (Esping-Andersen, 1990, 1999; Titmuss, 1963, 1987) – is chosen to serve as a proxy for prospective modelling (Daatland and Motel-Klingebiel, 2007). To do so, three hypotheses will be discussed and tested empirically: the hypothesis of (relative) levels, the distribution hypothesis and the social structure hypothesis:

The 'hypothesis of (relative) levels' states that levels of QoL in later life are highest in a well-developed Scandinavian social-democratic system and should be lowest in a liberal system that is highly privatised and offers relatively poor resources to older people. The 'distribution hypothesis' expects variation of QoL to be highest under a liberal welfare regime, while it should be more moderate in highly redistributive systems like the Norwegian social-democratic regime or even the German conservative-corporatist one. The 'social structure hypothesis' predicts the relevance for inequality positions to be highest in England, while redistribution by the welfare state should level out the relative impact of social strata, educational levels, or gender in Northern and Central Europe.

METHODOLOGY – DATABASE AND INDICATORS

The empirical analyses are based on the data-set of the European research project, Old Age and Autonomy: The Role of Service Systems and Intergenerational Family Solidarity (OASIS). It is a five-country study involving Norway, England, Germany, Spain, and Israel (Lowenstein and Ogg, 2003). OASIS was funded within the 5th Framework Programme, European Community Contract N°QLK6-CT-1999-02182). It focuses conceptually on analyses of QoL in old age (Tesch-Roemer et al., 2003) and on the relevance and meaning of service systems, family structures, and social support. The overall goal of OASIS is to determine the relation between private support and formal service systems and how this mix of support relates to the QoL. The project combines qualitative and quantitative approaches and, hence, allows a triangulation of research methods (Kelle and Erzberger, 1999; Kluge and Kelle, 2001). (For details on the quantitative methodology, see Motel-Klingebiel et al., 2003b; for further information on the qualitative study, see Phillips and Ray, 2003.) In the present case data from the quantitative survey will be analysed.

The survey sample was drawn as a representative, stratified sample of the urban population of age 25 and older living in private households in the participating countries. In Norway all three available urban areas were included, while in England and Germany a selection of such urban areas was made (England: selection of six major regions with 120 wards which were considered as representative for the English urban areas, Germany: random selection of 31 urban regions within 16 states). Sampling strategies in respect to participants differed in the participating countries. The goal was to optimise the sampling according to national best practice (Norway: mixture of random route and register sampling; England: use of electoral registers; Germany: random sampling based on municipality registries).

Interviews took place between September 2000 and May 2001. Analyses of sample selectivity were conducted (Motel-Klingebiel and Gilberg, 2002; Motel-Klingebiel et al., 2003b), but did not show special problems with the OASIS sample. As is known from many other gerontological studies, healthy older people have a higher opportunity to participate than frailer people (Lindenberger et al., 1996; Kühn and Porst, 1999; Künemund, 2000; Motel-Klingebiel et al., 2003; Hauser and Willis, 2005). People of age 75 and older were structurally overrepresented in the survey sample as a result of a disproportional sample design. This was employed to retain an adequate number of cases including for age-specific analyses of retired people or even the very old in each of the countries. If parameters for the entire reference population are presented, special post-stratification weights are applied to adjust for this overrepresentation. These weights are based on official population statistics for the urban population of a particular country. The analyses presented here are restricted to the comparison between Norway, England, and Germany according to traditional regime typologies and include only people in the post-retirement stage of their life course; in order to level out national differences in retirement policies, being retired was operationalised as being of age 65 and over. People younger than 65 were excluded from the analyses. Table 1 gives an overview of national and overall

TABLE 1. Sample Structure – Unweighted and Weighted Number of Cases (OASIS, 2000)

Unweighted/weighted	Norway	England	Germany	Total
65–74	100/261	226/424	154/386	480/1071
75+	413/252	398/200	499/266	1309/718
Total	513/513	624/624	652/652	1789/1789

samples. The final overall number of cases for the 65+ sample in the three-country data-set was $n = 1789$.

The OASIS questionnaire covers a variety of topics, among them sociodemographic data (including education, occupational status, and income), subjective health and functional ability, use of services, family structure and relations (including mutual support), norms and preferences, and subjective QoL (Lowenstein et al., 2002). The analyses presented in this chapter will mainly focus on subjective representations by applying the instrument WHOQoL (WHOQoL Group, 1994) in the short format of the WHOQoL scale (WHOQoL-Bref) as it is used in the OASIS study (Tesch-Römer et al., 2000). QoL is here seen as the individual's interpretation of the current living situation under the condition of their respective culture, norms, and values compared with their goals, expectations, standards, and interests.

The WHOQoL-Bref instrument basically concentrates on four major life domains that are defined to cover the entire living situation of the respondents. Domain scores are computed as individual means in the areas 'physical health', 'psychological health', 'social relationships', and 'environment' that are substantially correlated with the complete WHOQoL measures (World Health Organization, 1996; WHOQoL Group, 1998; for the use of WHOQoL-Bref in the OASIS study see Motel-Klingebiel et al., 2003a; Lowenstein et al., 2002). In addition, there are two single items on overall QoL and life satisfaction measured on a five-point Likert scale, used as a semantic differential. The WHOQoL and WHOQoL-Bref instruments are composed to be employed in international comparative research.

The analyses presented in this chapter apply the complex structure of the WHOQoL-Bref instrument (Table 2). Firstly, the analyses employ a factor 'QoL' that is statistically computed from a principal component analyses based on the four domains of the WHOQoL-Bref (Motel-Klingebiel et al., 2005; Daatland et al., 2002). Secondly, the WHOQoL-Bref single indicators on 'overall life satisfaction' ('In general, how satisfied are you with your life?'), on 'resources and economic needs' ('Have you enough money to meet your needs?'), and on 'satisfaction with financial situation' ('How satisfied are you with your financial situation?') are brought into the investigation. Thirdly, the individuals' relative income position (net equivalent income per capita computed with 'new' OECD scale) is applied. This indicator is not part of the WHOQoL. The goal is to contrast the subjective evaluations with an objective socio-economic indicator (Table 2) and offer an extended perspective for the empirical analyses that should not simply become dependent on decisions on certain perspectives of QoL indicators.

TABLE 2. Quality of Life Indicators Employed in this Study

Overall QoL (WHOQoL-Bref)
Factor solution (PCA) based on the four main domains of QoL
(1) Physical health, (2) psychological health, (3) social relations, (4) environment
– domains are equally weighted, additive indices

Overall life satisfaction (WHOQoL)
'In general, how satisfied are you with your life?'

Resources and economic needs (WHOQoL-Bref)
'Have you enough money to meet your needs?'

Satisfaction with financial situation (WHOQoL)
'How satisfied are you with your financial situation?'

Relative income position
Net equivalent income per capita (new OECD scale)

TABLE 3. Levels of Income, QoL, and Subjective Evaluations (65+) (OASIS, 2000)

65+	Germany	Norway	England	$p<$
Overall QoL (OQoL)				
Mean	0.12	0.02	−0.13	0.01
Std. dev.	0.99	0.90	1.06	
Overall life satisfaction (OLS)				
Mean	3.98	4.06	4.01	n.s.
Standard deviation	0.77	0.72	0.84	
Resources and needs (SUBNEED)				
Mean	3.74	3.70	3.12	0.01
Standard deviation	0.90	0.88	0.91	
Satisfaction with financial situation (SUBINC)				
Mean	3.83	3.91	3.63	0.01
Standard deviation	0.80	0.83	0.96	
Equivalent income (AEQINC)				
Mean in euro[*]	€1265	€1701	€802	0.01
Standard deviation	581	830	475	

[*] Exchange rates as on 1 January 2002. weighted.

The distribution of these indicators is given in Table 3. As can be seen, levels of all indicators differ significantly between societies, with the exception of the single indicator on overall life satisfaction. A more detailed discussion will be presented in the following section.

EMPIRICAL RESULTS

The subsequent analyses will be conducted to test the different hypotheses described above. Firstly, absolute and relative levels of older people's QoL as measured by the different indicators described above will be discussed. This will be done by analysing absolute values for those aged 65 and over as shown in Table 3 as well as relative

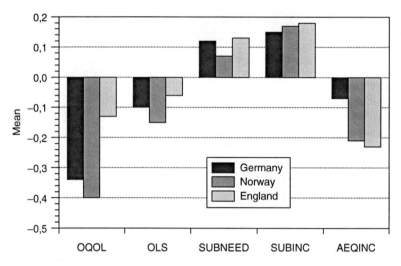

Values: Age group specific means of country specific z-scores for the overall OASIS sample.
Source: OASIS 2000, weighted.

Figure 1. Relative levels of QoL, subjective evaluations, and income (65+).
Values: age group–specific means of country-specific z-scores for the overall OASIS sample. (OASIS, 2000, weighted.)

levels indicated by age group–specific means of z-scores, that show older people's QoL relative to overall adult population of age 25 and older (Figure 1). Secondly, the variation of different dimensions of QoL in the countries will be analysed by testing intercountry differences of standard deviations, coefficients of variance, and Gini-coefficients (Figure 2). Thirdly, ε-coefficients will be studied to analyse the country-specific associations between QoL and categorical indicators of social structure such as gender, education, and social class (Figure 3 – for the definition of these indicators see Motel-Klingebiel *et al.*, 2003a). In addition, multiple OLS-regression models on QoL indicators will be estimated to define standardised net effects of particular social structure indicators on QoL in old age within the countries analysed (Table 4).

As can be seen in Table 3, levels of older people's QoL differ considerably between societies, but the cross-country differences depend to a certain extent on the indicators analysed. While overall QoL measured as a factor based on all WHOQoL-Bref domains is highest in Germany and lowest in England, there are no cross-country differences in overall life satisfaction. There are similar values for Germany and Norway in resources and economic needs as well as in satisfaction with the financial situation, with England showing significantly lower values than the other countries. Absolute income levels are obviously highest in Norway and lowest in England, with Germany somewhere in between. But comparing absolute levels of QoL may, to a degree, be a risky business, as indicators may be affected by country-specific answering behaviour, cultural norms, and diverging references for evaluation. Nevertheless, existing data on indicators of subjective well-being prove that such differences between societies are substantive and not merely an artefact (Diener and Suh, 1999).

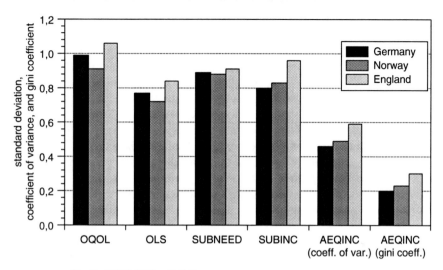

Source: OASIS 2000, weighted.

Figure 2. Distribution of income, QoL, and subjective evaluations (65+). (OASIS, 2000, weighted.)

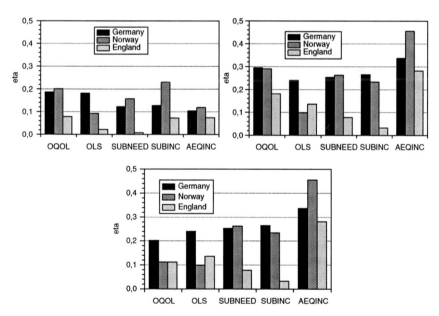

Source: OASIS 2000, weighted.

Figure 3. Income, QoL, and subjective evaluations (65+) by gender, education, and social class. (OASIS, 2000, weighted.)

However, it makes a considerable difference whether absolute values are analysed or if relative levels are examined that relate older people's QoL to levels of QoL in the countries' overall populations (Figure 1): as absolute income levels are obviously highest in Norway and lowest in England, the analysis of age specific means of z-scores shows that – in a relative perspective – the retirees still do best in Germany (even if the level is still, to some extent, lower in later life than at earlier ages) while there are only minor differences between Norway and England, where retirees have significantly lower incomes than their younger neighbours. Reverse effects can even be shown for other relative measures. While England basically shows low or intermediate levels for absolute measures for the older population, it seems to be the other way round when it comes to relative indicators. Relative to overall levels and subjective indicators, English older people seem to be better off than older Norwegians or Germans – as previously mentioned, income positions are an exception.

With respect to the level hypothesis, this means that on the one hand, we can find some evidence for lower levels of QoL in the case of England, the liberal welfare model, but only weak evidence for highest levels in Norway. On the other hand, the level hypothesis seems to be partly rejected by the finding that while in absolute terms and in a relative perspective on objective resources older people do worst in England compared to the two other countries, this is not the case for the important subjective indicators of QoL.

To test the 'distribution hypothesis' the variation of different dimensions of QoL in the three countries will be analysed. Figure 2 shows standard deviations, coefficients of variance as well as Gini-coefficients and their differences between societies. As can be seen, England shows the highest variability on all indicators – with the exception of the subjective evaluation of economic resources and needs, where no relevant difference can be proven. For overall QoL as well as overall life satisfaction, we find lowest values for older Norwegians and intermediate values in Germany. When it comes to subjective economic situations, there are no differences between both countries and with reference to the objective economic conditions measured by the income situation of older people, a somehow higher variability in Norway in comparison to Germany is evident regardless of the inequality measure (coefficients of variance or Gini-coefficients) chosen for analysis. However, the difference is only minor. Again the hypothesis is partially rejected but principally confirmed: as expected, England with its liberal welfare regime shows the highest variability of QoL among older people. The difference between Germany and Norway is not great and is dependent on the indicators chosen for the analyses. Overall subjective measures point in the expected direction as set by the distribution hypothesis, while the objective income measure unexpectedly shows lowest variability in the conservative-corporatist model.

To test the 'social structure hypothesis' on the relevance of traditional indicators of social inequality, ε-coefficients for gender, level of education and social class were computed. The results basically show that the relative impact of these social inequality indicators is lowest in England, while there is an ambivalent relationship between Norway and Germany. Surprisingly, the explanatory power of social structure indicators such as gender, education and strata does not seem to be higher under the English liberal welfare regime than in the social-democratic Scandinavian and the conservative-corporatist German system. On the contrary, the absolute explanatory power of gender, education and social class is lower. This seems to contradict the social structure hypothesis as discussed before. This needs to be controlled for before conclusions are drawn as, firstly, the indicators applied may be correlated and, secondly, the lower relative impact is based on a higher variability of QoL measures among older people in England,.

To do so, multiple ordinary least square regression models (Fahrmeir *et al.*, 1996; Gujarati, 1995) of QoL measures on these social structure indicators were estimated to analyse standardised net effects on QoL. To control for differences in the particular age structures of the three countries, chronological age was also controlled for but not displayed in the models presented in Table 4.

Analysing standardised measures leads to an ambivalent picture (Table 4). It shows high relevance of gender for subjective economic indicators in Norway (which may be connected to earlier expectations of gender equality in such societies as significant gender differences in objective dimensions like income positions cannot be observed), substantial effects of educational levels in the German and in the

TABLE 4. Gender, Education, and Social Class in Multivariate Perspective
(Multiple OLS Regression Models) (OASIS, 2000)

		Overall QoL (OQoL)	Overall life satisfaction (OLS)	Subjective needs (SUBNEED)	Subjective income (SUBINC)	Equivalent income (AEQINC)
Gender	Germany	0.090^*	0.063	0.047	0.034	0.013
	Norway	0.057	0.95	0.169^{**}	0.225^{**}	0.070
	England	0.052	0.019	0.040	−0.056	−0.039
Education	Germany	0.166^{**}	0.108^*	0.195^{**}	0.160^{**}	0.106^*
	Norway	0.202^{**}	0.022	0.124^*	0.153^{**}	0.218^{**}
	England	0.049	0.043	−0.003	−0.038	0.020
Social class	Germany	0.131^{**}	0.082	0.198^{**}	0.200^{**}	0.275^{**}
	Norway	0.141^{**}	0.122^*	0.255^{**}	0.184^{**}	0.315^{**}
	England	0.191^{**}	0.149^{**}	0.202^{**}	0.146^{**}	0.265^{**}

Standardised β-coefficients; statistically controlled for age of respondent.
As in these models independent variables are considered as metric for the reason of clarity, the effects of that simplification needed to be tested. There is no effect for gender and education, while social class performs like a metric variable in Germany and Norway. However, for most dependent variables there is a remarkable cut off in England between lower middle classes, on the one hand, and middle and higher classes, on the other hand – with strong discrepancies between lower and middle classes but only minor differences between the upper parts of the class structure.
$^*p < 0.05$; $^{**}p < 0.01$.

Norwegian case (in Germany this may result from the well-documented inequality effects of the German education system: Farkas, 2003; Hradil, 2004) and the great importance of social class in all the countries (with the strongest effects on income situations in all three countries). This again does not simply confirm the hypothesis as mentioned above. Instead a more complex picture needs to be drawn with differential effects of different welfare models on different sets of indicators and areas.

CONCLUSION

In the debates on provision for old age – in Germany as well as in numerous other modern welfare states, there is a shift in the paradigms of redistribution. Generational equity becomes predominant while intragenerational distribution recedes into the background. This shift is widely associated with a preference for increased privatisation of old age security and social security in general – an increasing relevance of insurance principles and the share of private provision even in the so-called conservative-corporatist and social-democratic regimes. Among other things, this leads to the question of the effects of such shifts on social inequality in later life. The international comparative perspective as chosen in this chapter allows an approximate testing of hypotheses in a cross-sectional perspective as a substitute for prospective or retrospective modelling.

Briefly, descriptive analyses as well as multivariate models prove that levels as well as variation of QoL among older people are significantly influenced by welfare systems and that higher inequality in privatised systems cannot be satisfactorily explained by structural indicators. Instead inequality among older people in a liberal system expresses a higher degree of structural uncertainty – with potential effects for individual life planning and the biographical perspectives of later life that need to be tested in future research on QoL and social inequality among older people.

The comparison between Norway, England, and Germany shows that in a liberal system combining public basic provision with a high incidence of private pensions, low levels of resources and absolute outcomes can be observed, while older people do not see themselves as disadvantaged compared to younger members of society. In England widespread social problems can be found, as indicated by objective and subjective undersupply and an increased diversity in all the indicators analysed, while the picture is more complex for the relationship between the Norwegian and the German examples as well as for the interconnection between traditional measures of social structure and QoL in later life.

The level hypothesis was partly supported, as England shows the lowest absolute levels but (a) older people did best in the German system but not in Norway and (b) England's older people do not describe themselves as marginalised if asked for an evaluation of their QoL – even if they are obviously deprived when it comes to objective measures such as income. The distribution hypothesis is supported, as there is highest variation and a high spread of economic problems in the liberal system with limited public old age provision and high relevance of private provision. The social structure hypothesis was not supported: gender, education, and partly strata are not more intensely linked with resources and outcomes in a more liberal system. The increased inequality that is associated with a liberalised system cannot be explained by an increase in the impact of conventional indicators of social inequality (gender, education, and strata) but needs to be analysed by more elaborate modelling.

From a social policy perspective, the conclusion can be drawn that a liberalisation of welfare systems is likely to lead to an increase in diversity and social inequality among older people – a tendency that can already be observed in the case of Germany, where rates of old age poverty increased during the last decade and specific groups of older people no longer continue to benefit from increases in household incomes. This increased inequality cannot simply be explained by a renaissance of the impact of traditional social structure indicators such as education, gender, or social class. Instead, there may be growing uncertainty regarding what can be expected in later life or a relevance of unobserved predictors such as region, industry segment or type of old age security. While absolute levels of QoL and its distribution currently seem to be effected by differences in welfare systems and in living conditions, older people seem subjectively to compensate (or even overcompensate) for changing environments and resources, as indicated by the relative levels.

REFERENCES

Allardt, E. (1975) *Dimensions of Welfare in a Comparative Scandinavian Study*, Helsinki, Research Group for Comparative Sociology, Research reports.

Atkinson, A.B. (1983) *The Economics of Inequality*, Oxford, Clarendon.

Bäcker, G. and Koch, A. (2003) 'Die Jungen als Verlierer? Alterssicherung und Generationengerechtigkeit', *WSI-Mitteilungen*, 2, 111–117.

Bolte, K.M., Kappe, D., and Neidhardt, F. (1975) *Soziale Ungleichheit*, Opladen, Leske and Budrich.

Bommier, A., Lee, R., Miller, T., and Zuber, S. (2004) *Who Wins and Who Loses? Public Transfer Accounts for US Generations Born 1850 to 2090*, NBER Working Paper No. w10969, Cambridge, MA, National Bureau of Economic Research.

Bulmahn, T. (2000) 'Zur Entwicklung der Lebensqualität im vereinten Deutschland', *Aus Politik und Zeitgeschichte* (B40), pp.30–38.

Campbell, A., Converse, P.E., and Rogers, W.L. (1976) *The Quality of American Life*, New York, Sage.

Clasen, J. and van Oorschot, W. (2002) *Changing Principles in European Security*, Presentation at the 2nd COST A15 Conference 'Welfare Reforms for the 21st Century', Oslo, April 5–6, 2002.

Daatland, S.O. and Motel-Klingebiel, A. (2007) 'Separating the local and the general in cross-cultural aging research' in H.-W. Wahl, C. Tesch-Römer, and A. Hoff (eds) *New Dynamics in Old Age: Individual, Environmental and Societal Perspectives*, Amityville, NY, Baywood, pp.343–359.

Daatland, S.O., Herlofson, K., and Motel-Klingebiel, A. (2002) 'Methoden und Perspektiven international vergleichender Alter(n)sforschung (Perspectives on comparative research on ageing in Europe)' in A. Motel-Klingebiel and U. Kelle (eds) *Perspektiven der empirischen Alter(n)ssoziologie*, Opladen, Leske and Budrich, pp.221–248.

Diener, E. (2000) 'Subjective well-being', *American Psychologist*, 55, 34–43.

Diener, E. and Suh, E.M. (1998) 'Age and subjective well-being: an international analysis' in K.W. Schaie and L.M. Powell (eds) *Annual Review of Gerontology and Geriatrics*, Vol. 17: *Focus on Emotion and Adult Development*, New York, Springer, pp.304–324.

Diener, E. and Suh, E.M. (1999) 'National Differences in Subjective Well-being' in D. Kahnemann, E. Diener, and N. Schwarz (eds) *Well-Being: The Foundations of Hedonic Psychology*, New York, Sage, pp.434–450.

Diener, E., Suh, E.M., Lucas, R.E., and Smith, H.L. (1999) 'Subjective Well-being: Three Decades of Progress', *Psychological Bulletin*, 125, 276–302.

Easterlin, R.A. (1987) 'The New Age Structure of Poverty in America: Permanent or Transient?', *Population and Development Review*, 13, 195–208.

Erikson, R. (1974) 'Welfare as a planning goal', *Acta Sociologica*, 17, 273–278.

Esping-Andersen, G. (1990) *The Three Worlds of Welfare Capitalism*, Princeton, NJ, Princeton University Press.

Esping-Andersen, G. (1999) *Social Foundations of Postindustrial Economies*, Oxford, Oxford University Press.

Estes, C.L., Linkins, K.W., and Binney, E.A. (1996) 'The Political Economy of Aging' in R.H. Binstock and L.K. George (eds) *Handbook of Aging and the Social Sciences*, San Diego, CA, Academic Press, pp.346–361.

Fahey, T. and Smyth, E. (2004) 'Do Subjective Indicators Measure Welfare?', *European Societies*, 6, 5–27.

Fahrmeir, L., Hamerle, A., and Tutz, G. (eds) (1996) *Multivariate statistische Verfahren*, Berlin, de Gruyter.

Farkas, G. (2003), 'Human Capital and the Long-Term Effects of Education on Late-Life Inequality' in S. Crystal and D. Shea (eds) *Annual Review of Gerontology and Geriatrics*, New York, Springer, pp.138–154.

Filipp, S.-H. (2002) 'Gesundheitsbezogene Lebensqualität alter und hochbetagter Frauen und Männer' in Deutsches Zentrum für Altersfragen (ed.), *Expertisen zum vierten Altenbericht der Bundesregierung, Band I, Das hohe Alter – Konzepte, Forschungsfelder, Lebensqualität*, Hannover, Vincentz, 315–414.

Gujarati, D.N. (1995) *Basic Econometrics*, New York, McGraw-Hill.

Hauser, R.M. and Willis, R.J. (2005) 'Survey Design and Methodology in the Health and Retirement Study and the Wisconsin Longitudinal Study' in L.J. Waite (ed.) *Aging, Health and Public Policy: Demographic and Economic Perspectives*, New York, Population Council, pp.209–235.

Hradil, S. (2004) *Die Sozialstruktur Deutschlands im internationalen Vergleich*, Wiesbaden, VS Verlag für Sozialwissenschaft.

Kelle, U. and Erzberger, C. (1999) 'Integration qualitativer und quantitativer Methoden. Methodologische Modelle und ihre Bedeutung für die Forschungspraxis', *Kölner Zeitschrift für Soziologie und Sozialpsychologie*, 51, 509–531.

Kluge, S. and Kelle, U. (eds) (2001) *Methodeninnovation in der Lebenslaufforschung. Integration qualitativer und quantitativer Verfahren in der Lebenslauf-und Biographieforschung*, Weinheim, Juventa.

Kohli, M. (1985) 'Die Institutionalisierung des Lebenslaufs – Historische Befunde und theoretische Argumente', *Kölner Zeitschrift für Soziologie und Sozialpsychologie*, 37, 1–29.

Kreckel, R. (1992) *Politische Soziologie der sozialen Ungleichheit*, Frankfurt/M., New York, Campus.

Kühn, K. and Porst, R. (1999) Befragung alter und sehr alter Menschen: Besonderheiten, Schwierigkeiten und methodische Konsequenzen. Ein Literaturbericht. ZUMA Arbeitsbericht 99/03, Mannheim, Zentrum für Umfragen, Methoden und Analysen.

Künemund, H. (2000) 'Datengrundlage und Methoden' in M. Kohli and H. Künemund (eds) *Die zweite Lebenshälfte – Gesellschaftliche Lage und Partizipation im Spiegel des Alters-Survey*, Opladen, Leske and Budrich, pp.33–40.

Lindenberger, U., Gilberg, R., Pötter, U., Little, T.D., and Baltes, P.B. (1996) 'Stichprobenselektivität und Generalisierbarkeit der Ergebnisse in der Berliner Altersstudie' in K.U. Mayer and P.B. Baltes (eds) *Die Berliner Altersstudie*, Berlin, Akademie Verlag, pp.85–108.

Lowenstein, A. and Ogg, J. (eds) (2003) *OASIS – Old Age and Autonomy: The Role of Service Systems and Intergenerational Family Solidarity*. Final Report (http://www.dza.de/forschung/oasis_report.pdf. Haifa: University of Haifa).

Lowenstein, A., Katz, R., Mehlhausen-Hassoen, D., and Prilutzky, D. (2002) *The Research Instruments in the OASIS Project – Old Age and Autonomy: The Role of Service Systems and Intergenerational Family Solidarity*, Haifa, The Center for Research and Study of Aging – The Faculty for Welfare and Health Studies, University of Haifa.

Motel-Klingebiel, A. (2004) 'Quality of Life and Social Inequality in Old Age' in S.O. Daatland and S. Biggs (eds) *Ageing and Diversity. Multiple Pathways in Later Life*, Bristol, Policy Press, pp.189–205.

Motel-Klingebiel, A. and Gilberg, R. (2002) 'Zielsetzungen, Perspektiven und Probleme bei Surveybefragungen mit alten Menschen' in A. Motel-Klingebiel and U. Kelle (eds), *Perspektiven der empirischen Alter(n)ssoziologie*, Opladen, Leske and Budrich, pp.133–153.

Motel-Klingebiel, A., Hoff, A., Christmann, S., and Hämel, K. (2003) *Altersstudien und Studien mit alter(n)swissenschaftlichem Analysepotential. Eine vergleichende Kurzübersicht. Diskussionspapiere*, Nr. 39, Berlin, Deutsches Zentrum für Altersfragen.

Motel-Klingebiel, A., Tesch-Römer, C., and von Kondratowitz, H.-J. (2003a) 'The Role of Family for Quality of Life in Old Age – A Comparative Perspective' in V.L. Bengtson and A. Lowenstein (eds) *Global Aging and Challenges to Families*, New York, de Gruyter, pp.323–354.

Motel-Klingebiel, A., Tesch-Römer, C., and von Kondratowitz, H.-J. (2003b) 'The Quantitative Survey' in A. Lowenstein and J. Ogg (eds) *OASIS – Old Age and Autonomy: The Role of Service Systems and Intergenerational Family Solidarity*. Final Report, Haifa, Haifa University, pp.63–101 (http://www.dza.de/forschung/oasis_report.pdf).

Motel-Klingebiel, A., Tesch-Römer, C., and von Kondratowitz, H.-J. (2005) 'Welfare States Do Not Crowd Out the Family: Evidence for Mixed Responsibility from Comparative Analyses', *Ageing and Society*, 25, 863–882.

Noll, H.-H. (1999) Konzepte der Wohlfahrtsentwicklung: Lebensqualität und "neue" Wohlfahrts-konzepte. Project "Towards a European System of Social Reporting and Welfare Measurement", Subproject "European System of Social Indicators", EuReporting Working Paper No. 3, Mannheim, ZUMA.

Noll, H.-H. and Schöb, A. (2002) 'Lebenqualität im Alter' in Deutsches Zentrum für Altersfragen (eds) *Das hohe Alter – Konzepte, Forschungsfelder, Lebensqualität. Expertisen zum Vierten Altenbericht der Bundesregierung, Band 1*, Hannover, Vincentz, pp.229–313.

Phillips, J. and Ray, M. (2003) 'The Qualitative Phase' in A. Lowenstein and J. Ogg (eds) *OASIS – Old Age and Autonomy: The Role of Service Systems and Intergenerational Family Solidarity*, Final Report, Haifa, Haifa University, pp.103–126.

Preston, S.H. (1984) 'Children and the Elderly – Divergent Paths for America's Dependents', *Demography*, 21, 435–357.

Price, D. and Ginn, J. (2003) 'Sharing the Crust. Gender, Partnership Status and Inequalities in Pension Accumulation' in S. Arber, K. Davidson, and J. Ginn (eds) *Gender and Ageing*, Maidenhead, UK, Open University Press, pp.127–147.

Schmähl, W. (2004) *"Generationengerechtigkeit" als Begründung für eine Strategie "nachhaltiger" Alterssicherung in Deutschland*, Bremen, Zentrum für Sozialpolitik.

Sen, A.K. (1973) *On Economic Inequality*, Oxford, Oxford University Press.

Smith, J., Fleeson, W., Geiselmann, B., Settersten, R., and Kunzmann, U. (1996) 'Wohlbefinden im hohen Alter: Vorhersagen aufgrund objektiver Lebensbedingungen und subjektiver Bewertung' in P.B. Baltes (ed.) *Die Berliner Altersstudie*, Berlin, Akademie Verlag, pp.497–523.

Smith, J., Fleeson, W., Geiselmann, B., Settersten, R., and Kunzmann, U. (1999) 'Sources of Well-being in Very Old Age' in K.U. Mayer and P.B. Baltes (eds) *The Berlin Aging Study Aging from 70 to 100*, A research project of the Berlin-Brandenburg Academy of Sciences, Berlin, Akademie Verlag, pp.450–471.

Tesch-Roemer, C. and Motel-Klingebiel, A. (2004) 'Gesellschaftliche Herausforderungen des demographischen Wandels' in A. Kruse and M. Martin (eds) *Enzyklopädie der Gerontologie*, Bern, Hans Huber, pp.561–575.

Tesch-Roemer, C., Motel-Klingebiel, A., and von Kondratowitz, H.-J. (2003) 'Quality of Life' in A. Lowenstein and J. Ogg (eds) *OASIS – Old Age and Autonomy: The Role of Service Systems and Intergenerational Family Solidarity*. Final Report, Haifa, Haifa University, pp.259–284 (http://www.dza.de/forschung/oasis_report.pdf).

Tesch-Römer, C., von Kondratowitz, H.-J., Motel-Klingebiel, A., and Spangler, D. (2000) *OASIS – Old Age and Autonomy: The Role of Service Systems and Intergenerational Family Solidarity. Erhebungsdesign und Instrumente des deutschen Survey. Diskussionspapiere Nr. 32*. Berlin, Deutsches Zentrum für Altersfragen.

Titmuss, R.M. (1963) *Essays on 'the Welfare State'*, London, Unwin University Books.

Titmuss, R.M. (ed.) (1987) *The Philosophy of Welfare. Selected Writings of Richard Titmuss,* London, Allen & Unwin.

WHOQoL Group (1994) 'Development of the WHOQOL: Rationales and Current Status', *International Journal of Mental Health,* 23, 24–56.

WHOQoL Group (1998) 'Development of the World Health Organization WHOQOL-Bref Quality of Life Assessment', in *Psychological Medicine,* 28, 551–558.

World Health Organization (1996) *WHOQOL-Bref – Introduction, Administration, Scoring and Generic Version of the Assessment,* Field Trial Version, December 1996, Genf, World Health Organization – Programme on Mental Health.

HANS-WERNER WAHL, HEIDRUN MOLLENKOPF, FRANK OSWALD,
AND CHRISTIANE CLAUS

7. ENVIRONMENTAL ASPECTS OF QUALITY OF LIFE IN OLD AGE

Conceptual and empirical issues

INTRODUCTION

This chapter is based on the insight that person–environment relationships are essential components of the quality of life (QoL) of older people (Lawton, 1991; Walker, 2005). In addition to the social environment, the maintenance and improvement of autonomy and well-being of ageing individuals have much to do with the utilisation and optimisation of environmental resources such as housing and local amenities. Nevertheless, the mainstream of QoL research concerned with older people seldom provides in-depth consideration of the role of the environment as people age and, if it does, it is the social environment that is given primacy most frequently.

In our own treatment of environmental aspects of QoL, in this chapter, we will concentrate – in line with what frequently has been coined environmental gerontology (Lawton, 1999; Wahl, 2001; Wahl and Gitlin, 2007) – on the physical-spatial and material part of the environment and its relationship to outcomes like autonomy, well-being, and societal participation. The physical environment refers to the totality of the diverse range of phenomena, events, and forces that exist outside the ageing individual and that are directly linked to the material and spatial sphere. Lawton (1999, p.106) has been even more rigorous in his definition of the physical environment as 'all that lies outside the skin of the participant, is animate, and may be specified by counting or by measuring in centimeters, grams, or seconds'. Furthermore, a necessary distinction within the realm of the physical environment is the one between the home environment and the out-of-home environment. The home environment includes the physical structure of all spatial components available as a potential resource or constraining force for a given inhabitant and eventual co-inhabitants in a given house or apartment. When it comes to out-of-home environments, it is difficult to provide a more clear-cut definition in addition to the very general approach offered by Lawton (1999). We will concentrate on outdoor mobility and transportation issues and their implications for QoL. The use of public areas has many direct relationships to autonomy and well-being, for example, via the accessibility of parks, shopping areas, or public buildings. Also, outdoor mobility is a major sphere that is able to support or undermine independence and well-being and thus to force or prevent disability.

Although we primarily refer to 'the physical environment' and use this term throughout the chapter, all the social, societal and cultural accoutrements, and implications of the physical environment are also acknowledged (see Wahl and Lang, 2004; for an in-depth discussion of this issue). It should also be emphasised that we will concentrate on person–physical environment interactions in the realm of

H. Mollenkopf and A. Walker (eds.), Quality of Life in Old Age, 101–122.
© 2007 *Springer.*

community dwelling older people, because differences in ageing in the private household setting and ageing in institutional environments are such that it would be too much to consider both in depth in one chapter.

DEFINING QoL DRIVEN BY A PERSON–ENVIRONMENT PERSPECTIVE

Lawton (1991, p.6) has provided the field with a definition of QoL that directly addresses the environment: 'Quality of Life is the multidimensional evaluation, by both intrapersonal and social-normative criteria, of the person–environment system of an individual in time past, current, and anticipated'. Most importantly, Lawton uses the term person–environment system as an indication that personal life and ageing are always embedded in given environmental conditions able to shape the overall QoL for better or worse. However, not all environments are equal when it comes to the QoL these environments offer. An important facet of this view is that the physical environment provides a major context and a major outcome of the QoL of older people. Another related idea is that older people should be regarded as capable of proactively selecting and shaping their physical environments in accordance with their needs and functional capacities (Lawton, 1989). This should not exclude, however, the obvious fact that there are limits to such proactivity in the case of major competence loss such as with dementia-related disorders.

Furthermore, the intrapersonal component points to the role of the internal standards of an older person, which are also critical, when it comes to the evaluation of the physical environment. On the one hand, studies on housing and neighbourhood satisfaction have shown significantly lower satisfaction scores in areas of poor quality (Mollenkopf et al., 2004a; Mollenkopf and Kaspar, 2005). On the other hand, there is often a paradox to be found, underpinning the observation that high residential satisfaction can be observed in situations of rather low objective housing standard (Oswald et al., 2003a). It thus seems of limited value to focus only on subjective or only objective criteria when it comes to the physical environment.

In addition, what Lawton (1991) has coined as social-normative criteria gains importance regarding home and out-of-home environments. For example, legally based regulations and guidelines drive community understanding and inform social policy as to what 'good environments' in the realm of housing or out-of-home environments are for ageing individuals and what kind of adaptations are needed to achieve given objective standards. Conflicts between intrapersonal and social-normative criteria are rather common in ageing people's relationships to their physical environments and thus frequently make the definition of QoL a difficult task. A classic example is barrier-free environments, using generally accepted and/or legally based criteria, which may suggest profound housing modifications in a given person–environment system with clearly expected positive QoL outcomes, such as prevention of falls and support of long-term independence. However, it is not rare for such externally identified environmental modification needs to find acceptance by the older person, because there is high residential satisfaction and no subjectively experienced need towards environmental improvement.

Going further, the temporal aspect of QoL addressed in Lawton's definition points to the dynamic and ongoing stability vs. change characteristic inherent in the transactions of ageing individuals with their physical environments. It is a fundamental feature of the person–environment system in ageing that vulnerability increases, particularly in the transition to the period of very old age. Thus, a given housing environment or out-of-home context may add significantly to the QoL of a given older person at T1 in terms of allowing for the involvement in preferred leisure activities or desired social interaction. At T2, which may happen in only weeks due to occurrences such as a stroke or a fall, the same environment may turn to a major stress factor and prevent basic independence due to barriers or underdeveloped public transportation.

Another important insight on QoL touched upon in Lawton's definition is multidimensionality. Such multidimensionality may be understood as the need to consider a full range of domains able to add to QoL such as health, wealth, housing, or social relationships. However, one also may apply the issue of multidimensionality to the additional qualification of person–physical environment systems. A variety of attributes has been suggested as cross-cutting or characteristic of any type of environment (Lawton, 1999). *Safety* of environments is among the most important prerequisites for users of physical environments. *Accessibility* of environments is of vital importance to all citizens', but particularly to old and disabled people's participation in society. *Orientation* addresses the potential of the environment to guide navigation or way finding. *Privacy* refers to the possibility provided by the environment to retreat and have a place of pronounced and undisturbed personal intimacy. *Control* of environmental conditions addresses the possibility of exerting self-directed changes and achieving personal needs or life goals (Oswald *et al.*, 2003b). Finally, *stimulation for novel experiences* may become a major QoL attribute for redesigning public spaces, park areas, or a new housing complex aimed to enhance social interaction among inhabitants, for instance, between younger and older people.

In sum, the evaluation of QoL related to person–physical environment systems presumes a convergent approach driven by a multidimensional set of criteria, which should be additionally defined based on intrapersonal as well as social-normative criteria.

QoL AND THE PHYSICAL ENVIRONMENT: MAJOR THEORETICAL APPROACHES

The in-depth understanding of person–environment processes and outcomes has so far not been achieved by one major theory; rather, a multitude of conceptual approaches, which pluralistically augment and build on each other, infuse the field.

Interaction with the Objective Physical Environment

Major theoretical efforts to better understand the relation of the ageing individual with his or her physical environment have been provided within the field of environmental gerontology or the ecology of ageing (Wahl, 2001). The Competence-Press

Model originally suggested by Lawton and Nahemow (1973; see also an update provided by Scheidt and Norris-Baker, 2004) provides a broad, overarching framework allowing different types and levels of competence such as sensory loss, physical mobility loss, or cognitive decline and environmental factors including housing standards, neighbourhood conditions, or public transport to be considered. Perhaps the most important element of the Competence-Press Model is its fundamental assumption that for each older person there is an optimal combination of remaining competence and environmental circumstances leading to the relative highest possible behavioural and emotional functioning for that person. A term frequently used for this in environmental gerontology is person–environment fit (Carp and Carp, 1984; Iwarsson, 2004). The model also suggests that it is at the lower levels of competence that older people become the most susceptible to their environment such that low competence in conjunction with high 'environmental press' negatively impacts autonomy, affect, and well-being. The competence-environmental press framework continues to provide the basic mechanism of person–environment relationships as people age and has been supported by a considerable body of empirical research (Gitlin, 2003; Lawton, 1989; Scheidt and Norris-Baker, 2004). For example, research on the impact of physical distances on social interaction patterns of older people in institutional settings shows that longer distances undermine social relationships, thus highlighting the 'environmental docility' as people age. Furthermore, the Competence-Press Model has become a major driver in the practical world of designing and optimising environments for older people.

Besides putting emphasis on competence as is done in the Competence-Press Model, other conceptions of person–environment fit as suggested by Carp and Carp (1984) and Kahana (1982) emphasise the role of motivation in person–environment relationships. The basic assumption is that misfits between given needs and given environmental options to fulfil these needs are linked with lowered behavioural functioning and well-being. Empirical support for this assumption has been reported from studies conducted in institutional settings as well as in private home environments. For example, retaining autonomy in visually impaired older people was found to strongly depend on how much the physical layout of home environments adds to the compensation of the vision loss (Wahl et al., 1999). Furthermore, a distinction between an older person's basic vs. higher-order needs has been suggested by Carp and Carp (1984). Basic needs are related to issues of maintaining competence (e.g. maintaining independent self-care), while higher-order needs address issues such as privacy and affiliation. While person–environment misfit in the basic need domain predominantly results in reduced behavioural autonomy, misfit in the realm of higher-order needs may undermine, predominantly, emotional well-being and mental health.

Role of Perceived Physical Environment

In addition to the foregoing approaches pointing mostly to the role of the objective environment, the set of processes operating when ageing individuals form cognitive and affective ties to their physical-spatial environments have also found much conceptual and empirical consideration. One widely acknowledged approach suggested by Rowles (1983) as well as Rowles and Watkins (2003) has focused on the many person-environment-related faces of what may be coined 'insideness.' According to Rowles, different types of insideness, all speaking to the transition of 'spaces into places', can be found as the result of empirical analysis. Whereas *social insideness* arises from everyday social exchange over long periods of time, *physical insideness* is characterised by familiarities and routines within given settings such as the home environment. A third element of insideness has been labeled *autobiographical insideness*, i.e. places that carry a rich collection of memories and thus support the ageing individual's sense of place identity. Empirical research has also shown that physical insideness is particularly important for ageing individuals with chronically disabling conditions.

Rubinstein's (1989) conception of the *meaning of home* relies on the assumption that the active management of the environment in itself represents a major source of well-being as people age, especially for those who are frail or living alone. Rubinstein identifies three classes of psychosocial processes that give meaning to the home environment, namely *social-centred*, defined as ordering of the home environment based on a person's version of sociocultural rules for domestic order, *person-centred*, defined as expression of one's life course in features of the home, and *body-centred* processes, defined as the ongoing relationship of the body to the environmental features that surround it.

Furthermore *place attachment* has been defined as reflecting feelings about a geographic location and emotional binding of a person to places (Oswald and Wahl, 2005). Empirical evidence shows that place attachment seems to grow steadily across the life course, reaching its culminating point in very old age. This also underlines the notion that forced relocation in very old age due to a chronically disabling condition is a critical life event and a profound challenge for maintaining well-being and a sense of identity.

Physical Environments on the Macro Level of Analysis

All the aforementioned conceptual approaches predominantly consider person–environment interactions on the micro level of analysis; i.e. the focus is on obtaining an in-depth understanding of the role of the immediate environment within which people age. However, as Bronfenbrenner's (1999) ecological theory of human development and other social science approaches related to QoL (Glatzer and Zapf, 1984) suggest, in targeting person–environment relationships, the micro environmental level is undoubtedly influenced and shaped by macro environmental level contexts including the neighbourhood at large, housing policies, economic standards, cultural values and beliefs, and social and long-term care service provision and policies. These factors all form the environmental context of ageing with varying immediate and distal impacts on everyday life of people as they age. Also, issues of equality or inequality emerge from such macro-level societal conditions, in that differences in QoL of older people are greatly influenced by uneven access to resources such as housing or transportation. From this vantage point, person–physical environment systems represent an important array of what have been coined social indicators, allowing to set a given ageing individual in a hierarchy of available environmental resources as well as to help characterise whole countries in terms of how their treatment of ageing people is expressed on the level of concrete physical environments.

An important point is also the consideration of a life course perspective. It could well be that people initially live in high quality neighbourhoods, which decline over time, placing people in the later years of their life at a disadvantage. Another classic research question is the urban–rural distinction and how this affects ageing. For example, use of the outside environment may depend upon public transportation possibilities and these may be less developed in rural than urban regions (Mollenkopf *et al.*, 2004a). On the other hand, the macro-level perspective also considers the diversity in cultural norms and it could well be that disadvantages in person–environment options in rural areas are compensated for because of stronger intergenerational bonding and higher prevalence of cohabitation of older and younger family members. At yet another level, the consideration of person–environment interrelationships is worth considering across countries, i.e. as a cross-cultural research challenge. For example, the definition of barrier-free environments and respective legislative norms and regulations likely depend on the economic situation as well as on different cultural conceptions of autonomy and the role of physical environments compared to social environments such as the family system.

EMPIRICAL ILLUSTRATION I: HOUSING, QOL, AND AGEING

To exemplify evidence on the relationship between housing and facets of QoL, we first present findings on the variety of meanings of home in the face of competence loss. Next, we focus on the interaction of both perceived and objective housing aspects, in relation to autonomy and well-being, indicating QoL in old age. Finally, we consider relocation in later life as a multimotive process to maintain independence and life quality.

Perceived Housing, Competence Loss, and QoL

The focus of our work in this regard was on the relation between perceived housing and competence loss in later life. Data were drawn from qualitative in-depth semi-structured interviews with 126 older people, aged 61–92. One-third of the participants were in good health, one-third suffered from severe mobility impairment, and one-third were blind (Oswald and Wahl, 2005). During a multiphase coding procedure, different meaning categories were established with satisfying reliability (Cohen's Kappa: 0.77–0.83). The identified categories were: (1) 'physical', focusing on the experience of housing conditions such as experience of the residential area and furnishing; (2) 'behavioural', related to the everyday behaviour of the person at home and to ways of rearranging items in the home; (3) 'cognitive', representing statements of cognition, especially biographical bonding to the home, such as the experience of familiarity and insideness; (4) 'emotional', expressing the experience of privacy, safety, pleasure, and stimulation; and (5) 'social', expressing

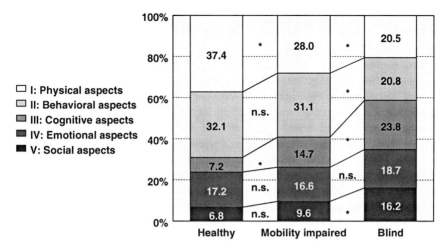

Figure 1. Relative frequencies of meaning of home domains for healthy, mobility-impaired, and blind older adults.

Note: Evaluation of verbal transcripts and tapes, based on 1,804 statements of $N = 126$ subjects. Mean number of statements per person amounted to 14.3. Multivariate MANOVA procedure for five domains was conducted to test differences between subgroups (Wilk's Lambda = 0.707; $F = 4.40$ (10, 232); *$p < 0.05$; explained variance: 29.3%). To show differences in detail, univariate simple contrasts were computed for each category between healthy and mobility-impaired, as well as betweenmobility-impaired and blind subjects. (Reproduced from Oswald, 2003, p.138)

relationships with fellow-lodgers, neighbours, or visitors. The findings are depicted in Figure 1.

Concerning group differences, healthy participants were more appreciative of the physical location, access, and amenity aspects of the home. Impaired participants emphasised the cognitive and biographical significance of the home. Concerning behavioural and social aspects, blind participants concentrated more on their social and cognitive sphere and less on behavioural and physical aspects of the home, while the meaning patterns for the mobility-impaired participants included behavioural aspects to a greater extent. About the same share of statements were made with regard to emotional themes in all three groups.

In sum, cognitive and emotional aspects of home are important 'non-observable' resources of QoL and there is variation depending on competence loss (Oswald and Wahl, 2005).

Perceived and Objective Housing – Implications For QoL

To address housing with regard to the relationship of processes of perceived and objective housing on one side and outcomes related to QoL on the other, selected findings from the European ENABLE-AGE project are presented (Iwarsson *et al.*, 2004). The overarching aim of this study was to explore the home environment as a determinant for healthy ageing in very old age in Germany, Sweden, the UK, Hungary, and Latvia. Findings are based on data drawn from the Swedish, British, and German subsamples of the survey at T1 as the target sample in each country was comparable in terms of the age-range, i.e. a total of 1,223 80- to 89-year-old adults living alone in urban districts.

Methodological issues in terms of measures and data-analytic strategy are only briefly introduced here: perceived housing is addressed by seven indicators, covering housing satisfaction (single-item self-evaluation), meaning of home (four subscale sum-scores addressing behavioural, physical, cognitive/emotional, and social aspects), housing-related control beliefs (one external control beliefs subscale sum-score; Oswald et al., 2003b) and usability in the home (two subscale sum-scores; Fänge and Iwarsson, 2003). Objective housing was assessed by an observation measure on p–e fit, indicating both the number of environmental barriers at home as well as the individual number of housing accessibility problems based on environmental barriers and functional limitations (two sum-scores; Iwarsson and Slaug, 2001). Measures on housing-related outcomes of life quality cover independence in daily activities, assessed by means of the activities of daily living (ADL) staircase sum-score (Iwarsson and Isacsson, 1997) on basic and instrumental activities of daily living. Perceived functional independence and life satisfaction were each administered with a global single-item self-evaluation rating. Psychological well-being was assessed by means of the subscale on perceived environmental mastery from the Psychological Well-being Questionnaire (Ryff, 1989), addressing sense of mastery and competence in managing the environment. Affects were assessed by means of the Positive and Negative Affect Schedule (PANAS) sum-scores (Watson et al., 1988). Depression was assessed with the short version of the Geriatric Depression Scale (GDS) sum-score (Yesavage et al., 1983). To statistically address the relationships between housing indicators and outcomes of QoL in old age, canonical correlations were assessed for the same subsample (Germany, Sweden, and UK). Indicators of objective and perceived housing were put together in one set of variables, juxtaposed to indicators of autonomy and well-being, indicating QoL (Table 1).

The findings show that both indicators of objective and perceived housing are related to outcomes of autonomy and well-being. It was not the mere number of barriers in the home environment, but housing accessibility that proved important for housing-related autonomy in terms of independence in daily life and well-being. Also several aspects of perceived usability and meaning of home correlated with autonomy and well-being. Those participants with good accessibility at home, who perceived their home as useful and meaningful (high scores in physical and behavioural meaning as well as in usability), and who thought that others or fate were only a little responsible for their housing situation (low external control beliefs), had a better autonomy in daily life/instrumental activities of daily living (ADL/IADL, perceived independence), a better sense of well-being (life satisfaction, environmental mastery), were in a better mood (positive affect) and suffered less from depressive symptoms. Although the level of accessibility, life satisfaction, or depression was different at different sites, there was a tendency towards partially comparable patterns of relationships between housing-related processes of p–e exchange and housing-related autonomy and well-being.

TABLE 1. Correlations of Aspects on Housing and Healthy Ageing (Canonical Variates) (Adapted from Oswald *et al.*, in press)

	Sweden	Germany	UK
	(n = 346)	(n = 343)	(n = 350)
Eigenvalues	1.2***	1.3***	1.6***
Canonical correlations	0.74 (73%)	0.75 (79%)	0.78 (74%)
Housing variable set			
Environmental barriers	−0.03 (0.15)	−0.08 (−0.01)	−0.09 (0.07)
Magnitude of p–e fit/accessibility problems	**−0.73 (-0.48)**	**−0.61 (−0.30)**	**−0.67 (−0.32)**
Usability in the home			
Physical environmental aspects	**0.45 (0.05)**	**0.42 (0.03)**	**0.58 (0.09)**
Activity aspects	**0.64 (0.24)**	**0.71 (0.35)**	**0.55 (0.08)**
Meaning of home			
Behavioural aspects	**0.81 (0.45)**	**0.74 (0.38)**	**0.85 (0.45)**
Physical aspects	0.17 (−0.08)	0.57 (0.17)	0.68 (0.15)
Cognitive/emotional aspects	0.34 (0.18)	0.35 (0.03)	0.45 (0.10)
Social aspects	0.30 (−0.03)	0.13 (−0.08)	0.35 (−0.09)
Housing-related external control beliefs	**−0.53 (−0.21)**	**−0.58 (−0.20)**	**−0.64 (−0.33)**
Housing satisfaction	0.05 (0.06)	0.16 (−0.03)	0.15 (−0.09)
Autonomy and well-being variable set			
Independence in daily activities (ADL)	**0.83 (0.52)**	**0.68 (0.37)**	**0.75 (0.34)**
Perceived functional independence	**0.80 (0.41)**	**0.76 (0.38)**	**0.82 (0.37)**
Life satisfaction	**0.36 (0.04)**	**0.50 (0.13)**	**0.47 (−0.01)**
Environmental mastery (Ryff)	**0.59 (0.23)**	**0.76 (0.45)**	**0.66 (0.20)**
Depression (GDS)	**−0.55 (−0.12)**	**−0.53 (0.01)**	**−0.76 (−0.42)**
Positive affect (PANAS)	0.33 (0.05)	**0.43 (0.10)**	**0.39 (0.02)**
Negative affect (PANAS)	−0.22 (−0.06)	−0.32 (0.01)	−0.28 (0.06)

Note: ***p < 0.001 Sub-samples: are reduced due to listwise deletion in canonical correlation procedures. Standardized canonical coefficients in brackets. Correlations > 0.35 are in bold type.

In sum, the data provide empirical evidence for the assumption that meaningful bonding to the home, high usability and low external housing-related control beliefs (reflecting perceived housing), as well as high levels of accessibility (reflecting objective housing or p–e fit, respectively) are linked to the achievement and maintenance of independence in daily living (objective and perceived ADL) and well-being (environmental mastery, positive affect, and low depression) in very old age. Such findings demonstrate complex links between housing and important outcomes of autonomy and well-being, indicating QoL in a group of vulnerable very old people. In detail, daily autonomy and well-being are not just related to environmental barriers at home, but to housing accessibility, usability and housing-related control beliefs, which supports the need to simultaneously refer to concepts such as the p–e fit model (Kahana, 1982) and assumptions on place attachment (Rubinstein, 1989; Rubinstein and Parmelee, 1992; Rowles, 1983; Rowles and Watkins, 2003).

Relocation and Quality of Life

The argument that different cohorts age differently underpins the role of the environment in relation to increasing numbers of new housing options and the subsequent process of relocation in later life (Oswald and Rowles, 2006). In order to address relocation not only as a stressful life event, but also from a perspective of its potentials to improve the QoL, we emphasised moving from home to home as an active and goal-directed process of p–e regulation. The main objectives of this study were to explore what kind of relocation motivations are prevalent and what objective environmental changes are associated with relocation. Data were drawn from telephone-based interviews with 217 older adults (60–89 years old) who moved from one home to another within the 3-year period immediately prior to being interviewed. Using a combined qualitative and quantitative methodology, data on reasons for moving were obtained by assessing responses to open-ended questions supplemented by more in-depth probing of each of the motivating factors identified by the participants (Oswald et al., 2002).

The study differentiated between content (e.g. person, physical, social environment) and level of need-related motivations (basic vs. higher-order needs; Carp and Carp, 1984). Basic needs reflect maintenance of personal autonomy with respect to necessary activities of daily living and competencies in everyday life. Higher-order needs reflect more development-oriented domains including privacy, comfort, stimulation or favoured personal style and preferences. Furthermore, four domains of objective change (household amenities, stimulation, availability of resources, and social network) were addressed. The major assumption was that relocation contributes to housing-related autonomy and thus, serves to maintain or even enhance QoL. Hence, addressing the motivation to move could detect the extent to which older people move in order to cope with increasing environmental press or if they proactively benefit from environmental richness (Lawton, 1989).

Evidence emerged showing that participants had multiple reasons for moving, about half of which related to the satisfaction of basic needs to maintain autonomy and half to higher-order needs. They mentioned, on average, four different reasons for moving. With respect to the content level of motives, physical environment-related motivations were the most prevalent (Figure 2). These included basic housing needs ('I found the apartment was too large to do my daily work') as well as higher-order needs ('We wanted to have a balcony and a view'). In the domain of motives pertaining to the social environment, the differentiation between basic ('My daughter can do the shopping for me now, because she lives just around the corner') and higher-order needs ('I wanted to spend more time with my grandchildren') can also be shown. We conclude that there was rarely one single reason for moving, but rather a set of needs that, in conjunction, led to relocation. Although the participants varied in health status and basic needs, most reported that higher-order needs were increasingly significant elements in terms of relocation decision-making in old age.

Another aim of the study was to analyse objective improvement in the social and physical home environment after relocation. Without going into much detail on measurement in this regard we may note that the amount of stability in most of the

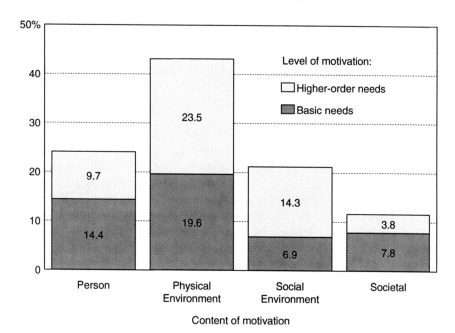

Figure 2. Reasons for moving cross-classified by level and content of motivation.
Note: The relative frequencies of varying move motives are shown in Figure 2. A total of 961 reasons for the move were coded from 217 participants, which represents a mean of 4.4 reasons for moving per person. (Reproducedfrom Oswald, 2003, p. 141)

environmental conditions was remarkably high. However, participants neverthe-less generally improved several domains of the home as a consequence of the move. Not only household amenities (e.g. barriers at home) of or proximity to family, but in many cases, stimulation (in terms of scenic landscape and lighting) was improved.

In sum, our results support the assumption that older movers do not always react passively to environmental press but appeared to have proactively optimised their environment and enhanced their environmental richness (Lawton, 1989). Findings on relocation from home to home in old age show that environmental regulation processes in later life may be triggered by basic needs to maintain autonomy (e.g. to reduce physical barriers) as well as by higher-order needs (e.g. to fulfil personal goals), which contributes to the maintenance and enhancement of QoL.

EMPIRICAL ILLUSTRATION II: OUT-OF-HOME ENVIRONMENTS, QoL, AND AGEING

Simultaneously with the dwelling place and the equipment of the home, the out-of-home environment with its various physical-spatial and social components is a substantial prerequisite for QoL in old age.

The Close Relationship Between Out-of-home Environments and QoL

As noted in previous research, the presence of service facilities for daily necessities and appropriate health care nearby, areas of recreation and stimulation, cultural facilities, natural characteristics, and feeling safe and secure in the area are all essential factors linked to autonomy and personal well-being (Golant, 1986; Krause, 1993; Zimmer and Chappell, 1997).

Also, urban–rural comparisons have repeatedly shown disadvantages for rural older persons. Facilities and leisure opportunities presented to older people as well as adequate health services are more widespread in urban areas than in the countryside (Golant, 2004; Krout, 1998). In most European cities, elderly people can reach essential facilities mainly on foot or by public transit. In rural areas, however, they often depend on a car or on other (family, formal) transportation support, because not only are shops and services usually less abundant, but public transportation is less readily accessible (for the USA see, e.g. Rosenbloom, 2004).

Regional conditions, especially those relating to the degree of urbanisation and distances between places of residence, can also clearly affect the frequency and nature (personal, technically mediated) of contacts with important reference people and participation in social activities that are regarded as crucial elements for the QoL of elderly people (Krause, 2004; Takahashi et al., 1997; Thomése and van Tilburg, 2000; Wenger, 1997). Older people in rural areas live generally less far away from their most important confidants than older urbanites, while the transportation times needed in, for example, large urban areas, increase the difficulties of maintaining social relationships (Baas et al., 2005; Moser et al., 2002).

Such structural characteristics refer to the significance of out-of-home mobility as an increasingly important precondition for maintaining an independent lifestyle and keeping up social relationships and the connection with one's community. Improvements in transport technologies, accompanied by changes in the constitution and nature of urban and rural settlements, have resulted in a growing dispersal of travel origins and destinations and the necessity of bridging the widening gap between functional areas (Transportation Research Board, 2001; World Business Council for Sustainable Development, 2002). Apart from these functional aspects of getting to places, mobility can be performed for its own sake (Mokhtarian, 2005; Mollenkopf, 2003). Moreover, in modern society, mobility is associated with highly valued societal goals such as freedom, autonomy, and flexibility (Coughlin, 2001; Handy et al., 2005; Lash and Urry, 1994; Mollenkopf et al., 2004c; Rammler, 2001). The automobile, in particular, symbolises such modern values and among the whole range of currently available transportation modes, it opens up the most independent and flexible options for moving about.

The importance of mobility has often been emphasised in research on mobility in old age (e.g. Carp, 1980, 1988; Coughlin, 2001; Marottoli et al., 1997; Owsley, 2002; Rosenbloom and Ståhl, 2002; Schaie, 2003), but has largely been neglected in satisfaction research. However, there is increasing empirical evidence that older adults' diminishing actual mobility as it was observed in many studies (usually measured as frequency of trips made) does not correspond to their needs and wishes. More than 25 years ago Carp (1980, p.140) stressed the importance of mobility in old age: 'In modern society old people, like those of other ages, must go outside their homes to provide for maintenance needs as well as for sociability and recreation'. A series of recent studies showed that the possibility to move about contributes significantly to older people's subjective QoL (Banister and Bowling, 2004; Cvitkovich and Wister, 2001; Farquhar, 1995).

Contributions of the MOBILATE Study

In our own work, several research projects enabled us to shed some light upon the complex interactions between older adults and the various environments they are living in and, as a result, to demonstrate the impact of the diverging conditions on their QoL. In a study on outdoor mobility in four European cities, older adults participating in case studies were asked what out-of-home mobility means to them. Statements like 'Joy', 'It's nourishing being with others', 'You feel free to get around as you like', or 'It's everything, it's life', demonstrate the high value they place on the ability to move about outside their homes (Mollenkopf et al., 2004b, pp.126–127). In the USA, Coughlin (2001) got similar answers with respect to the meaning of transportation, albeit mostly related to being able to drive a car.

Data collected in the European project 'MOBILATE. Enhancing Outdoor Mobility in Later Life' (Mollenkopf et al., 2005), funded by the European Commission, and conducted in 2000 in urban and rural regions of five European countries (from North to South Finland, the Netherlands, Germany, Hungary, and Italy) with diverging geographical, economic, and cultural conditions, made it possible to compare older adults' living conditions with respect to personal resources and environmental conditions prevailing in these areas. The randomly chosen sample included $N = 3,950$ community dwelling people aged 55 years or older, disproportionately stratified according to gender, age, and region. Standardised questionnaires were used to assess the essential features of the community and various aspects of life that are important for autonomy and well-being. In view of the diverging national conditions of the participant countries with respect to spatial extension, settlement structure, and population density, middle-sized cities and villages or rural areas were chosen in proportion to each country's characteristics.

We presumed that regional differences would substantially affect the QoL of people. In fact, objective social and structural factors, as well as the older people's respective subjective evaluations, differ substantially between the urban and rural areas both between and within the five countries. In correspondence with past research (cf. Golant, 2004; Krause, 2004) we can state that economic resources, on the one hand, and environmental conditions in terms of housing amenities, variety of transport options, and availability of services and cultural amenities, on the other, tend to be better in urban than in rural areas, while social conditions such as living together and proximity of important confidants as well as natural environmental conditions are more favorable in rural areas (Mollenkopf et al., 2004a). Apart from these general tendencies pointing to structural urban–rural differences prevailing in all countries, each of the areas seems to provide its own advantages and disadvantages, because the results concerning general satisfaction with the living area were mixed: in western Germany and the Netherlands, this appraisal was significantly higher in the rural than in the urban areas studied. In Hungary, it was just the opposite. In Finland, eastern Germany, and Italy, neighbourhood satisfaction was almost the same in both settlement types. Besides, within the countries studied, satisfaction with life in general did not differ very much between older adults living in urban areas and people of the same age living in the countryside. Only in the regions studied in the former socialist countries eastern Germany and, in particular, Hungary were the older study participants generally less satisfied than their contemporaries in the other European regions, and satisfaction with life was even lower among the rural older person (Table 2). This finding is also in accordance with macro-level research comparing life satisfaction between countries (Veenhoven, 1996a). With respect to emotional well-being, Italian urban older person tended to be highest and Finnish rural older person tended to be lowest in Positive Affect (PANAS; Watson et al., 1988).

Similarly, multiple regression analyses performed for each region showed that, apart from the well-known and highly significant impact of basic human conditions like objectively and subjectively assessed health and fundamental economic preconditions (e.g. Barresi et al., 1983–1984; Diener et al., 2003; Lamb, 1996; Fernández-Ballesteros et al., 2001; Veenhoven, 1996a; see also Chapter 12), no single indicator revealed to be equally important in all regions studied. The only aspect of relevance in at least half the regions (rural Finland, the West German city, and both research regions each in East Germany and Italy) was the indicator 'satisfaction with the possibility to pursue leisure time activities' (not shown in Table 2). These findings show that as different as regional conditions are, so are the components of the social, built, and natural environments that contribute to older people's subjective QoL in terms of satisfaction with life in general.

Our main research interest was then whether or not aspects of mobility contribute to the QoL of older people in such diverging regional conditions. In case the assumption of the high significance of out-of-home mobility in modern society would hold true, this should be reflected in a higher subjective QoL if they would be provided with a broad range of transport options and the possibility to pursue a great diversity of out-of-home activities in addition to basic human conditions such as health and income. As indicators for subjective QoL, we used a one-item scale assessing satisfaction with life in general (Veenhoven, 1996b; Zapf and Habich, 1996) as a measurement of the cognitive dimension of QoL, and the positive or negative affect scale (PANAS) (Watson et al., 1988) as a measurement of emotional well-being. Multiple regression analyses were carried out separately for each country. Tables 3 and 4 show the variables included and the respective findings. For a better overview, only the standardised β values of variables reaching statistical significance of at least $p < 0.05$ and the semi-partial r^2 values are documented in the tables.

TABLE 2. Comparisons of satisfaction with living area, satisfaction with life in general, and emotional well-being in urban and rural areas

	Finland		The Netherlands		Germany West		Germany East		Hungary		Italy	
	Urban	Rural	Urban	Rural	Urban	Rural	Urban	Rural	Urban	Rural	Urban	Rural
Satisfaction with living area[a] (0–10; M; SD)	8.7 (1.3) n.s.	8.6 (1.4)	7.4 (1.7) ***	8.1 (1.3)	7.7 (2.3) ***	9.0 (1.4)	8.0 (1.8) n.s.	8.1 (2.2)	7.9 (2.5) ***	7.0 (2.6) *	8.3 (2.0)	8.0 (2.1)
Satisfaction with life[a] (0–10; M; SD)	8.5 (1.2) n.s.	8.4 (1.5)	7.8 (1.4) *	8.0 (1.1)	7.4 (2.1) n.s.	7.7 (1.9)	7.3 (1.8) *	7.0 (2.3)	6.5 (2.4) *	6.0 (2.5)	7.7 (2.1) n.s.	7.4 (1.9)
Emotional well-being (Positive affect; 1–5; M; SD)	3.1 (0.7) ***	2.9 (0.7)	3.4 (0.6) n.s.	3.4 (0.5)	3.3 (0.7) ***	3.0 (0.9)	3.4 (0.6) ***	3.2 (0.7)	3.3 (0.7) n.s.	3.2 (0.6)	3.5 (0.7) ***	3.2 (0.7)

Note: MOBILATE Survey 2000.

Weighted data are given to correct for disproportional sampling with respect to age and gender.

* $p < 0.05$; ** $p < 0.01$; *** $p < 0.001$; urban–rural differences were tested using t-statistics.

[a] Self-evaluation rating on an 11-point scale (range 0–10), higher scores indicating higher satisfaction.

TABLE 3. Predictors of satisfaction with life (Adapted from Mollenkopf *et al.*, 2006)

	Finland Stand. β-weights	The Netherlands Stand. β-weights	Germany Stand. β-weights	Hungary Stand. β-weights	Italy Stand. β-weights
Region (1 = urban; 2 = rural)					0.12**
East/West (1 = East; 2 = West)					
Age			0.05*		
Gender (1 = male; 2 = female)	0.11***				
Diversity of network				0.15***	0.09*
Satisfaction with health[a]	0.20***	0.34***	0.23***	0.29***	0.29***
Satisfaction with income[a]	0.48***	0.20***	0.34***	0.28***	0.30***
Satisfaction with leisure activities[a]	0.21***	0.21***	0.23	0.15***	0.22***
Wish for more outdoor activities (0 = no; 1 = yes)					
Importance of being out[b]		0.15***	0.10***	0.09*	0.02
Options of transport modes[c]					
Options of outdoor activities[d]			0.09***		0.11*
Actual mobility[e]					
N	584	571	1476	548	599
Model r^2	0.4755	0.3815	0.4790	0.3218	0.3573

Note: MOBILATE Survey 2000; MOBILATE Diary 2000; $p < 0.05$; $p < 0.01$; $p < 0.001$

[a] Self-evaluation rating on an 11-point rating scale, higher scores indicating higher satisfaction

[b] Self-evaluation rating on an 11-point rating scale, higher scores indicating higher importance

[c] Sumscore ranging from 0 (no transport mode used, respondent is immobile) to 13 (all transport modes are used)

[d] Standardized sumscore ranging from 0 (no outdoor activity purposed) to 1 (all asked outdoor activities pursued)

[e] Mean number of trips per person and day.

The findings confirm approved knowledge but also offer new perspectives on older people's QoL. As expected and in accordance with previous studies (see Diener *et al.*, 1999 for an overview), satisfactions with one's health and income were revealed to be the most important predictors of satisfaction with life in general. These aspects contributed substantially to emotional well-being as well, albeit with clearly less impact. The negative impact of age also observed in other studies (Baltes *et al.*, 1999) turned out among the Finnish, German, and Italian older person. In Germany, age affected satisfaction with life as well.

When comparing further sociodemographic and structural factors, region and gender are not generally decisive for QoL. Only in Finland were women significantly more satisfied with life than men, and only in the Netherlands and Hungary

TABLE 4. Predictors of subjective well-being (positive affect) (Adapted from Mollenkopf *et al.*, 2006)

	Finland Standard β-weights	The Netherlands Standard β-weights	Germany Standard β-weights	Hungary Standard β-weights	Italy Standard β-weights
Region (1 = urban; 2 = rural)			-0.23^{***}	0.12^{**}	
East/West (1 = East; 2 = West)					
Age	-0.30^{***}		-0.09^{***}		-0.23^{***}
Gender (1 = male; 2 = female)		0.10^{**}		0.08^{*}	
Diversity of network			0.15^{***}	0.13^{**}	
Satisfaction with health[a]	0.13^{**}		0.11^{***}	0.13^{**}	0.11^{**}
Satisfaction with income[a]		0.09^{*}			
Satisfaction with leisure activities[a]	0.09^{*}	0.09^{*}		0.17^{***}	0.28^{***}
Wish for more outdoor activities (0 = no; 1 = yes)					0.09^{*}
Importance of being out[b]	0.09^{*}		0.09^{***}	0.18^{***}	
Options of transport modes[c]	0.13^{**}				
Options of outdoor activities[d]	0.19^{***}	0.19^{**}	0.32^{***}	0.18^{***}	0.19^{***}
Actual mobility[e]			0.13^{***}	0.12^{**}	
N	565	535	1438	491	597
Model r²	0.4101	0.1569	0.3817	0.3077	0.4184

Note: MOBILATE Survey 2000; MOBILATE Diary 2000; * $p < 0.05$; ** $p < 0.01$; *** $p < 0.001$

[a] Self-evaluation rating on an 11-point rating scale, higher scores indicating higher satisfaction

[b] Self-evaluation rating on an 11-point rating scale, higher scores indicating higher importance

[c] Sum-score ranging from 0 (no transport mode used, respondent is immobile) to 13 (all transport modes are used)

[d] Standardized sumscore ranging from 0 (no outdoor activity purposed) to 1 (all asked outdoor activities pursued)

[e] Mean number of trips per person and day

women reported slightly more positive affect than their male counterparts. However, the influence was rather small in each case. Older people living in the rural areas of Italy and Hungary reported higher life satisfaction and well-being, respectively, than their urban compatriots, but the opposite holds for Germany. In addition, living in the eastern or western Germany did not impact on the German older people's subjective QoL. A big social network outside one's household is an important prerequisite for general life satisfaction for older people in Italy and, in particular, Hungary, where it is also associated with greater emotional well-being. The latter holds true for German older people as well.

With respect to the impact of mobility and activity factors for QoL, we have to differentiate between objective and subjective predictors and between cognitive and emotional aspects. The three objective components of mobility (options of transport modes, options of outdoor activities, and actual mobility) did not contribute substantially to the older people's satisfaction with life in general. Only in Germany and Italy did a large diversity of activities play a positive role with respect to the cognitive evaluation of life quality. However, subjective mobility-related aspects, such as the importance assigned to going out (in three out of the five participating countries) and, in particular, satisfaction with the possibilities to pursue leisure time activities (in all regions studied), were significant predictors of satisfaction with life, thus supporting our notion of a comprehensive understanding of mobility.

When considering the emotional aspects of QoL, the impact of the objective indicators used for a comprehensive understanding of mobility increases. A great diversity of outdoor activity options, in particular, contributes significantly to positive affect in all countries, even more so than satisfaction with health and income. Actual mobility in terms of trips made plays a positive role in the German and Hungarian regions, and a broad range of transport options is an important additional predictor of well-being in Finland. As in the case of general life satisfaction, subjective aspects such as outdoor orientation and satisfaction with one's possibilities for leisure activities explain emotional well-being as well, in at least three of the countries.

In sum, the findings confirm that out-of-home mobility factors contribute substantially to older people's QoL. The strong impact of both objective and subjective aspects of being able to pursue activities for satisfaction with life and emotional well-being supports the hypothesis that in modern society older adults' QoL is largely affected by these mobility aspects. Interestingly, subjective factors seem to be more important for the cognitive dimension of life quality while the impact of objective conditions equalises or even exceeds subjective appraisals in the emotional dimension. Furthermore, the findings point to the necessity of considering both regional peculiarities on the macro level and cognitive and emotional aspects on the micro level in order to fully understand this complex relation.

CONCLUSION

The aim of this chapter is to underline that a person–physical environment perspective is able to substantially add to the understanding of QoL. Our aim to define QoL in terms of person–environment relations has been drawn mainly from the now classic proposal of Lawton (1991). Theoretical accounts such as the Competence-Press Model as well as the concept of person–environment fit are ongoing drivers to better understand the objective component of physical environments for QoL outcomes such as autonomy. Subjective and experiential aspects provide major additions to the consideration of objective physical environments and person–environment QoL outcomes. The extension of such individualised person–environment system approaches to the macro level of analysis, as taken in a social science perspective on ageing, is important for understanding equality and inequality in access and use of objective environmental resources as people age. A range of empirical findings was unfolded in order to support the resource character of the indoor and outdoor environments for the QoL of older people.

REFERENCES

Baas, S., Kucsera, C., Mollenkopf, H., and Széman, Z. (2005) 'Social Relations and Mobility' in H. Mollenkopf, F. Marcellini, I. Ruoppila, Z. Széman, and M. Tacken (eds) *Enhancing Mobility in Later Life. Personal Coping, Environmental Resources and Technical Support. The Out-of-Home Mobility of Older Adults in Urban and Rural Regions of Five European Countries*, Amsterdam, IOS Press, pp.195–220.

Baltes, M.M., Freund, A.M., and Horgas, A.L. (1999) 'Men and Women in the Berlin Aging Study' in P.B. Baltes and K.U. Mayer (eds) *The Berlin Aging Study*, Cambridge, Cambridge University Press, pp.259–281.

Banister, D. and Bowling, A. (2004) 'Quality of Life for the Elderly: The Transport Dimension', *Transport Policy* 11(2), 105–115.

Barresi, C.M., Ferraro, K.F., and Hobey, L.L. (1983–1984) 'Environmental Satisfaction, Sociability, and Well-being among Urban Elderly', *International Journal of Aging and Human Development*, 18(4), 277–293.

Bronfenbrenner, U. (1999) 'Environments in Developmental Perspective: Theoretical and Operational Models' in S.L. Friedman and T.D. Wachs (eds) *Measuring Environment Across the Life Span*, Washington, DC, American Psychological Association, pp.3–28.

Carp, F.M. (1980) 'Environmental Effects upon the Mobility of Older People', *Environment and Behavior*, 12(2), 139–156.

Carp, F.M. (1988) 'Significance of Mobility for the Well-being of the Elderly' in Committee for the Study on Improving Mobility and Safety for Older People, *Transportation in an Aging Society*, TRB Special Report 218, Washington, DC, National Academy Press, p.13.

Carp, F.M. and Carp, A. (1984) 'A Complementary/Congruence Model of Well-being or Mental Health for the Community Elderly' in I. Altman, M.P. Lawton, and J.F. Wohlwill (eds) *Human Behavior and Environment: Elderly People and the Environment*, Vol. 7, New York, Plenum Press, pp.279–336.

Coughlin, J. (2001) *Transportation and Older Persons: Perceptions and Preferences*, Washington, DC, AARP.

Cvitkovich, Y. and Wister, A. (2001) 'The Importance of Transportation and Prioritization of Environmental Needs to Sustain Well-being among Older Adults', *Environment and Behavior*, 33(6), 809–829.

Diener, E., Oishi, S., and Lucas, R.E. (2003) 'Personality, Culture, and Subjective Well-being: Emotional and Cognitive Evaluations of Life', *Annual Review of Psychology*, 54, 403–425.

Diener, E., Suh, E.M., Lucas, R.E., and Smith, H.L. (1999) 'Subjective Well-being: Three Decades of Progress', *Psychological Bulletin*, 125, 276–302.

Fänge, A. and Iwarsson, S. (2003) 'Accessibility and Usability in Housing. Construct Validity and Implications for Research and Practice', *Disability and Rehabilitation*, 25, 316–325.

Farquhar, M. (1995) 'Elder People's Definitions of Quality of Life', *Social Science and Medicine*, 41(10), 1439–1446.

Fernández-Ballesteros, R., Zamarrón, M.D., and Ruíz, M.A. (2001) 'The Contribution of Socio-demographic and Psychosocial Factors to Life Satisfaction', *Ageing and Society*, 21, 25–43.

Gitlin, L.N. (2003) 'M. Powell Lawton's Vision of the Environment in Aging Processes and Outcomes: A Glance Backwards to Move us Forward' in K.W. Schaie, H.-W. Wahl, H. Mollenkopf, and F. Oswald (eds) *Aging Independently. Living Arrangements and Mobility*, New York, Springer, pp.62–76.

Glatzer, W. and Zapf, W. (1984) *Lebensqualität in der Bundesrepublik. Objektive Lebensbedingungen und subjektives Wohlbefinden [Quality of life in the Federal Republic of Germany. Objective living conditions and subjective well-being]*, Frankfurt, Campus.

Golant, S.M. (1986) 'The Influence of the Experienced Residential Environment on Old People's Life Satisfaction', *Journal of Housing for the Elderly*, 3, 3–4, 23–49.

Golant, S.M. (2004) 'The Urban–Rural Distinction in Gerontology: An Update of Research' in H.-W. Wahl, R.J. Scheidt, and P.G. Windley (eds) *Aging in Context: Socio-Physical Environments. Annual Review of Gerontology and Geriatrics, 2003*, New York, Springer, pp.280–312.

Handy, S., Weston, L., and Mokhtarian, P.L. (2005) 'Driving by Choice or Necessity?', *Transportation Research A*, 39(2–3), 183–203.

Iwarsson, S. (2004) 'Assessing the Fit Between Older People and Their Physical Home Environments: An Occupational Therapy Research Perspective', in H.-W. Wahl, R.J. Scheidt, and P.G. Windley (eds) *Aging in Context: Socio-Physical Environments (Annual Review of Gerontology and Geriatrics, 2003)* (S. 85–109), New York, Springer.

Iwarsson, S. and Isacsson, Å. (1997) 'Quality of Life in the Elderly Population: An Example Exploring Interrelationships among Subjective Well-being, ADL Dependence, and Housing Accessibility', *Archives of Gerontology and Geriatrics*, 26(1), 71–83.

Iwarsson, S. and Slaug, B. (2001) *Housing Enabler. An Instrument for Assessing and Analyzing Accessibility Problems in Housing*, Lund, Sweden, Studentlitteratur.

Iwarsson, S., Wahl, H.-W., and Nygren, C. (2004) 'Challenges of Cross-National Housing Research with Older Persons: Lessons from the ENABLE-AGE Project', *European Journal of Ageing*, 1(1), 79–88. DOI: 10.1007/s10433-004-0010-5.

Kahana, E. (1982) 'A Congruence Model of Person--Environment Interaction' in M.P. Lawton, P.G. Windley, and T.O. Byerts (eds) *Aging and the Environment. Theoretical Approaches*, New York, Springer, pp.97–121.

Krause, N. (1993) 'Neighborhood Deterioration and Social Isolation in Later Life', *International Journal of Aging and Human Development*, 36(1), 9–38.

Krause, N. (2004) 'Neighborhoods, Health and Well-being in Late Life' in H.-W. Wahl, R.J. Scheidt, and P.G. Windley (eds) *Focus on Aging in Context: Socio-Physical Environments (Annual Review of Gerontology and Geriatrics*, Vol. 23), 5th edn, New York, Springer, pp.272–294.

Krout, J.A. (1998) 'Services and Service Delivery in Rural Environments' in R.T. Coward and J.A. Krout (eds) *Aging in Rural Settings. Life Circumstances and Distinctive Features*, New York, Springer.

Lamb, V.L. (1996) 'A Cross-National Study of Quality of Life Factors Associated with Patterns of Elderly Disablement', *Social Sciience & Medicine*, 42(3), 363–377.

Lash, S. and Urry, J. (1994) *Economics of Signs and Space*, London, Sage.

Lawton, M.P. (1989) 'Environmental Proactivity in Older People' in V.L. Bengtson and K.W. Schaie (eds) *The Course of Later Life*, New York, Springer, pp.15–23.

Lawton, M.P. (1991) 'A Multidimensional View of Quality of Life in Frail Elders' in J.E. Birren, J.E. Lubben, J.C. Rowe, and D.E. Deutchman (eds) *The Concept and Measurement of Quality of Life in the Frail Elderly*, San Diego, CA, Academic Press, pp.3–27.

Lawton, M.P. (1999) 'Environmental Taxonomy: Generalizations from Research with Older Adults' in S.L. Friedman and T.D. Wachs (eds) *Measuring Environment across the Life Span*, Washington, DC, American Psychological Association, pp.91–124.

Lawton, M.P. and Nahemow, L. (1973) 'Ecology and the Aging Process' in C. Eisdorfer and M.P. Lawton (eds) *The Psychology of Adult Development and Aging*, Washington, DC, American Psychological Association, pp.619–674.

Marottoli, R., Mendes de Leon, C.F., Glass, T.A., Williams, C.S., Cooney, L.M., Berkman, L., and Tinetti, M.E. (1997) 'Driving Cessation and Increased Depressive Symptoms: Prospective Evidence from the New Haven EPESE', *Journal of the American Geriatrics Society*, 45, 202–206.

Mokhtarian, P.L. (2005) 'Travel as a Desired End, not Just a Means', *Transportation Research A*, 39(2–3), 93–96.

Mollenkopf, H. (2003) 'Outdoor Mobility in Old Age' in J.R. Miller, R.M. Lerner, L.B. Schiamberg, and P.M. Anderson (eds) *The Encyclopedia of Human Ecology*, Vol. 2, New York, ABC-Clio, pp.537–542.

Mollenkopf, H., Kaspar, R., Marcellini, F., Ruoppila, I., Széman, Z., Tacken, M., and Wahl, H.-W. (2004a) 'Quality of Life in Urban and Rural Areas of Five European Countries: Similarities and Differences', *Hallym International Journal of Aging*, 6(1), 1–36.

Mollenkopf, H. and Kaspar, R. (2005) 'Ageing in Rural Areas of East and West Germany: Increasing Similarities and Remaining Differences', *European Journal of Ageing*, 2(2), 120–130. DOI: 10.1007/s10433-005-0029-2.

Mollenkopf, H., Marcellini, F., Ruoppila, I., and Tacken, M. (eds) (2004b) *Ageing and Outdoor Mobility. A European Study*, Amsterdam, IOS Press.

Mollenkopf, H., Marcellini, F., Ruoppila, I., Széman, Z., Tacken, M., and Wahl, H.-W. (2004c) 'Social and Behavioural Science Perspectives on Out-of-Home Mobility in Later Life: Findings from the

European Project MOBILATE', *European Journal of Ageing*, 1, 45–53. DOI: 10.1007/s10433-004-0004-3.

Mollenkopf, H., Marcellini, F., Ruoppila, I., Széman, Z., and Tacken, M. (eds) (2005) *Enhancing Mobility in Later Life – Personal Coping, Environmental Resources and Technical Support: The Out-of-Home Mobility of Older Adults in Urban and Rural Regions of Five European Countries*. Amsterdam, IOS Press.

Mollenkopf, H., Baas, S., Kaspar, R., Oswald, F., and Wahl, H.-W. (2006) 'Outdoor Mobility in Late Life: Persons, Environments and Society' in H.-W. Wahl, H. Brenner, H. Mollenkopf, D. Rothenbacher, and C. Rott (eds) *The Many Faces of Health, Competence and Well-Being in Old Age: Integrating Epidemiological, Psychological and Social Perspectives*, Dordrecht, The Netherlands, Springer, p.42.

Moser, G., Ratiu, E., and Fleury-Bahi, G. (2002) 'Appropriation and Interpersonal Relationships. From Dwelling to City through the Neighborhood', *Environment and Behavior*, 34(1), 122–136.

Oswald, F. (2003). 'Linking Subjective Housing Needs to Objective Living Conditions among Older Adults in Germany' in K.W. Schaie, H.-W. Wahl, H. Mollenkopf, and F. Oswald (eds) *Aging Independently: Living Arrangements and Mobility*, New York, Springer, pp.130–147.

Oswald, F. and Rowles, G.D. (2006) 'Beyond the Relocation Trauma in Old Age: New Trends in Today's Elders' Residential Decisions' in H.-W. Wahl, C. Tesch-Römer, and A. Hoff (eds) *New Dynamics in Old Age: Environmental and Societal Perspectives*, Amityville, NY, Baywood, pp.127–152.

Oswald, F., Schilling, O., Wahl, H.-W., and Gäng, K. (2002) 'Trouble in Paradise? Reasons to Relocate and Objective Environmental Changes among Well-off Older Adults', *Journal of Environmental Psychology*, 22(3), 273–288.

Oswald, F. and Wahl, H.-W. (2005) 'Dimensions of the Meaning of Home' in G.D. Rowles and H. Chaudhury (eds) *Home and Identity in Late Life: International Perspectives*, New York, Springer, pp.21–45.

Oswald, F., Wahl, H.-W., Mollenkopf, H., and Schilling, O. (2003a) 'Housing and life-satisfaction of older adults in two rural regions in Germany', *Research on Aging*, 25(2), 122–143.

Oswald, F., Wahl, H.-W., Martin, M., and Mollenkopf, H. (2003b) 'Toward Measuring Proactivity in Person–Environment Transactions in late Adulthood: The Housing-related Control Beliefs Questionnaire', *Journal of Housing for the Elderly*, 17(1/2), 135–152.

Oswald, F., Wahl, H.-W., Schilling, O., Nygren, C., Iwarsson, S., Sixsmith, A., Sixsmith, J., Széman, S., and Tomsone, S. (in press). Relationships between housing and healthy aging in very old age. *The Gerontologist*.

Owsley, C. (2002) 'Driving Mobility, Older Adults and Quality of Life', *Gerontechnology*, 1(4), 220–230.

Rammler, S. (2001) *Mobilität in der Moderne. Geschichte und Theorie der Verkehrssoziologie* [Mobility in modern society. History and theory of traffic sociology], edition sigma, Berlin.

Rosenbloom, S. (2004) 'Mobility of the Elderly: Good News and Bad News', *Transportation in an Aging Society. A Decade of Experience*, Washington, DC, Transportation Research Board (TRB), pp.3–21.

Rosenbloom, S. and Ståhl, A. (2002) 'Automobility among the Elderly; the Convergence of Environmental, Safety, Mobility and Community Design', *EJTIR*, 2(3/4), 197–213.

Rowles, G.D. (1983) 'Geographical Dimensions of Social Support in Rural Appalachia' in G.D. Rowles and R.J. Ohta (eds) *Aging and Milieu. Environmental Perspectives on Growing Old*, New York, Academic Press, pp.111–129.

Rowles, G.D. and Watkins, J.F. (2003) 'History, Habit, Heart and Hearth: On Making Spaces into Places' in K.W. Schaie, H.-W. Wahl, H. Mollenkopf, and F. Oswald (eds) *Aging Independently: Living Arrangements and Mobility*, New York, Springer, pp.77–96.

Rubinstein, R.L. (1989) 'The Home Environments of Older People: A Description of the Psychological Process Linking Person to Place', *Journal of Gerontology*, 44(2), 45–53.

Rubinstein, R.L. and Parmelee, P.A. (1992) 'Attachment to Place and Representation of Life Course by the Elderly' in I. Altman and S.M. Low (eds) *Human Behavior and Environment*, Vol. 12: *Place Attachment*, New York, Plenum Press, pp.139–163.

Ryff, C.D. (1989) 'Beyond Ponce de Leon and Life Satisfaction: New Directions in Quest of Successful Ageing', *International Journal of Behavioral Development*, 12(1), 35–55.

Schaie, K.W. (2003) 'Mobility for what' in K.W. Schaie, H.-W. Wahl, H. Mollenkopf, and F. Oswald (eds) *Aging Independently: Living Arrangements and Mobility*, New York, Springer, pp.18–27.

Scheidt, R.J. and Norris-Baker, C. (2004) 'The General Ecological Model Revisited: Evolution, Current Status, Continuing Challenges' in H.-W. Wahl, R.J. Scheidt, and P.G. Windley (eds) *Aging in Context: Socio-Physical Environments (Annual Review of Gerontology and Geriatrics 2003)*, Vol. 23, New York, Springer, pp.34–58.

Takahashi, K., Tamura, J., and Tokoro, M. (1997) 'Patterns of Social Relationships and Psychological Well-being among the Elderly', *International Journal of Behavioral Development*, 21(3), 417–430.

Thomése, F. and vanTilburg, T. (2000) 'Neighbouring Networks and Environmental Dependency. Differential Effects of Neighbourhood Characteristics on the Relative Size and Composition of Neighbouring Networks of Older Adults in the Netherlands', *Ageing and Society*, 20, 55–78.

Transportation Research Board. National Research Council (TRB) (2001) *Making Transit Work, Insight from Western Europe, Canada and the United States*. TRB Special Report 257. Washington, DC, National Academy Press.

Veenhoven, R. (1996a) 'Average Level of Satisfaction in 10 European Countries: Explanation of Differences' in W.E. Saris, R. Veenhoven, A.C. Scherpenzeel, and B. Bunting (ed.) *A Comparative Study of Satisfaction with Life in Europe*, Budapest, Eotvos University Press, pp.243–253.

Veenhoven, R. (1996b) 'Developments in Satisfaction-Research', *Social Indicators Research*, 37, 1–46.

Wahl, H.-W. (2001) 'Environmental Influences on Aging and Behavior' in J.E. Birren and K.W. Schaie (eds) *Handbook of the Psychology of Aging*, 5th edn, San Diego, CA, Academic Press, pp.215–237.

Wahl, H.-W. and Gitlin, L.N. (2007) 'Environmental Gerontology' in J.E. Birren (ed.) *Encyclopedia of Gerontology: Age, Aging, and the Aged* (2nd ed., pp.494–501) Oxford, Elsevier.

Wahl, H.-W. and Lang, F.R. (2004) 'Aging in Context Across the Adult Life: Integrating Physical and Social Research Perspectives' in H.-W. Wahl, R. Scheidt, and P.G. Windley (eds) *Aging in Context: Socio-Physical Environments (Annual Review of Gerontology and Geriatrics, 2003)*, New York, Springer, pp.1–33.

Wahl, H.-W., Oswald, F., and Zimprich, D. (1999) 'Everyday Competence in Visually Impaired Older Adults: A case for Person–Environment Perspectives', *The Gerontologist*, 39, 140–149.

Walker, A. (2005) 'A European Perspective on Quality of Life in Old Age', *European Journal of Ageing*, 2(1), 2–12.

Watson, D., Clark, L.A., and Tellegen, A. (1988) 'Development and Validation of Brief Measures of Positive and Negative Affect: The PANAS Scales', *Journal of Personality and Social Psychology*, 54(6), 1063–1070.

Wenger, G.C. (1997) 'Social Networks and the Prediction of Elderly People at Risk', *Aging and Mental Health*, 1(4), 311–320.

World Business Council for Sustainable Development (WBCSD) (2002) *Mobility 2001. World Mobility at the End of the Twentieth Century and Its Sustainability* (www.wbcsdmobility.org), 2001.

Yesavage, J.A., Brink, T.L., Rose, T.L., Lum, O., Huang, V., Adey, M., and Leirer, V.O. (1983) 'Development and Validation of a Geriatric Depression Screening Scale: A Preliminary Report', *Journal of Psychiatric Research*, 17(1), 37–49.

Zapf, W. and Habich, R. (eds) (1996) *Wohlfahrtsentwicklung im vereinten Deutschland. Sozialstruktur, sozialer Wandel und Lebensqualität (Welfare development in unified Germany. Social structure, social change and quality of life)*. Berlin, Edition Sigma.

Zimmer, Z. and Chappell, N. (1997) 'Rural–Urban Differences in Seniors' Neighbourhood Preferences', *Journal of Housing for the Elderly*, 12(1/2), 105–124.

FERMINA ROJO-PÉREZ, GLORIA FERNÁNDEZ-MAYORALAS,
VICENTE RODRÍGUEZ-RODRÍGUEZ AND JOSÉ-MANUEL ROJO-ABUÍN

8. THE ENVIRONMENTS OF AGEING IN THE CONTEXT OF THE GLOBAL QUALITY OF LIFE AMONG OLDER PEOPLE LIVING IN FAMILY HOUSING

INTRODUCTION

Quality of Life and Older People: A Holistic Approach

Quality of life (QoL) has been defined in numerous ways from different disciplinary viewpoints and analysis objectives. Recent scientific work has again underscored the idea of the complexity of the concept, the multiple approaches to its study, definitions, and methods of measurement (Gilhooly *et al.*, 2005; Martínez, 2006). One of the most widely considered definitions is the one established by the WHOQOL Group (1995), which emphasises both the subjective–objective dichotomous approach and the variety of domains to be considered.

Nor does there seem to be a consensus about the domains that influence older people's living conditions. The element considered most universally has been people's health, health status, and functional ability (Fernández-Mayoralas and Rojo, 2005). Yet ageing is also related to a deterioration in other aspects of living conditions (a smaller income after retirement, a smaller network of relationships, inadequate residential environment, etc.) and a broad variety of care needs, so more recent research has analysed a wide range of explanatory factors in seeking to define the QoL of older people (Walker, 2005b; Walker and Hagan Hennessy, 2004), although more often from a domain-specific viewpoint.

In order to go beyond this unitary approach to QoL domains, recent research is contributing to an analysis of the factors that influence individual QoL at old age from a broader approach and for different contexts within Europe (Walker, 2005a), as well as from a global or holistic approach (Fernández-Mayoralas *et al.*, 2005). All the aspects that form part of individual QoL are important to the living conditions of older people, and both the list of domains and the importance or preference of each one of them has been the subject of different research projects, a structured synthesis of which may be seen in Brown *et al.* (2004). According to these authors, there is no single list of dimensions, nor does their preference in a global QoL model in old age follow an established order, for many reasons, such as the research goals sought in each case, the age and level of competence of the demographic group analysed, the instruments used, the cultural tradition as well as the differences between the geographical places studied.

A QoL evaluation from an individual viewpoint is an issue that is being researched among the older population of Madrid through a semi-structured questionnaire including the Schedule for the Evaluation of Individual Quality of Life – Direct Weighting (SEIQoL-DW) (Browne *et al.*, 1994), to establish the important domains

123

H. Mollenkopf and A. Walker (eds.), Quality of Life in Old Age, 123–150.
© 2007 *Springer.*

of the QoL of the individual and its relative weight, as well as a long list of questions to closely examine the explanatory factors of individual QoL.[1]

Applying the SEIQoL instrument demonstrates that health, family network, economic situation, social network, and free time or leisure activities are the five dimensions mentioned most often by older people (Fernández-Mayoralas et al., 2005). The residential environment, with its housing, neighbourhood, and neighbour components, ranks last of all, even though the global average QoL score in these domains is high, especially the neighbour network and the neighbourhood.

In this context, this study aims to find out about older people's living space and how they rate the conditions of their residential environment in relation to their needs and expectations and establish which elements or predictors weigh most in their satisfaction with their residential environment, in order to further ascertain the impact that this dimension of QoL has on the population as it ages.

Ageing at Home: The Residential Environment as a Domain of Quality in Later Life

Residential environment studies play an important role in gerontology (Wahl and Weisman, 2003; Chapter 7), but the term 'ageing at home' is relatively new to gerontological nomenclature (Callahan, 1993). For Pastalan (1990) it means not having to move from one's habitual residence and delaying admittance to an institution for as long as possible. Houben (2001) analyses the innovations carried out in the field of housing and care for older people and conducts an international comparative study describing how European Union (EU) countries are tackling the coordination of housing and social policies on ageing.

In Spain, according to the 2001 Census, 97.7% of the population aged 65 and over live in their own house, as opposed to living with others, either in homes for older people or other types of institutional accommodation. Also ageing at home is the strategy chosen by 82% of the population aged 60 and over in Madrid, as opposed to the rest who had moved (11%) or would move (6%) from their home to another relative's home elsewhere, while a minimum proportion of older people would move into a residential home (Abellán and Rojo, 1997).

This is backed up by the fact that, for some time, public policies have been advocating having older people stay in their habitual residence, at home, in their environment, in order to 'ensure that elderly people's homes are as accessible and habitable as necessary, and equipped with enough services to allow them to lead their daily lives' (INSERSO, 1993). The aim is to 'avoid and delay, as long as possible, institutionalization ... the last refuge when people's other self-sufficiency and defence mechanisms have failed, either abruptly or naturally' (Comunidad de Madrid, 1998). Through schemes designed to make it easier for older people to continue living in their homes, by adapting housing and the environment to their needs, the Spanish 2003–2007 Action Plan seeks to 'foster elderly people's autonomy, through integrated accessibility plans and the use of technical aids and new technologies' (IMSERSO, 2003), in line with the enhancement in technologies for home care (Wiles, 2005).

The low level of residential mobility (Puga, 2004; Rojo et al., 2002) and sense of attachment to the place of residence are related to the home and environment as places where older people spend most time and to which they feel most closely attached. And this could be one of the reasons for the high mean value of the overall SEIQoL index score with which the people of Madrid rate their neighbours, neighbourhood, and home (82%, 77%, and 72%, respectively, in the set of important areas of their QoL), even though they are the facets least nominated spontaneously by older people among the 19 recorded. Regardless of the order in which they are mentioned, the residential environment is only cited by 12% of older people, as against nearly 38% who mention leisure activities, 57% social network, 76% financial situation, 82% family network, and 96% health; these five areas or dimensions of QoL attain an average SEIQoL index score of 71%, 72%, 70%, 71%, and 71% respectively, as opposed to an overall average index of 71%. These results emphasise the importance of the residential environment in older people's lives and the high level of satisfaction with each one of the elements of the habitat where they live. However, it is a well-known fact that older people live in old, poorly equipped homes, and, *a priori*, one would expect them to rate their residential environment lower, even though their homes tend to be in consolidated urban spaces (Rodríguez *et al.*, 2005). These are further reasons for the need to study and analyse this dimension of QoL in greater depth.

In this research, the conceptualisation of the residential environment is broad, yet converges with the physical, social, and psychological facets, according to Peace *et al.* (2006). For instance, the physical aspects considered include the built environment, the home and the building, its characteristics and amenities. The social environment would affect the outdoor space where older people enter into contact with the rest of population, and the psychological environment is expressed through their experience of the physical and social environment, and what it means to them.

The approach taken in this work follows other studies (Rojo *et al.*, 2005) which analyse the residential environment of Spanish older people who live in family housing, differentiating between private space (indoor space or to the home and the building or property where it is located) and public space (the broader, surrounding outer environment, formed by the neighbourhood or community where one lives), under the premise of the importance of the nearest private ambit, in evaluating satisfaction with the residential environment.

When evaluating all these aspects of the residential context of ageing, it is essential to bring into play objective and subjective indicators that measure what can be objectified consistently for all its users and what is exclusive to each person's perception. Both the objective aspects of the home and the environment, and the subjective ones are filtered by the users' demographic, social, and cultural factors. Spaces are not neutral, they do not mean the same for everyone. Even objective aspects can be modified by subjective assessments (Pacione, 2003). Age, gender, and size and structure of the household, could be regarded as decisive factors in how people, specially older citizens, rate their homes and the urban environment (Rojo *et al.*, 2005), and the level of income, or its expression through social class, stands out as a factor of greater impact on how people rate their home and the environment, especially in the population with fewer economic resources (Bratt, 2002; Fernández-Mayoralas *et al.*, 2004).

MATERIAL AND METHODS

Data Collection

The data come from face-to-face interviews in the project QoL of the Elderly Living in Family Housing in the Region of Madrid (CadeViMa), conducted in 2005, with a sample of 499 (4% of permitted error, $p \leq 0.05$ for a confidence level of 95%) interviewees aged 65 and over, representing a total of 959,993 people (Instituto de Estadística, 2004). The survey was designed as a semi-structured questionnaire containing sociodemographic characteristics and objective and subjective questions about the following QoL dimensions: health; household and living arrangements; family and social relationships; formal and informal support; leisure activities; residential environment; emotional well-being; values and beliefs; and economic resources. Moreover, an instrument of individual QoL, SEIQoL-DW, was applied to explore the dimensions of QoL and their weights.

The residential features, their quality and the resident's perceptions have been measured from multiple attributes. The items selected include housing and neighbourhood characteristics, older people's evaluations of them as well as their perceptions of neighbourhood problems. Data on household and living arrangements; economic resources; residential environment according to housing, neighbourhood, and neighbours; sociodemographic features; and size of the residential area have been used to explore the residential satisfaction as an attribute of older people's QoL. All these variables have been processed in an ordinal scale in numerical order.

With regard to the sociodemographic variables, a social classification was generated based on the level of education and socio-economic status, by applying Correspondence Analysis (CorA) to establish an association among them and define social class groups taking into account the retained dimensions and contribution of each variable in each dimension. The self-values exceed 1 and the discrimination values are high in both dimensions (82% for first dimension and 65% for the second). Seventy-eight per cent of the variance of the first dimension is explained and rather less, 70%, of the second one, and the two-dimensional solution jointly solves a high variance percentage, 74%.

The standardised scores (mean 0 and variance 1) of each subject, in each dimension of the correspondence table, constitute the input for sorting the subjects and grouping those with homogeneous profiles through Cluster Analysis under the non-hierarchical clustering procedure; the results display five social class groups (Table 1): high social, upper middle, middle, lower middle, and lower (Fernández-Mayoralas et al., forthcoming).

Exploring Factors and Relationships

The space where older people normally spend their lives is their home, neighbourhood, and the network of neighbourly relationships, and is obviously closely related to ageing-in-place. The built environment offers a series of amenities (buildings, activities, social services, etc.) that the residents use as and when they need them and that are analysed as expression of residential satisfaction.

TABLE 1. Social Class: Final Cluster Centres

| | Cluster | | | | | |
Dimension	1: Lower	2: Lower middle	3: Middle	4: Upper middle	5: Upper	
Dimension 1	−0.45	−0.32	0.19	2.04	4.21	
Dimension 2	−0.99	0.07	1.78	0.07	−1.29	
Number of cases in each cluster						
Total 139		236	83	22	19	499
% 27.86		47.29	16.63	4.41	3.81	100.00

Taking into account the multidimensionality, of residential satisfaction, a Factor Analysis (FA), under the principal component method, is applied to explore the relationship among the variables and to condense it into a smaller set of components or factors. Results show that the data are adequate to run FA: the determinant of the correlation matrix is low, chi-square value is high and it is associated with a significance level of $p \leq 0.0001$, and values of the sampling adequacy Kaiser–Meyer–Olkin (KMO) measure > 0.50 (Table 2). The communalities, or the total amount of variance each variable shares with all other variables, are high.

With regard to sociodemographic and household characteristics, four factors accounted for 73% of the variance: social class, perceived economic resources, ageing and household structure (the smaller the size of the household, the higher the age of their members), and living independently.

For home and building or state, the analysis of 13 selected variables resulted in a factor solution of seven principal components that accounted for 69% of the cumulative variance: availability of space in the house, ageing of the house vs. quality of house, maintenance work needed in the house vs. necessary and comfort amenities in the building, basic amenities in the house, barriers inside the house (the greater the perception of mobility-hindering barriers inside the home, the less likely they are to have an independent bedroom, or a bedroom shared with somebody else other than the spouse), home ownership, and basic amenities in the building.

With regard to district and neighbourhood: seven components condense the information of 23 variables explaining 64.6% of the total variance (distance to neighbourhood services or community level,[2] distance to municipality services, distance to cultural and sports centres in the district vs. number of perceived problems in the residential area, agreement with positive image of the neighbours, disagreement with negative descriptions of neighbours' behaviour, relatives living in the neighbourhood and length of stay).

Dependent Variable: Residential Satisfaction

Residential satisfaction is expressed by evaluating the conditions of the physical habitat lived-in, based on the expectations and achievements of the individuals with regard to that habitat. Therefore the quality of the living space is evaluated from the

TABLE 2. Principal Components of the Explanatory Variables

Variables	Communalities	Factor loadings	Principal components
(1) Sociodemographic and household characteristics			
Total variance explained: 72.9			
KMO measure of sampling adequacy: 0.584; Bartlett's test of sphericity: chi-square 463.1; sig.: 0.000			
Educational level	0.875	0.923	Social class
Social class of the interviewee	0.851	0.910	
Social class of the household head (excluding interviewee) (1)	0.280	0.364	
Feelings of safety in relation with the own economic future	0.820	0.898	Perceived economic resources
Satisfaction with personal economic situation	0.822	0.893	
Self-perceived socio-economic status of the household	0.501	0.516	
Mean age of the household members	0.886	0.914	Ageing and household structure
Age of the interviewee	0.790	0.740	
Household size (number of members)	0.739	-0.682	
Place of living (own house, with others in their house)	0.731	-0.837	Living independently
(2) Home and building or estate			
Total variance explained: 68.9			
KMO measure of sampling adequacy: 0.607; Bartlett's test of sphericity: chi-square 492.4; sig.: 0.000			
Home crowded index (surface/household size)	0.686	0.810	Availability of space in the house
Home surface (m²)	0.724	0.782	
Ageing of the house	0.602	-0.760	Quality of the house vs ageing of the house
Number of necessary amenities in the home	0.679	0.679	Maintenance work in the house vs. necessary and comfort amenities in the building
Number of comfort amenities in the home	0.509	0.519	
Maintenance works in the home, housing adaptations	0.646	0.749	
Number of necessary amenities in the building	0.505	-0.583	
Number of comfort amenities in the building	0.647	-0.500	
Number of basic amenities in the house	0.734	0.825	Basic amenities in the house
Barriers inside the house	0.798	0.831	Barriers inside the house
Having independent bedroom	0.628	-0.514	
Home tenure patterns	0.875	0.926	Home ownership
Number of basic amenities in the building	0.937	0.960	Basic amenities in the building

(3) District and neighbourhood
Total variance explained: 64.6
KMO measure of sampling adequacy: 0.785; Bartlett's test of sphericity: chi-square 2890.7; sig.: 0.000

Item			
Grocer's shops	0.712	0.825	Distance to neighbourhood or community services
Bars, cafeterias, restaurants	0.591	0.719	
Educational centres: primary schools, secondary schools	0.545	0.698	
Means of transport: bus, underground, taxi	0.577	0.688	
Gardens, green parks	0.516	0.658	
Primary health care centre	0.447	0.638	
Parish church, other religious services	0.485	0.589	
Centre of social services	0.814	0.892	Distance to municipality services
Cultural centres	0.680	0.800	
Elderly care – day centres, elderly social centres	0.675	0.727	
Cultural amenities: theaters, cinemas, exhibition centres	0.697	0.713	Distance to cultural and sports centres of district level
Number of problems perceived in the neighbourhood of residence	0.577	-0.704	
Sports centres	0.590	0.597	
Health centre of medical specialties	0.572	0.526	
Generally speaking, the neighbours are polite and friendly people	0.746	0.833	Agreement with positive image of the neighbours
The neighbours are people who you can ask for help in case of need	0.543	0.723	
The neighbours are people like oneself or who share similar interests	0.687	0.723	Disagreement with negative descriptions of neighbours' behavior
The neighbours do not meddle in other people's private lives	0.853	0.918	
The neighbours are not troublesome, noisy or dirty people	0.831	0.861	
Relatives (siblings, grandchildren, other relatives) living in the same neighbourhood or in the same village	0.778	0.874	Relatives living nearby
Relatives (children, parents, parents-in-law) living in the same neighbourhood or in the same village	0.775	0.872	
Length of time living in the neighbourhood	0.629	0.733	Length of stay in the same neighbourhood
Neighbour relations	0.549	0.701	

Extraction method: principal component analysis. Rotation method: Varimax with Kaiser normalization.
(1) 136 cases of household head; in factor analysis process cases with missing values have been substituted by mean value (meansub).

perspective of the contents offered by the habitat (seen as services or amenities) and how its residents use them in terms of their needs. When a balance exists between individual needs and environmental contents residential satisfaction will be high and vice versa.

Thus overall residential satisfaction has been considered as the dependent variable. Three items in the questionnaire ask for the level of satisfaction with each aspect of the residential environment (house of residence, neighbourhood, and neighbours). Level of satisfaction was measured by a high to low five-point scale: very satisfied, quite satisfied, fairly satisfied, not very satisfied, and not at all satisfied.

Applying the FA to these variables (Table 3) reveals the existence of a factor (60% of the total cumulative variance) relative to satisfaction with the residential environment, with a positive correlation of the variables of partial satisfaction with the home, the residential area, or neighbourhood and relationships with neighbours.

Analysis

The description of the characteristics of the environment of older people in Madrid and the expression of their residential satisfaction are based on analysis of contingency tables and mean tests, with probability parameters. Multiple Linear Regression Analysis (MLRA) is used to achieve another of the goals of this study, that is to say, the identification of the overall residential satisfaction predictors, taking into account the elements of the environment in which older people live, and the importance of each explanatory factor. MLRA is used to study relationships between a series of predictors or explanatory variables and a dependent variable or criterion. The extent to which the dependent variable is explained in the regression model is measured by the coefficient of determination (R^2).

The method of including variables in the regression model is the stepwise selection, with a criteria of probability of F-to-enter ≤ 0.05 and F-to-remove ≥ 0.1. The explanatory variables are the 18 factors obtained from the factor analysis (Table 2) of the sociodemographic and household features, characteristics and amenities of the home and the community, and neighbourly relationship network.

TABLE 3. Principal Components Analysis of the Partial Residential Satisfaction Variables

Variables	Communalities	Factor loadings	Principal components
Total variance explained: 59.5			
KMO measure of sampling adequacy: 0.638; Bartlett's test of sphericity: chi-square 208.5; sig.: 0.000			
Satisfaction with neighbour relations	0.667	0.817	Residential satisfaction
Satisfaction with the neighbourhood	0.621	0.788	
Satisfaction with house of residence	0.499	0.706	

Extraction method: principal component analysis. Rotation method: Varimax with Kaiser normalization.

The dependent variable is the main factor or component with which the home, neighbourhood, and neighbour satisfaction variables are correlated (Table 3). Statistical association between global residential satisfaction and variables as gender, marital status, and size of the residential area has been observed by means of the ANOVA statistical technique.

RESULTS OF THE ANALYSES

Sociodemographic Characterisation of Older People: The Feminisation of Ageing

As a representative sample of the population aged 65 and over living in the Madrid region, the most outstanding demographic characteristics of the interviewed population are the proportion of those older than 75, the majority presence of women and the significant proportion of widows and widowers, most of them being interviewees either married or living with a partner. A few of these characteristics increase with age and gender: widowhood is significantly more common among women over 74 and, in contrast, men predominate among the younger elderly, being married or living with a partner. The average age is quite high (75), and 1 year higher among women. This age, gender, and civil status structure is very similar to that observed for the Spanish older population (Rojo et al., 2005). In Madrid, this group seems to be people with a relatively low level of education albeit higher than the national average for the same age. For instance, nearly one-third completed secondary schooling, but four out of every ten only finished primary school and 2.6% are illiterate. There are no differences observed by gender, but there are in terms of age, with the younger elderly having the highest levels of education.

One aspect where the gender-based differences are most significant relates to employment. More than half of the older people are retired and a quarter have a widow's or disability or non-subsidised pension, while only 18% do or help out with household tasks and less than 2% are still working. Most of the men receive an employment-based pension, while most of the women receive a widow's pension, which tends to be smaller, or else they are in charge of the household chores. According to the social class levels, as explained above, less than 10% of older people would be classified in the upper-middle or upper class. The gender- and age-based differences are significant: more than one-third of men fall into the middle–middle and upper classes, whereas only 18% of women do; furthermore, an inverse relationship is observed between social class and age. As tends to occur when information is gathered via a questionnaire, one-third of the interviewees did not give any information about their monthly level of income. Of those who did answer, nearly 8 out of every 10 said that their monthly income (including their pension and other items) ranges from € 300 to € 900, 4% receive less than € 300, and 17% receive € 900 or more. The significance of these values can be appreciated if one takes into account the average retirement pension, which in Spain amounted to € 686.61 in 2005 (http://www.mtas.es/estadisticas/BEL/PEN/index.htm);

consequently, nearly half the older people in the Madrid region receive a below-average retirement pension. Although no statistically significant gender-based differences are observed, women declare a lower level of income; and, in terms of age, there is an inverse statistical association. The perceived economic status of the household in which the old person lives fits a normal curve, with 70% of the interviewees placing themselves towards the middle; there are no significant gender or age-based differences in this variable.

The average size of households is almost two people, because more than half of this population lives with somebody else and nearly one-third live alone, as a result of the majority presence of married people and widows, respectively. There are fewer households formed by an older person living alone with children and/or grandchildren, or those formed by an older person with his/her spouse or partner who also live with children, grandchildren or other relatives. Consequently, the average age of these households is around 71 years old. In other words, most older people live in old households, and only 18% live in households with an average age of less than 65 years old. The majority of older people who live alone are women, mostly widows, but also the oldest people, as a result of their civil status having changed to widowhood.

Ageing at home implies independence, based on the principle of privacy (Pynoos and Regnier, 1991), explained as the fact of being alone, not being bothered by others, having control over where one goes, feeling free of company or observed by anyone. Privacy could be altered if it becomes necessary to share the house or even the room with other people, or even vulnerability if the older person has to go and live with his or her children. Despite their advanced age most older people live in their own home (93.5%) and the remainder live with other relatives, mainly children, or every so often have to move from one relative's house to another's.

One trait of the lifestyle that undeniably influences other conditions of older people's lives, especially in accessibility to amenities and services, is that around 95% of the population aged 65 and over in the Madrid region live in urban areas, especially in the capital but also in its metropolitan area.

Older People Live in Large, Old Houses that they Own

Nine out of ten of the target population live in their own house, which has been paid for in full (Table 4), chiefly due to the current stage of their life cycle, because they have had enough time to buy their houses, which is a common form of tenancy among the Spanish population (Rojo et al., 2005).

Home ownership seems to afford individuals the means through which they can achieve a sense of 'onthological security' in their daily lives, in a world that is sometimes perceived as threatening and uncontrollable (Dupuis and Thorns, 1998). Ownership represents financial security, in as much as the house can constitute a value of change, a guarantee that offsets the drop in their level of income upon retirement, in addition to being something to pass on to the next generation; it also gives them freedom of choice when it comes to deciding where to spend their old age; and confers a certain measure of control over the home, in so far as they can make any repairs, modifications or improvements necessary to adapt the home to the

TABLE 4. Summary of Older People's Housing Status According to Sociodemographic and Household Characteristics

Variables (499 valid cases) Categories	Number of cases	Owned house (row %)	Mean age of the house	Mean surface of the house (m²)	Surface (m²) by person (mean)	Mean number of amenities in the house Basic(*)	Necessary(*)	Comfort(*)
		96.9	41.3	83.9	51.6	3.0	7.5	0.5
Gender		(NS)	(NS)	(NS)	(NS)	(NS)	(NS)	(NS)
Male	204	96.5	41.4	84.0	48.5	3.0	7.5	0.6
Female	295	97.2	41.2	83.9	53.9	2.9	7.4	0.5
Age (mean age = 75.0)		(NS)	(NS)	(NS)	(0.008)	(NS)	(0.009)	(0.0001)
65–74 years old	247	96.3	40.2	86.4	48.2	2.9	7.6	0.7
75–84 years old	179	98.3	41.1	83.0	53.8	3.0	7.4	0.5
85 and more	73	95.8	45.4	77.9	58.2	3.0	7.3	0.3
Social class		(0.039)	(NS)	(0.0001)	(0.001)	(NS)	(NS)	(0.0001)
Lower	139	93.4	39.6	76.1	48.9	3.0	7.3	0.4
Lower middle	236	97.8	42.6	82.4	49.3	2.9	7.5	0.5
Middle	83	100.0	42.7	90.6	57.9	3.0	7.5	0.7
Upper middle	22	100.0	35.6	100.7	52.1	3.0	7.5	0.8
Upper	19	94.7	37.9	112.3	72.6	3.1	7.9	1.0
Household size (mean size = 1.9)		(NS)	(0.0001)	(0.001)	(0.0001)	(NS)	(NS)	(0.013)
1 person	160	96.8	44.9	80.8	80.8	2.9	7.3	0.4
2 persons	259	96.9	40.8	82.6	41.3	3.0	7.5	0.5
3+	80	97.4	35.3	94.8	27.2	3.0	7.5	0.7
Size of the residential area		(0.002)	(0.0001)	(0.0001)	(0.014)	(0.028)	(0.001)	(0.018)
Madrid municipality	343	97.6	42.7	79.4	49.4	2.9	7.5	0.5
Metropolitan municipalities	128	97.6	35.8	94.3	55.6	3.0	7.6	0.7
Non-metropolitan municipalities	28	85.7	50.1	92.0	61.0	3.3	6.9	0.3

(NS) Non-significant $p > 0.05$; (*) these concepts are explained in text, pages 134–135.

characteristics of the people occupying the spaces (Phillips *et al.*, 2005). One-third of older people reported a need to adapt or improve their homes, and 12% reported a very serious need for improvements to avoid mobility barriers inside the home, or to adapt it for a disabled person.

Home ownership is liable to vary in line with sociodemographic features and can affect the quality of the housing and how it is self-rated. Thus, the proportion of older people owning a home is smaller among those with lower levels of education, social class, monthly income, or perceived socio-economic status, who live in non-metropolitan municipalities, and, as was to be expected, among those who live in the home of their children or another relative.

The age of the home and the building in which it is located is another characteristic that could influence the quality of the housing, limiting the presence of amenities. Or they might have become inadequate for the new needs and demands that emerge as one ages. The average age of the houses where the respondents live is 41 years, and nearly half of them live in houses that are older than that. The most modern homes are those of older people who live in the metropolitan area, which has been developed in recent years, and who live with their children or in larger families. The wealthier older people also live in relatively more modern houses.

The size of the home is another element to be considered when assessing whether it is suited to their needs and its members' ability to cope on their own every day. It is important to bear in mind the present stage of the older person's lifecycle, his or her age, and other demographic characteristics, in order to ascertain whether the size of the home meets his or her needs. Madrid's older people live in homes with an average size of 84 m^2, making for an average of nearly 52 m^2 per person (crowding index). The people who live in larger and usable houses are outside the capital, mainly in the metropolitan area, and generally they enjoy more advantageous demographic conditions, i.e. the young-old, with higher levels of education, social class, income, or perception of the economic position of their household. Other people who live in larger homes include those who live accompanied, in large families living with other relatives, or when the house does not belong to them but to their children, being in these cases smaller in average surface area per person. Households with the smallest number of members, where the older person lives alone or with his or her spouse or partner, can enjoy more space in terms of square metres per member.

The characteristic of the housing and building is the presence and type of amenities in order to ascertain the objective level of well being of older people (Fernández-Mayoralas *et al.*, 2002), which have been grouped into three levels: basic, necessary, and comfort. The most basic ones, regarded as elementary for living a dignified life, such as running water or one's own toilet, are nearly universal (more than 97%); two-thirds of the interviewees report that they have another basic amenity, namely a shower; however, it must be stressed that more than one-third say that they only have some kind of moveable or fixed heating appliance to heat their home. Consequently, the average for this type of appliance is three out of four, without any significant differences in terms of demographic characteristics.

Necessary amenities seem to be very widespread (99% of those interviewed), with no gender or age-based differences being observed with respect to electrical appliances such as TV, refrigerator, or washing machine. A similar proportion says they have a telephone or hot water with no gender-based differences, though there are aged-based differences, such that, despite being very necessary, the older the people are, the smaller the proportion of people who have them. Nearly 82% of the interviewees report having other necessary amenities, such as a bath or piped gas for cooking or heating water; but a smaller proportion for heating installation. Therefore the average number of these amenities is high, 7.5 out of 8. Even though the differences are not very significant, this average is higher among the interviewees who live in the metropolitan area, and also among those who live, in addition to their spouse, with other relatives, or else in their children's home, and even higher among younger older people, with a higher level of education, social class, monthly income, and perception of the economic status of their household.

The amenities least reported are those which relate to comfort, such as air conditioning (mentioned by only 19% of interviewees), or dishwashers (by one-third). The average count is barely 0.5 amenities out of 2, and is higher among men, young-old people and, broadly speaking, among older people whose sociodemographic conditions are also more comfortable, as well as those who live in metropolitan municipalities, in households that are larger and not so old, because they live with other relatives, or in one of their children's homes.

Home ownership, the age of the home, size, and its amenities are parameters that are interrelated, in a variety of situations that make an older person's home a fairly comfortable place to live. Those living in rented accommodation are more likely to live in older and smaller houses and have fewer amenities, while older people who own their houses, especially if they are still paying for them, are likely to live in houses that are more modern, larger and well equipped. The age of the house also seems to be associated with its size, in that those who live in more modern houses are likely to enjoy a larger surface area. Finally, the presence of amenities of one kind or another seems to be associated with and reinforce the objective conditions of houses that are not so old and larger in size.

As with the amenities in the house, three levels have also been considered with regard to the amenities of the building where the house is located: the basic level, which includes having one toilet shared by several houses, it being noteworthy that this still applies to more than 2% of older people; the necessary level, with amenities such as a lift, which 67% of interviewees report having; and the comfort level, which includes a caretaker (43%), garden (37%), garage (20%), or swimming pool (6%). With regard to having devices that help to avoid access barriers both to the building and the house where the older people live, three out of every ten interviewees state some type of obstacle that limits their access, either due to the lack of a lift to save them having to climb the stairs or the lack of a wheelchair ramp.

The Outer Residential Environment

For older people, the area around their house is perhaps just as important for their well-being as the house (Carp and Carp, 1982). Not only are the spatial characteristics of the neighbourhood fundamental but also the social neighbour-related characteristics. It is interesting to examine the physical accessibility to the neighbourhood's resources and also their integration in community life, their relationships with their neighbours, as well as the closeness of their relatives, as aspects that benefit their personal development and improve their QoL (Bowling *et al.*, 1993; Baxter *et al.*, 1998).

The relationship between proximity to goods and services, and satisfaction with the neighbourhood/neighbours has been proven by some studies, although sometimes accessibility to amenities and services goes hand-in-hand with adverse living conditions (Jirovec *et al.*, 1985), that it is necessary to consider how older people perceive their closest spatial and personal environment and what problems they report. How personal characteristics are strong predictors of satisfaction with the home have also been described, while the residential characteristics (sense of security, having friends in the neighbourhood, etc.) are strong predictors of satisfaction with the neighbourhood (Bruin and Cook, 1997). If an individual has a favourable attitude towards the residential environment, if the person is satisfied, his/her behaviour will be consistent with that attitude in terms of staying in his/her home and the neighbourhood, good neighbourly relationships, and participation in activities (Amérigo and Aragonés, 1997).

How long they have lived in the neighbourhood, the relatives who live nearby, and the nature of the community problems that they perceive could be good indicators of older people's integration in their residential environment. This population reports having lived in the same residential habitat for an average of 38 years, without significant differences by gender or social class, though certain differences are observed in the other parameters analysed, such that a longer stay in the same neighbourhood, town, or community is associated with older age, lower income, living alone, in aged households and with a low perceived economic status, as well as in non-metropolitan areas (Table 5). Nearly 8 out of every 10 older people have relatives living nearby, especially those who live in larger, less aged households.

As for problems in the neighbourhood, 1 out of 10 people reports none, and on average 3.9 problems are mentioned. Although no significant gender or age-based differences are observed, those most critical of their community seem to be older people from higher social classes, not only due to the proportion of interviewees who refer problems but also because they report a larger number of them. In accordance with living arrangements and place of residence, those who live on their own tend to be less dissatisfied, as do those who live in non-metropolitan, or rural areas.

The most significant problems for older people (Table 6) have to do with the lack of a pleasant atmosphere and safety in the environment, so 14% of the answers refer to muggings, burglaries, gangs, not enough police, graffiti, etc.; 8% to suspicion of strangers, including foreigners, gypsies, squatters, prostitutes, etc.; and, finally, a few have to do with unemployment, poverty, low pensions, the lack of young people, the lack of communication with people and loneliness. The second group of most

TABLE 5. Number of Years Living in Same Place, Relatives Living Nearby, and Number of Problems in the Neighbourhood

Characteristics of the population (499 valid cases)				Perception of problems in the neighbourhood (first and second mentions)	
Categories	Number of cases	Mean number of years living in the same neighbourhood (mean = 38.1)	Relatives living nearby (same building, neighbourhood, village) (Row %) (76%)	None (Row %) (11.3%)	Mean number (excluded none) (3.9)
Gender		(NS)	(NS)	(NS)	(NS)
Male	204	38.0	75.5	14.7	4.0
Female	295	38.2	76.3	9.0	3.9
Age (mean = 75.0)		(0.023)	(NS)	(NS)	(NS)
65–74 years old	247	36.8	77.7	11.7	4.0
75–84 years old	179	38.0	74.9	11.2	4.1
85 and more	73	42.8	72.6	10.6	3.3
Social class		(NS)	(NS)	(0.0001)	(0.006)
Lower	139	39.5	75.5	16.9	4.6
Lower middle	236	38.7	78.0	6.9	3.7
Middle	83	37.6	74.7	16.4	3.3
Upper middle	22	32.8	81.8	11.8	4.0
Upper	19	29.3	52.6	0.0	5.9
Household size (mean = 1.9)		(0.0001)	(0.0001)	(0.025)	(NS)
1 person	160	41.1	69.4	15.2	3.7
2 persons	259	38.7	74.9	8.2	4.1
3+	80	30.5	92.5	13.7	3.8
Size of the residential area		(0.032)	(NS)	(0.0001)	(0.015)
Madrid municipality	343	37.2	76.7	6.9	3.7
Metropolitan municipalities	128	38.9	73.4	13.8	4.6
Non-metropolitan municipalities	28	45.4	78.6	61.9	2.5

(NS) Non-significant $p > 0.05$

important problems is related to deterioration of the environment, such as the widespread noise and pollution, all the building and roadworks that make it hard to move around the neighbourhood, and dirt in the form of rubbish, animal droppings, and so on. With regard to the lack of city amenities (13.4% of the answers), interviewees state to the shortage of green spaces, untidy parks and gardens, the fact that the area is poorly served by public transport and inaccessible from the rest of the city, poorly lit and badly paved streets, narrow pavements, and poor signposting. Labelled as urban pressure problems they report heavy traffic, the lack of car parks and, subsequently, the fact that there are cars parked on the pavements and zebra crossings, thereby hindering their mobility. Moreover, and in a smaller proportion of answers, older people refer to problems related to deficient urban morphology (poorly built housing, old housing, and unfinished streets) and negative social behaviour (drug dealers, beggars). The lack of services (not enough neighbourhood shops and stores, many traditional shops closing, shortage of social, health, cultural and educational services, as well as services for older people) was barely mentioned in 7% of the answers.

The nature of the problems reported, typical of large, developed cities with huge inflows of immigrants that could have a negative effect on perceived living conditions (Fokkema, 1996), confirms, however, that accessibility to community services is not a drawback that older people mention. The amenities perceived as closest are grocers' shops, transport, bars, cafeterias, and restaurants; religious services; and green spaces. In contrast, the amenities perceived as furthest away includes cultural ones such as libraries, cinemas, theatres, as well as sports centres or specialised health services (facilities of 'district level', less frequent in terms of demand and supply),[3] although they find that other community social services, such as clubs and day centres for older people, are relatively far away.

A High Level of Residential Satisfaction

The descriptive results show that older people in Madrid have a high level of residential satisfaction, thus 88% of the interviewees report being very and quite satisfied with the different elements of their living environment (Figure 1). The home,

TABLE 6. Type of Problems Perceived by Older People as Most Important in their Residential Environment (Neighbourhood or Village)

List of problems	Multiple response	
	Number of responses	% of responses
Lack of pleasant and safe atmosphere	480	30.4
Environmental degradation	374	23.7
Lack of urban amenities	211	13.4
Urban pressure	185	11.7
Deficient urban morphology	128	8.1
Not enough services	114	7.2
Negative social behaviour	82	5.2
Other, unspecified problems	4	0.3
Total responses	1.578	100.0
Total cases	399	
Without problems	51 (11.3% over 450 valid cases)	

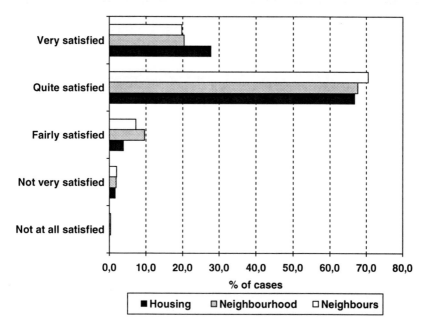

Figure 1. Level of residential satisfaction.

first of all, and perception of neighbourly relations, are the highest scoring elements; nearly 3 out of every 10 older people report being very satisfied with their house, and 2 out of every 10 with their neighbours. Men are more satisfied with their neighbourhood and house, while women seem to be more satisfied with the neighbours and their relationships with them. Those aged between 65 and 74 years old say that they are both very and quite satisfied with their home and neighbours, while the oldest tend to be very and quite satisfied with the neighbourhood or village where they live, perhaps directly related to the number of years spent living in the same place.

Overall residential satisfaction shows no significant statistical differences in terms of gender, age, and marital status, as revealed by other research (Diener, 1999), although a tendency towards higher satisfaction is observed among men, single people, and the younger interviewees. However, there are significant differences in terms of social class; being included in a higher class, as an expression of level of education and socio-economic status, does seem to influence the higher residential satisfaction ($p \leq 0.005$). Thus, the lower-middle and middle class interviewees are the most satisfied with their residential environment. Statistically significant differences exist between residential satisfaction and the size of the residential habitat ($p \leq 0.003$), such that it is older people who live in the city of Madrid, followed by those who live in rural areas, who express the highest level of satisfaction, with a big difference with those who live in the metropolitan area.

THE OVERALL RESIDENTIAL SATISFACTION PREDICTORS

The Importance of Subjective Variables

Older people in the Madrid region say that they are relatively satisfied with their residential environment, according to other research (Golant, 1984; Ginsberg, 1987; Rojo et al., 2002). However, just as interesting as the descriptive knowledge of the levels of satisfaction, is working out the environmental elements which best explain residential satisfaction, so as to address the fundamental research objective as well as providing information to establish the forms of action to face up to the ideal situation of growing old at home. MLRA has been used to achieve this – a method which has also provided appropriate solutions in other research projects. In Spain, it has been used to study the residential satisfaction of the population as a whole in the cities of Madrid (Amérigo, 1995) and La Coruña (García, 1997). This statistical technique has been used to discover the predictors of residential satisfaction with the house and neighbourhood attributes among older women (Carp and Christensen, 1986), the Perceived Environmental Quality Index (PEQI) predictors among the older female (Christensen and Carp, 1987), satisfaction with life among older women (Cutler Riddick, 1985), housing satisfaction among people aged 60 and over (Golant, 1982), the residential satisfaction of people who live in the city centre (Ginsberg, 1987), residential satisfaction with the housing, neighbourhood, and neighbour attributes of older people in the city of Madrid (Rojo et al., 2000), and overall residential satisfaction (Rojo et al., 2001).

Of all the information used in the regression model nine variables have turned out to be statistically significant ($p \leq 0.05$) in predicting overall residential satisfaction (Table 7). The strength of the connection between the dependent variable and the explanatory variables is 0.60, with a determination coefficient $R^2 = 0.36$, indicating that 36% of the variability of overall residential satisfaction can be explained by all of the predictors variables selected; of them, the top six are subjective in nature, which, as a whole, account for one-third of the variance of the overall residential satisfaction.

The greatest predictive power is provided by the perception of economic resources, either of the person interviewed or the household, and it contributes 6.4% (column 6, Table 7) to the explanation of the overall model. This explanatory variable is defined by the satisfaction perceived by an older person in relation to their general financial situation (taking into consideration all of their income and savings), their confidence in their future financial situation, and also to the economic self-positioning of their household in classes ranging from extremely low to extremely high. How does the interaction of these variables work in the overall residential satisfaction model? The higher the valuation of their own financial situation or that of their household of residence and the greater their confidence in their future financial situation, the higher their residential satisfaction will be.

It has already been said that almost 5 out of every 10 interviewees say that their income is less than the 2005 national average (€ 686.61). In this financial situation, monthly income appears to affect the older person's satisfaction in relation to their

TABLE 7. Predictors of the Overall Residential Satisfaction Among Older People in Madrid Region

Criterion variable: overall residential satisfaction[a]	Multiple correlation coefficient (R) = 0.600; coefficient of determination (R^2) = 0.360; adjusted R^2 = 0.342; sig. F < 0.05					
Predictor variables	1 – Correlation between the dependent variable and each independent variable (R)	2 – Unstandardized coefficients (B)	3 – Significance level	4 – R^2 change	5 – Contribution coefficient 5 = 1*2	6 – Contribution coefficient (in %)
Constant		0.000				
Perceived economic resources	0.358	0.179	0.000	0.128	0.064	6.42
Agreement with positive image of the neighbours	0.296	0.272	0.000	0.077	0.081	8.06
Disagreement with negative descriptions of neighbours' behavior	0.304	0.262	0.000	0.060	0.080	7.97
Distance to cultural and sports centres of district level	0.230	0.204	0.000	0.026	0.047	4.70
Distance to municipality services	0.167	0.187	0.000	0.022	0.031	3.13
Distance to neighbourhood or community services	0.166	0.153	0.001	0.018	0.025	2.53
Basic amenities in the house	0.104	0.121	0.012	0.013	0.013	1.26
Availability of space in the home	0.096	0.105	0.032	0.008	0.010	1.01
Relatives living nearby	0.107	0.090	0.049	0.008	0.010	0.96
					0.360	36.04

[a] A factor component obtained from the Factor Analysis (FA) with variables of partial residential satisfaction (housing, neighbourhood, neighbours)

overall financial situation: the greater their income, the more positive their perception (Figure 2). Likewise, the valuation of their future financial situation is directly related to their satisfaction with their current financial position.

The financial situation is the first variable which explains residential satisfaction, while it is the third factor (in 76% of cases) of the five QoL elements spontaneously mentioned by older people as being the most important in their life (health, family network, economic situation, social network, and leisure activities), nevertheless reaching an average satisfaction of barely 60% and a weight of 16% (health 26%, family network 26%, social network 16%, and leisure activities 14%), representing 10% of the SEIQoL index, after family network and health (Fernández-Mayoralas *et al.*, forthcoming).

The second and third variables of the regression equation are related to a high appreciation of the opinion held by older people of their neighbours. The higher the level of agreement with positive descriptions of their neighbours as well as the disagreement with unfavourable descriptions, the higher the level of residential satisfaction. If then, the information concerning the perception of neighbours is included in the regression equation in second and third place, it contributes to the overall model with almost 16%. These two variables are based on different information. The first refers to aspects and opinions which describe the neighbours from a positive point of view, in that older people think that their neighbours are nice, educated people, perceive them to be similar to themselves, with the same interests, often helping others when required. Meanwhile, the second predictor variable shows

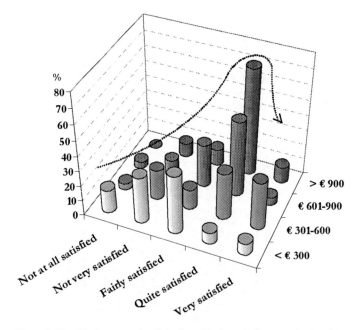

Figure 2. Monthly income and satisfaction with the overall economic situation.

a disagreement with pejorative characteristics defining the neighbour's behaviour, such that the neighbours are not seen as people who interfere in other people's private lives, or as people who cause a disturbance due to dirt or noise.

Another topic to which older people in the Madrid region attach a great deal of importance is the availability of amenities in the area where they live. The fourth, fifth, and sixth parameters of the regression equation confirm this. The overall residential satisfaction model attributes 10% (see column 6 of Table 7).

How does the perception of distance to the amenities affect residential satisfaction? A shorter distance to local or community amenities will cause residential satisfaction to increase. Moreover, according to the model, the parameter defined by the distance to sports facilities, specialist medical centres, and cultural places (theatres, cinemas, exhibition rooms) is the most important in the equation; not for nothing does it include the services which, since they are more specific, are known as 'district level services' as they serve geographical areas which are larger than the local community. This variable has a positive affect on the model and its inclusion explains why residential satisfaction is higher the closer the aforementioned amenities are perceived to be and the fewer perceived problems are reported in the area where they live.

The perception of the distance to municipal amenities (such as day centres and clubs for older people, cultural and social centres) and to amenities of a more local nature (falling under the typology of 'neighbourhood or community level', such as grocers' shops, bars, cafes and restaurants, parks and gardens, health centre and religious services) make up the fifth and sixth parameters of the regression equation, so the shorter the distance perceived by older people to the amenities the higher the residential satisfaction will be.

OBJECTIVE CONDITIONS

Home is considered to be the residential element which is closest to the individual, the most private (Marans, 2003), sometimes adapted to the needs and wishes of older residents (Bratt, 2002; Harrison, 2004), and sometimes restricted by environmental and personal circumstances (Abellán and Olivera, 2004). The more their residential needs are covered, the greater satisfaction the individual will derive; on the other hand, if the environment is far from meeting expectations due to there being problems and a lack of resources in the property (Slangen-De Kort et al., 1998; Schieman and Meersman, 2004; Struthers, 2005) there is a higher probability of residential dissatisfaction.

Home is the most basic area of the residential system and it forms part of a larger dimension, the local community and neighbours network, characterised by specific physical, non-residential, and social features, which do not belong to the person's sphere, although they use and enjoy it. It is then a multifaceted, public area, for general use but it is also an area for social and community relations which may end up being as important as private areas due to its contribution towards maintaining the person's identity and well being (Fernández-Mayoralas et al., 2004; Peace et al.,

2006). Sometimes the residential environment becomes oppressive and limits people's physical activity, reducing their QoL, making it necessary to undertake a series of improvements to overcome these limitations (Harrison, 2004).

Housing only affects residential satisfaction prediction by less than 3% of the contribution coefficient. Of all the housing elements, basic amenities and the space availability are retained in the model. Basic amenities such as running water and one's own toilet and shower are practically the same across the board in the housing of the population studied, but a third of the interviewees only has moveable or fixed heating appliance, but no central heating installation. For the most part, older people reside in spacious houses and since they are small households, the crowding index is relatively low (Victor, 1987).

The last parameter of the regression equation refers to the presence of relatives in the neighbourhood, whether they are first-degree (children, parents) or second-degree relatives (brothers and sisters, grandchildren, others). The request for this information was based on an assumption about the influence that a shorter distance to the family members' place of residence has on the QoL, and in this case on residential satisfaction, boosting personal development and improving QoL (Baxter et al., 1998; Bowling et al., 1993). Having family members in the local area is a relevant factor in old age; in fact it is one of the root causes of residential mobility at that age (Rojo et al., 2002).

CONCLUSION

QoL is developed by scientists from different disciplines and methodological approaches as an important research issue in Spain. Most of the research has focused on health and health-related QoL as the main dimensions (Martínez, 2006). The residential environment has attracted far less interest, and consequently it is not easy to find links between research into residential satisfaction and the public policy instruments to justify the decision-making regarding QoL of older people.

In research into the overall QoL evaluated by older people living in family housing in the Madrid region, the importance of the residential environment is not to be found in the proportion of people who nominate the living environment as one of the five dimensions of their QoL. Yet the interest in studying it lies in the fact that it is easier to understand a person's behaviour and state if the sociospatial context in which they live is known (Peace, 1987), reinforced by the high level of satisfaction with this dimension, and this will lead to the residential environment having a high relative weight in the overall QoL.

A high proportion of the overall residential satisfaction is explained in the regression model (36% of the variance). Although accurate comparisons cannot be made because different questionnaires have been used, the findings can be compared to other residential satisfaction research that has also applied the MLRA technique, with values similar to those found by other authors (Carp and Christensen, 1986; Christensen and Carp, 1987; Bruin and Cook, 1997). In previous studies (Rojo et al., 2001), 50% of the variance of the residential satisfaction of older people in the municipality of Madrid was explained by considering a questionnaire that addressed a wide range of issues; other studies with the same explanatory level include Weidemann *et al.* (1982) and García (1997); but there are also others with a smaller explanation of the dependent variable (Cutler Riddick, 1985; Mookherjee, 1992; Rohe and Basolo, 1997).

Therefore, it can be considered that the model used to predict overall residential satisfaction among older people in Madrid is a relevant model because it explains more than one-third of the criterion's variability; but it could also be indicating the existence of other predictor factors not contemplated in this study, together with the intrinsic variability of the individuals and their answers. Consequently, this issue should be examined more closely with qualitative methodology, in order to unearth other latent parameters.

According to the findings, 90% of the differential contribution in the explanation of the criterion is due to subjective factors, thereby underscoring, once more, the importance of these aspects in overall residential satisfaction as a dimension of QoL (Rojo *et al.*, 2001) with a small part being justified by objective elements; such that a deviation is found to exist between the characteristics of the objective environment and the perception reported by older people. As others point out, older people seem to be especially capable of adapting to different objective situations of the residential environment (Diener *et al.*, 1999; Oswald *et al.*, 2003).

Perception of the distance to community services and the evaluation of the neighbourly relationships network indicate that older people regard these elements from outside their homes as just as important as their homes *per se*, in line with the conclusions drawn by Carp and Carp (1982) in the study of the ideal residential area. Yet the heaviest explanatory parameter is due to the perception of their own economic resources and the household where they live, such that this result is related to the importance that older people attach to the economic resources dimension among the five most important areas of their lives. This subjective aspect would be supported by objective conditions in the following terms: people with better economic resources could choose where to live and afford a house according to their wishes – and consequently would be more satisfied with their residential environment.

Researchers and public policy managers alike have set their sights on QoL as a dynamic concept with many different faces. Researchers are committed to achieving a multidimensional conceptualisation of QoL and thereby define its essential domains as a basis for evaluating the needs of older people, as well as for validating methodological instruments. It is very important to evaluate the possible and foreseeable differences in the contrast of spontaneous results obtained with instruments such as SEIQoL, with such a small proportion of nominations of the residential environment as an overall QoL domain, with other types of measurement instruments that, based on researcher-defined lists of domains, also give interviewees the chance to rate them in terms of importance, satisfaction and weighting.

At the geographical level, significant comparative analyses of the urban vs. rural environment should be conducted, in as much as this study has discovered statistically significant differences in satisfaction depending on the size of the residential area. In Spain, no research has or is being conducted into the residential environment as an attribute of the QoL of older people in terms of the size of the residential habitat; or differentiating between broader spatial environments, as in Germany (Oswald *et al.*, 2003).

Further studies are required for the significance of the residential context among both older people who live on their own and those who live dependently in home care institutions. In the first case, what would be needed is a better understanding of the physical condition of the housing and of the environment that limit the person's QoL (Puga and Abellán, 2004). In the second, it would be a matter of shedding further light on the perception of the institutional residential environment of older people, according to their level of competence; recently, Chou *et al.* (2003) revealed the need to ascertain the interrelationship between the factors that influence residential satisfaction of older people in care centres, requesting information not only from the residents but also from the members of staff, and conclude that the staff's satisfaction plays a relevant role in shaping the residents' own satisfaction. In any case, studies of architectural barriers that limit mobility in the residential environment should be taken into consideration when evaluating independence and QoL (Aguado *et al.*, 2003), especially if a longitudinal study is taken into account, that is to say, if QoL is studied in an evolutionary sense on the basis of the continuous perception of a population cohort.

There also seems to be a gap in terms of studies that analyse the QoL of dependent people and their carers, whether such dependents live in housing or in a care centre. In this regard, research currently being conducted by the Spanish Epidemiological Ageing Studies Group[4] in several Spanish provinces will serve to fill this gap.

In Spain, with a regionally decentralised political and administrative structure, one must not forget the perspective of the courses of action that are being implemented at different decision-making levels. In order to properly structure older people's care schemes, the central, regional, and local authorities must collaborate in researching and designing a public policy capable of fostering the conditions required to improve the older population's QoL. To that end, the 2003–2007 Action Plan (IMSERSO, 2003) has built on the thoughts contributed by different international forums (recommendations of the Madrid International Action Plan on Ageing, World Forum of NGOs, Berlin Conference, and WHO) and Spanish forums to design its future courses of action for improving the QoL of older people.

In these circumstances and in order to attain a higher QoL, it is regarded as essential to keep older people in their family, community, and social environment, for which purpose three types of measures must be established: (a) to adapt housing to older people's functional capacity; (b) to supply social services in their own neighbourhood, such as day care centres; and (c) to provide a residential environment (supervised flats, residential homes) when specialised attention is required. Any of these three courses of action is likely to be evaluated in the service delivery context, but not so the QoL of the person being cared for, as a parameter assessed afterwards.

In this context, there seems to be a need to value QoL in relation to the residential environment, before and after the public policies are applied.

In short, every day brings further and more extensive research into the different components of QoL, and more consistent methods are applied; yet they remain to be integrated into public policy decision-making in relation to older people in Spain. It is time to do so.

NOTES

1. The research is being funded by two projects:
   - *Quality of Life Assessment among older people in Madrid Region*. Funded by the Government of Madrid (Regional R&D&I Plan). Headed by Dr. Gloria Fernández-Mayoralas (Ref. 06/HSE/0417/2004-2005).
   - *Quality of Life of non-institutionalizsed elderly in Spain. A contribution to the study of the Quality of Life from a holistic approach*. Funded by the Spanish Ministry of Education and Science (National R&D&I Plan). Headed by Dr. Fermina Rojo (Ref. BSO2003-00401 2004-2006).
2. In territorial administration terminology, several communities make up a district and several districts constitute a municipality. 'Neighbourhood or community level services' are those more frequent in terms of demand and supply, while 'district level services' are more specific as they serve geographical areas larger than the local community or neighbourhood areas.
3. See note 2.
4. Centro Nacional de Epidemiología – Instituto de Salud Carlos III, Madrid, Spain.

REFERENCES

Abellán, A. and Olivera, A. (2004) 'Dificultades en el entorno vivido', *Revista Multidisciplinar de Gerontología*, 14(3), 184–186.

Abellán, A. and Rojo, F. (1997) 'Migración y movilidad residencial de las personas de edad en Madrid', *Anales de Geografía de la Universidad Complutense*, 17, 175–193.

Aguado, A.L., Alcedo, M.Á., Fontanil, Y., Arias, B., and Verdugo, M.Á. (2003) *Calidad de vida y necesidades percibidas en el proceso de envejecimiento de las personas con discapacidad*, Madrid, IMSERSO, Estudios 20.

Amérigo, M. (1995) *Satisfacción residencial. Un análisis psicológico de la vivienda y su entorno*, Madrid, Alianza Editorial.

Amérigo, M. and Aragonés, J.I. (1997) 'A Theoretical and Methodological Approach to the Study of Residential Satisfaction', *Journal of Environmental Psychology*, 17(1), 47–57.

Baxter, J., Shetterly, S.M., Eby, C., Mason, L., Cortese, C.F., and Hamman, R.F. (1998) 'Social Network Factors Associated with Perceived Quality of Life. The San Luis Valley Health and Aging Study', *Journal of Aging and Health*, 10(3), 287–310.

Bowling, A., Farquhar, M., Grundy, E., and Formby, J. (1993) 'Changes in Life Satisfaction over a Two and a Half Year Period among Very Elderly People Living in London', *Social Science and Medicine*, 36(5), 641–655.

Bratt, R.G. (2002) 'Housing and Family Well-being', *Housing Studies*, 17(1), 13–26.

Brown, J., Bowling, A., and Flynn, T. (2004) *Models of Quality of Life: A Taxonomy, Overview and Systematic Review of the Literature*, European Forum on Population Ageing Research, 113 pp (http://www.shef.ac.uk/ageingresearch/pdf/qol_review_complete.pdf).

Browne, J.P., O'Boyle, C.A., McGee, H.M., Joyce, C.R.B., McDonald, N.J., O'Malley, K., and Hiltbrunner, B. (1994) 'Individual Quality of Life in the Healthy Elderly', *Quality of Life Research*, 3(4), 235–244.

Bruin, M.J. and Cook, C.C. (1997) 'Understanding Constraints and Residential Satisfaction among Low-income Single-parent Families', *Environment and Behavior*, 29(4), 532–553.

Callahan, J.J. (ed.) (1993) *Aging in Place*, New York, Baywood Publishing Company.

Carp, F.M. and Carp, A. (1982) 'The Ideal Residential Area', *Research on Aging*, 4(4), 411–439.

Carp, F.M. and Christensen, D.L. (1986) 'Technical Environmental Assessment Predictors of Residential Satisfaction: A Study of Elderly Women Living Alone', *Research on Aging*, 8(2), 269–287.

Chou, S.-C., Boldy, D.P., and Lee, A.H. (2003) 'Factors Influencing Residents' Satisfaction in Residential Aged Care', *Gerontologist*, 43(4), 459–472.

Christensen, D.L. and Carp, F.M. (1987) 'PEQUI-based Environmental Predictors of the Residential Satisfaction of Older Women', *Journal of Environmental Psychology*, 7, 45–64.

Comunidad de Madrid (1998) *Plan de Mayores*, Madrid, Consejería de Salud y Servicios Sociales, Comunidad de Madrid.

Cutler Riddick, C. (1985) 'Life Satisfaction for Older Female Homemakers, Retirees and Workers', *Research on Aging*, 7(3), 383–393.

Diener, E., Suh, E.M., Lucas, R.E., and Smith, H.L. (1999) 'Subjective Well-Being: Three Decades of Progress', *Psychological Bulletin*, 125(2), 276–302.

Dupuis, A. and Thorns, D.C. (1998) 'Home, Home Ownership and the Search for Ontological Security', *Sociological Review*, 46(1), 24–47.

Fernández-Mayoralas, G. and Rojo, F. (2005) 'Calidad de Vida y Salud: Planteamientos Conceptuales y Métodos de Investigación', *Territoris (Monográfico sobre Geografía de la Salud)*, 5, 117–135.

Fernández-Mayoralas, G., Rojo, F., Rodríguez, V., Prieto, M.E., Martínez, P., and Frades, B. (2005) *Research Projects on Quality of Life among the Older People in Spain*, Ostasiewicz, P.W., Third International Conference on Quality of Life Research 'Towards Quality of Live Improvement', Wroclaw, Poland, 14–16 September 2005.

Fernández-Mayoralas, G., Rojo, F., and Pozo, E. (2002) 'El entorno residencial de los mayores en Madrid', *Revista Estudios Geográficos*, LXIII, 248–249, pp.541–575.

Fernández-Mayoralas, G., Rojo, F., and Rojo, J.M. (2004) 'Components of the Residential Environment and Sociodemographic Characteristics of the Elderly', *Journal of Housing for the Elderly*, 18(1), 25–49.

Fernández-Mayoralas, G., Rojo, F., Prieto, M.E., León, B., Martínez, P., Forjaz, M., Frades, B., and García, C. (forthcoming) *El significado de la salud en la Calidad de Vida de la población mayor*, Granada, Caja Rural de Granada.

Fokkema, T. (1996) *Residential Moving Behaviour of the Elderly: An Explanatory Analysis for The Netherlands*, Amsterdam, Thesis Publishers.

García, R. (1997) *La ciudad percibida. Una psicología ambiental de los barrios de A Coruña*, A Coruña, Universidade da Coruña, Servicio de Publicacións.

Gilhooly, M., Gilhooly, K., and Bowling, A. (2005) 'Quality of Life: Meaning and Measurement' in A. Walker (ed.) *Understanding Quality of Life in Old Age*, Maidenhead, UK, Open University Press.

Ginsberg, Y. (1987) 'The Elderly in Central Tel Aviv' in W. Van Vliet, H. Choldin, W. Michelson, and D. Popeone (eds) *Housing and Neighborhoods. Theoretical and Empirical Contributions*, London, Greenwood Press.

Golant, S.M. (1982) 'Individual Differences Underlying the Dwelling Satisfaction of the Elderly', *Journal of Social Issues*, 38(3), 121–133.

Golant, S.M. (1984) 'The Effects of Residential and Activity Behaviors on Old People's Environmental Experiences' in I. Altman, M.P. Lawton, and J.F. Wohlwill (eds) *Elderly People and the Environment. Serie: Human Behavior and Environment. Advances in Therory and Research*, Vol. 7, New York, Plenum Press.

Harrison, M. (2004) 'Defining Housing Quality and Environment: Disability, Standards and Social Factors', *Housing Studies*, 19(5), 691–708.

Houben, P.P.J. (2001) 'Changing Housing for Elderly People and Co-ordination Issues in Europe', *Housing Studies*, 16(5), 651–674.

Instituto de Estadística (2004) *Padrón Continuo de Habitantes, diciembre de 2004*, Comunidad de Madrid, (electronic source).

IMSERSO (2003) *Plan de Acción para las personas mayores 2003–2007*, Madrid, Ministerio de Trabajo y Asuntos Sociales.

INSERSO (1993) *Plan Gerontológico*, Madrid, Ministerio de Asuntos Sociales.

Jirovec, R.L., Jirovec, M.M., and Bosse, R. (1985) 'Residential Satisfaction as a Function of Micro and Macro Environmental Conditions among Urban Elderly Men', *Research on Aging*, 7(4), 601–616.

Marans, R.W. (2003) 'Understanding Environmental Quality through Quality of Life Studies: the 2001 DAS and Its Use of Subjective and Objective Indicators', *Landscape and Urban Planning*, 65(1–2), 73–83.

Martínez, P. (ed.) (2006) *Calidad de vida en neurología*, Barcelona, Ars Medica.

Mookherjee, H.N. (1992) 'Perceptions of Well-being by Metropolitan and Non-metropolitan Populations in the United States', *Journal of Social Psychology*, 132(4), 513–524.

Oswald, F., Wahl, H.-W., Mollenkopf, H., and Schilling, O. (2003) 'Housing and Life Satisfaction of Older Adults in Two Rural Regions in Germany', *Research on Aging*, 25(2), 122–143.

Pacione, M. (2003) 'Urban environmental quality and human wellbeing – a social geographical perspective', *Landscape and Urban Planning*, 65(1–2), 19–30.

Pastalan, L.A. (1990) (ed.) *Aging in Place: The Role of Housing and Social Supports*, New York, The Haworth Press.

Peace, S.M. (1987) 'Residential Accommodation for Dependent Elderly People in Britain: the Relationship between Spatial Structure and Individual Lifestyle', *Espace, Populations, Sociétés*, 1, 281–290.

Peace, S.M., Holland, C., and Kellaher, L. (2006) *Environment and Identity in Late Life*, Maidenhead, UK, Open University Press, 182 pp.

Phillips, D.R., Siu, O.-L., Yeh, A.G.-O., and Cheng, K.H.G. (2005) 'Ageing and the Urban Environment' in G.J. Andrews and D.R. Phillips (eds) *Ageing and Place. Perspectives, Policy, Practice*, London, Routledge, pp.147–163.

Puga, M.D. (2004) *Estrategias residenciales de las personas de edad. Movilidad y curso de vida*, Barcelona, Fundación La Caixa.

Puga, M.D. and Abellán, A. (2004) *El proceso de discapacidad. Un análisis de la Encuesta sobre discapacidades, deficiencias y estado de salud*, Madrid, Pfizer.

Pynoos, J. and Regnier, V. (1991) 'Improving Residential Environments for Frail Elderly: Bridging the Gap between Theory and Application' in J.E. Birren, J.E. Lubben, J.C. Rowe, and D.E. Deutchman (eds) *The Concept and Measurement of Quality of Life in the Frail Elderly*, San Diego, CA, Academic Press, pp.91–119.

Rodríguez, V., Castro, T., Rojo, F., Fernández-Mayoralas, G., Vázquez, C., Puga, M.D., Rojo, J.M., and García, J.A. (2005) *Cambio demográfico y transformaciones económicas y sociales en el centro urbano de Madrid*, Madrid, Comunidad de Madrid, Consejo Económico y Social.

Rohe, W.M. and Basolo, V. (1997) 'Long-Term Effects of Homeownership on the Self-Perceptions and Social Interaction of Low-Income Persons', *Environment and Behavior*, 29(6), 793–819.

Rojo, F., Fernández-Mayoralas, G., and Pozo, E. (2000) 'Envejecer en casa: los predictores de la satisfacción con la casa, el barrio y el vecindario como componentes de la calidad de vida de los mayores en Madrid', *Revista Multidisciplinar de Gerontología*, 10(4), 222–233.

Rojo, F., Fernández-Mayoralas, G., Pozo, E., and Rojo, J.M. (2001) 'Ageing in Place: Predictors of Residential Satisfaction of Elderly', *Social Indicators Research*, 54(2), pp.173–208.

Rojo, F., Fernández-Mayoralas, G., Pozo, E., and Rojo, J.M. (2002) *Envejecer en casa: la satisfacción residencial de los mayores en Madrid como indicador de su calidad de vida*, Madrid, CSIC.

Rojo, F., Fernández-Mayoralas, G., Rodríguez, V., Prieto, M.E., and Rojo, J.M. (2005) *Condiciones residenciales de los mayores en España: espacio residencial privado vs. público*, in Asociación de Geógrafos Españoles (ed.) *Espacios públicos, espacios privados. Un debate sobre el territorio*. Santander, Cantabria, Universidad de Cantabria y Asociación de Geógrafos Españoles (CD-Rom:\PONENCIA 1\MESA 1\1_ROJO.pdf); (ISBN: 84-8102-982-5M).

Schieman, S. and Meersman, S.C. (2004) 'Neighborhood Problems and Health Among Older Adults: Received and Donated Social Support and the Sense of Mastery as Effect Modifiers', *Journals of Gerontology, Series B: Psychological Sciences and Social Sciences*, 59B(2), S89–S97.

Slangen-De Kort, Y.A.W., Midden, C.J.H., and Van Wagenberg, A.F. (1998) 'Predictors of the Adaptive Problem-solving of Older Persons in their Homes', *Journal of Environmental Psychology*, 18(2), 187–197.

Struthers, C.B. (2005) 'Housing Conditions and Housing Options for Older Residents: A Question of Need, a Question of Acceptable Alternative', *Journal of Housing for the Elderly*, 19, 1, pp.53–78.

Victor, C.R. (1987) *Old Age in Modern Society: A Textbook of Social Gerontology*, London, Croom Helm.

Wahl, H.-W. and Weisman, G.D. (2003) 'Environmental Gerontology at the Beginning of the New Millennium: Reflections on Its Historical, Empirical, and Theoretical Development', *Gerontologist*, 43(5), 616–627.

Walker, A. (ed.) (2005a) *Growing older in Europe*, Maidenhead, UK, Open University Press.

Walker, A. (ed.) (2005b) *Understanding Quality of Life in Old Age*, Maidenhead, UK, Open University Press.

Walker, A. and Hagan Hennessy, C. (eds) (2004) *Growing Older: Quality of Life in Old Age*, Maidenhead, UK, Open University Press.

Weidemann, S., Anderson, J.R., Butterfield, D.J., and O'Donell, P.M. (1982) 'Residents' Perception of Satisfaction and Safety. A Basis for Change in Multifamily Housing', *Environment and Behavior*, 14(6), 695–724.

The WHOQOL Group (1995) 'The World Health Organization Quality of Life assessment (WHOQOL): position paper from the World Health Organization', *Social Science and Medicine*, 41(10), 1403–1409.

Wiles, J. (2005) 'Home As a New Site of Care Provision and Consumption' in G.J. Andrews and D.R. Phillips (eds) *Ageing and Place. Perspectives, Policy, Practice*, London, Routledge, pp.79–97.

9. PERCEIVED ENVIRONMENTAL STRESS, DEPRESSION, AND QUALITY OF LIFE IN OLDER, LOW-INCOME, MINORITY URBAN ADULTS

INTRODUCTION

In this chapter we explore the predictors of perceived environmental stress (PES) and depression as indicators of quality of life (QoL) among older adults living in low-income subsidised senior housing in buildings with challenging social environments. We demonstrate that there is considerable variation in the ways that residents of these buildings respond to their environment that are closely linked to their mental health status and their perception of the stressors in their immediate environment.

Research has shown substantial effects of environmental stressors on mental and physical well-being, especially among residents of impoverished urban areas (Bullinger, 1989; Eriksson, 1999; Evans, 2003; Feldman and Steptoe, 2004; Garland, 1990; Krause, 1996; Levitt *et al.*, 1987; Shipp and Branch, 1999; Svanborg and Djurfeldt, 1987). These studies present evidence showing that factors including high levels of pollution, low socio-economic status, substandard living conditions, high potential for exposure to violence, limited access to social and health resources, and population density significantly contribute to QoL. In these studies, QoL is measured in a number of ways including indicators of mental and physical health distress. Relationships between contextual factors and these QoL indicators are often non-linear and very complex (Evans, 2003; Krause, 1996). The goal of this chapter is to discuss some of the major factors related to QoL as measured by PES and mental health status, in a sample of older, low-income, minority urban adults from a medium-sized city in north-eastern USA and to explain apparent gender differences.

ETHNIC AND RACIAL DISPARITIES AND QoL IN URBAN ENVIRONMENTS

Structural factors such as poverty, discrimination, and stigma, as well as environmental stressors such as noise and pollution, are more likely to be associated with reduced QoL (Bazargan and Hamm-Baugh, 1995; Bojrab *et al.*, 1988; Drewnowski and Evans, 2001; Evans, 2003; GSA, 2002; Krause, 1996; Tran *et al.*, 1996; Weinberger *et al.*, 1988) among lower income African Americans and Hispanics compared with middle-class suburban residents (Tran *et al.*, 1996; Fitzpatrick and LaGory, 2000; Krause, 1996; Schensul *et al.*, 2003). While demographic factors (ethnicity, gender, income, age) are significantly related to health and mortality in older adults (Morgan and Kunkel, 2001), characteristics of the physical or *built* environment such as stress, noise, sanitation, food, water, heat/cold, and toxins have

151

H. Mollenkopf and A. Walker (eds.), Quality of Life in Old Age, 151–165.
© 2007 *Springer.*

been shown to account for approximately 65% of the variance in life expectancy (Herskind *et al.*, 1996). Thus, the interaction of demographic and economic factors with stressful characteristics of inner city environments can be assumed to affect the mental health status and other indicators of QoL of low-income urban residents (Krause, 1996; Manton, 1982; Moody, 2002; Morgan and Kunkel, 2001).

THE CONTEXT OF PUBLIC AND SUBSIDISED HOUSING

The study we report on here was conducted in subsidised public or private senior housing in a small city in north-eastern USA. Conditions (and legal requirements) for residence in these buildings stipulate that individuals are able to live independently, i.e. they can manage the activities of daily living by themselves. Older adults obtain access to subsidised senior housing if they are aged 62 years or older and have incomes below a specified level, approximately \$732/month (HUD, 2003a, b). Recent Housing and Urban Development (HUD) and Americans with Disabilities (ADA) legislation has mandated that public housing, including housing for older adults, accept people under 62 years of age with governmentally certified disabilities who need housing subsidies (NAMI, 2003; Norquist, 2003). Many of the new residents in traditional senior housing arrangements are younger disabled individuals between the ages of 45 and 60. Most are male with a variety of disabilities including mental health problems, chronic physical health problems such as paralysis, deafness or the effects from serious accidents, or long-term drug addiction. Where these federal mandates have been implemented, a significant source of stress for the older residents is the feeling that their safety and security have been compromised because of the presence of younger adults in *their* buildings. We refer to this as 'perceived environmental stress' (PES). A significant proportion of this stress is attributed to the lifestyle differences and communication differences among residents attributed to both linguistic and cultural factors and age or developmental differences. As a result, both groups of residents complain that their QoL is compromised (HUD, 2003b, Schensul *et al.*, 2003; Ward *et al.*, 2004).

DEFINING PSYCHOLOGICAL QoL: PERCEIVED ENVIRONMENTAL STRESS, DEPRESSION, AND OTHER CORRELATES

This chapter uses components of a conceptual framework developed by Evans (2003) to help to explain links between PES, depression, and other measures of QoL. Evans suggests that direct dimensions of the external environment (e.g. air quality and other pollutants, toxic noise levels, safety), and aspects of external environment including building structures that indirectly affect or alter psychosocial processes (e.g. socially supportive relationships and sense of personal control) can affect mental health status. In addition, there is a relationship between burden of illness and social engagement and mental health status (Duggleby, 2002; Herzog *et al.*, 2002). We argue that personal control as measured by major life stressors, social support (from friends, family, and building staff), level of social engagement, and burden of illness affect QoL in this population of older adults, as measured by PES and depression. We also argue that the interaction of predictors will differ with respect to whether PES and depression are combined or independent, and differ by gender as well.

Personal control includes the ability to control one's surroundings (Evans, 2003). Lack of personal control is associated with feelings of powerlessness, help-lessness, fear, and insecurity that may contribute to PES and/or depression. Major life stressors (death of spouse, loved ones, and children; retirement) which can accumulate over time influence people's sense of personal control over their environments. Positive social support is associated with lower rates of depression, and has been shown to buffer depression (Bazargan and Hamm-Baugh, 1995; Hays et al., 2001). Many senior housing residents view other building residents as their primary source of social support (Schensul et al., 2003), but other building residents may not provide desired support, and relatives may make inappropriate demands on building residents. Similarly, positive social engagement or social cohesion can act as a buffer to reduce environmental or personal stress levels (Duggleby, 2002; Herzog et al., 2002; Powell et al., 1999). These relationships can be both supportive and stressful (Mutran et al., 2001). Finally, the structural characteristics of the building play an important role in the mental health and other indicators of QoL of the residents. High levels of crime in surrounding neighbourhoods, buildings that are ill-repaired, non- or poorly functioning security systems, little or no control over who enters the building, excessive noise especially at night, crowded conditions and small apartments, and disharmony among residents are all significant contextual stressors that can affect the mental and physical health of residents. These stressors can be measured independently or through the perception of residents. PES may act as an independent measure of QoL, or can be combined with a measure of depression. One could argue that coexistence of PES and depression indicates low QoL; the absence of both indicates high QoL and that PES and depression alone represent different and intermediate levels of QoL that interact differently with predictor variables.

HYPOTHESES

The goal of this chapter is to examine the predictors of QoL, defined as the combined effects of PES and depression, and to determine whether demographic factors coupled with perceived control, social support, health status, and social engagement predict variation in this distribution. We suggest that there is a continuum of QoL achieved by classifying the sample into four groups based on combining PES and depression, PES or depression alone, or neither. We explore two hypotheses: first, that the group classification can be justified and validated using results from analyses of variance and a set of dependent variables (DVs) related to QoL (physical health status, social support, major life stress, age, and involvement in leisure activities – variables known to be related to mental health and well-being in older adults); second, that the QoL predictors account differentially for variance in each group and that gender differences play a role as well.

METHOD

Data for these analyses were drawn from a 3-year study of the epidemiology of depression and anxiety in older low-income minority adults, and concurrent barriers to mental health evaluation and treatment. The study was conducted with a multi-ethnic population of older urban adults 50 years and over, residing in 13 public and subsidised senior housing buildings in Hartford, Connecticut. An interdisciplinary community-based mental health partnership methodology was employed that com-bined ethnography of emotions (Guarnaccia et al., 1989; Heurtin-Roberts et al., 1997) and social suffering (Kleinman et al., 1997) with survey research methods. Traditionally, studies that directly ask participants about their mental health problems are not necessarily well received and methodology must be adopted so that mental health topics are first introduced with a level of clarity for the participant so that participant interest and trust are optimised. Finding appropriate assessment techniques calls for the collaboration of community service providers, advocates, clinicians from within the community of interest, and well-trained researchers who are sensitive to the cultural, linguistic, socio-behavioural, and other issues relevant to this population.

The research methodology in this study synthesises psychology, anthropology, and medical disciplines (nursing and psychiatry), public health, gerontology, and experts from community-based agencies that have contextual experience in building trust-based relationships with older adults. Resident committees and tenant associa-tions, as well as building management and service staff, played an important role in supporting the project, promoting mental health as an important issue, and smooth-ing the way for the research team over the course of the study period.

Data collection involved: (1) ethnographic observations of buildings and resi-dents, with a focus on individual and social correlates to mental health and well-being; (2) administration of an epidemiologic survey of each building involving face-to-face interviews with residents which included a diagnostic mental health interview and related sociodemographic measures; (3) observations and formal interviews with residents, staff, and members of the community (health care providers, service agencies) to assess access and barriers to mental health care and indigenous definitions of depression and anxiety; and (4) referrals to mental health resources for participants who screened positive for depression and/or anxiety.

To obtain entry into the buildings and to gain resident support, the study team presented an initial formal interactive presentation describing the study simultane-ously in English and Spanish. Information was posted and residents received flyers before the formal presentation and again before survey recruitment and initiation. To qualify for the study, potential participants had to be at least 50 years old, residents of the building, living independently (without the assistance of a conservator), and successfully pass a screen for cognitive impairment. Residents were paid for their time, helped to construct a safety net of family and friends in the event of a mental health problem in the future, and were given a letter of referral to their primary care or mental health provider if their score on the study's mental health diagnostic tools indicated the likelihood of depression or anxiety.

Participants

The study sample included 635 residents, aged 50 years and over, from 13 public and subsidised buildings (see Table 1). Of these, 60% (384) were women; 42% (265) were Puerto Rican; and 37% (232) were black or African American.

TABLE 1. Demographic Information Overall and by Gender

	Variable overall (N = 635)	Men (n = 251)	Women (n = 384)
Age			
Mean (SD)	69.8 (9.2)	68.5 (8.6)	70.7 (9.5)
Median	70.0	68.0	71.0
Age group			
50–61 years	117 (18%)	49 (20%)	68 (18%)
62–74 years	317 (50%)	139 (55%)	178 (46%)
75+	201 (32%)	63 (25%)	138 (36%)
Ethnicity	232 (37%)	76 (30%)	156 (41%)
Black/African	265 (42%)	112 (45%)	153 (40%)
American	44 (7%)	17 (7%)	27 (7%)
Puerto Rican	33 (5%)	16 (6%)	17 (4%)
Other Latino	39 (6%)	22 (9%)	17 (4%)
West Indian/Caribbean	16 (2%)	3 (1%)	13 (3%)
White/Caucasian			
Other			
Length of time in the USA (years)			
Mean (SD)	49.5 (23.9)	48.6 (21.6)	50.0 (25.2)
Median	50.0	49.5	51.0
Length of time in Hartford (years)			
Mean (SD)	34.2 (21.0)	33.2 (20.1)	34.8 (21.6)
Median	35.0	34.0	35.5
Level of education			
Less than H.S. graduation	491 (77%)	199 (79%)	292 (76%)
Marital status			
Single	113 (18%)	68 (27%)	45 (12%)
Married/Cohabiting	93 (15%)	44 (17%)	49 (13%)
Separated	72 (11%)	34 (14%)	38 (10%)
Divorced	166 (26%)	64 (26%)	102 (26%)
Widowed	191 (30%)	41 (16%)	150 (39%)
Work status			
Not working	594 (94%)	232 (92%)	362 (94%)
Monthly income			
Less than $700	472 (74%)	173 (69%)	299 (78%)

Measures

Demographics. As shown in Table 1, selected demographic variables included age, age group cohort, gender, ethnicity, length of time in the USA, length of time in the city, education, relationship (civil) status, work status, and monthly income. Age was grouped into three categories: 50–61, 62–74, and 75 and over. Age groupings were defined based on those in pre-retirement (50–61), those who may or may not be retired (62–74), and retirees (aged 75 and over).

Perceived environmental stress index. The PES index consisted of a set of contextual stressors identified from a review of the literature, ethnographic observations in and around the buildings, and ethnographic interviews with older adults in typical residences. The index included three components (apartment, building, neighbourhood stressors). All items were binary. Examples of typical items include: 'Separation from family and friends', 'Theft and vandalism', 'Lack of information about programs and services', and items about perceived safety and security issues in the buildings. The question was introduced with 'here are some things that you may have encountered living here in the building. For each item, tell me if the situation has bothered you.' The α-coefficient for the apartment stress index was 0.76, and $\alpha = 0.86$ for both the building and neighbourhood item sets. Because the number of items for each of the three environmental components were not equal, the three indices were weighted (equal for apartment, building, and neighbourhood), and then combined for an overall environmental index score ($\alpha = 0.86$). Higher scores indicated higher levels of PES.

Composite International Diagnostic Interview. The Composite International Diagnostic Interview (CIDI) (WHO, 1990) is a well-used psychometrically validated diagnostic interview with excellent reliability and validity (Andrews and Peters, 1998; Kessler *et al.*, 1998; Wittchen, 1994). For this study, the CIDI-Short Form (CIDI-SF-D for depression, CIDI-SF-A for anxiety) (Kessler *et al.*, 1998) was used to screen for presence of depression and anxiety disorders over the last 12 months. The CIDI-SF is a comprehensive, fully standardised interview that can be used to assess mental disorders according to the definitions and criteria of the *International Classification of Diseases* Tenth Edition (ICD-10) and the *Diagnostic and Statistical Manual of Mental Disorders* Fourth Edition (DSM-IV) (Kessler *et al.*, 1998). The interview, designed for use by trained interviewers who are not clinicians, is the most widely used structured interview in the world. The CIDI has been designed for use in a variety of cultures and settings. It is primarily intended for use as an epidemiological tool, but can be used for other research and clinical tasks and has been shown to have acceptable validity (WHO, 1990). Participants above the standard cut-off point (coded as '1') were referred for clinical assessment and treatment to the most appropriate facility among those supporting the study. Non-referrals were coded as zero.

Environmental stress or depression classification. The primary DV for these analyses was group membership in one of four PES or depression status groups. Pearson product moment correlation results indicated a significant relationship between the continuous depression score (number of items endorsed) and the environmental stress score ($n = 635$, $r = 0.28$, $r^2 = 0.09$). Using the weighted PES score, those with scores of zero were placed in the *no PES* group and those with scores above zero were placed in the *PES* group. Using the results from the CIDI-SF-D scores for depression (0 and 1), four groups were created: no PES/no depression; PES/no depression; no PES/depression; and PES/depression.

Predictors were selected based on a review of the literature on QoL and environmental contextual factors related to older urban adults and results from current and past qualitative and quantitative research in these environments, and included participant age, gender, subjective health, and the composite indices. Subjective health, a single item (Ware *et al.*, 1995), asked participants to rate their overall physical health on a scale of 1 = Poor to 5 = Excellent. 'Diagnosed physical health problems' is an additive index that includes the number of diagnosed health problems each participant reported from among a total of 16. Problems included diabetes, arthritis, heart disease, high blood pressure, hearing and vision problems, and lung and breathing problems. Major life stress was assessed using an adapted version of the Holmes–Rahe Social Readjustment Rating Scale (Holmes and Rahe, 1967) and included items such as death of a spouse, retirement, personal injury, and illness. Participants responded 'no' or 'yes' (coded 1 and 2) to 17 items; then the number of affirmative responses were summed. Higher scores indicated increased life stress.

Leisure activities. This 16-item index includes spending time with friends, shopping, etc. (Cornoni-Huntly *et al.*, 1986). Affirmative answers were scored as 1 and summed. Higher scores indicated more participation in leisure activities.

Perceived social support index. This index was created by summing responses to four single items assessing instrumental, emotional, financial, and personal support (from close friends or family). Higher scores indicated increased social support.

Procedure

Following the initial formal study presentation in each building, residents were recruited for voluntary participation in the project. All participants provided written informed consent for study participation. All data were collected using survey or semi-structured interview formats. Community researchers who were diverse with respect to age, gender, and ethnicity, and trained in administration of the instruments, conducted the interviews. Participants were given the option of completing the interview in either English or Spanish. Bilingual and bicultural interviewers conducted interviews with those participants choosing to complete them. The study was approved by the Institutional Review Board (IRB) of the Institute for Community Research and Hartford Hospital.

Data were screened for assumption violations, as well as possible multicollinearity and singularity, and only the composite environmental stress score was transformed (square-root method) for better normality and distributional characteristic approximation (Tabachnick and Fidell, 2001). Missing data were less than 1% and were randomly scattered across the sample. Necessary replacements were made by case-by-case imputation using similar items highly related to the missing response.

Data Analyses

Data analyses included one-way analyses of variance (ANOVAs) to assess the mean group differences across four levels of QoL status (no PES/no depression, PES/no depression, no PES/depression, PES/depression) by the set of DVs and discriminant function analyses (DFA) to assess the best predictors of group (QoL status) membership. Because of the unequal gender distribution in the sample (60% of the sample was female) analyses were performed separately by gender.

RESULTS

Analyses of Variance

Analyses were performed by gender because of unequal sample size. Two sets of ANOVAs were performed to test the hypothesis that the four PES/depression groups could be statistically differentiated. All ANOVAs were statistically significant (see Table 2 for ANOVA summary statistics). Results from Tukey HSD (honestly significant difference) follow-up tests indicated that there were statistically significant mean differences between at least two of the PES/depression status groups for every DV for both men and women (Table 3). In addition, each of the four groups was significantly different from each of the other groups at least once, which supports the creation and differential effect potential for the four PES/depression groups.

Discriminant Function Analyses

Two discriminant analyses were performed to test whether six predictors (age, diagnosed physical health problems, participation in leisure activity, major life stress, subjective health, and social support) could accurately classify participants

TABLE 2. Summary of Uncorrected One-way Independent Groups ANOVAs by Gender of Six Dependent Variables (DVs) by Four Levels of Perceived Environmental Stress (PES)/depression Status

Source	F	η_p	P
Men ($df = 3{,}247$)			
Age	4.58	0.05	< 0.01
Diagnosed physical problems	4.82	0.06	< 0.01
Leisure activity	3.29	0.04	0.02
Major life stress	13.01	0.14	< 0.01
Subjective health status	10.51	0.11	< 0.01
Social support	16.93	0.17	< 0.01
Woman ($df = 3{,}380$)			
Age	14.60	0.10	< 0.01
Diagnosed physical problems	16.70	0.12	< 0.01
Leisure activity	10.78	0.08	< 0.01
Major life stress	7.12	0.05	< 0.01
Subjective health status	17.58	0.12	< 0.01
Social support	11.56	0.08	< 0.01

TABLE 3. Means and Standard Deviations by Gender of Six Dependent Variables (DVs) by Four Levels of Perceived Environmental Stress (PES)/depression Status

Variable	No PES No depression 1	PES No depression 2	No PES Depression 3	PES Depression 4	Significance
Men					
Age	70.2 (7.9)	69.0 (8.9)	65.8 (9.8)	64.9 (7.9)	(1,2 < 4)
Diagnosed physical problems	2.8 (2.4)	2.7 (2.0)	4.3 (2.3)	3.8 (2.5)	(2 < 3,4/1 < 3)
Leisure activity	9.2 (2.9)	9.6 (2.8)	9.1 (2.3)	7.9 (3.1)	(2 > 4)
Major life stress	4.8 (2.3)	5.2 (2.2)	6.3 (1.8)	7.3 (2.3)	(1,2 < 4)
Subjective health status	2.7 (1.0)	2.7 (0.9)	1.8 (0.7)	1.9 (1.0)	(1,2 > 3/1,2 > 4)
Social support	3.5 (0.9)	3.4 (1.1)	2.9 (1.3)	2.1 (1.5)	(1,2,3 > 4)
Women					
Age	73.2 (9.4)	71.8 (8.6)	68.0 (9.1)	65.1 (9.3)	(1 > 3,4/2 > 4)
Diagnosed physical problems	3.1 (1.8)	3.9 (2.1)	5.3 (2.0)	4.9 (2.3)	(1 < 2,3,4/2 < 3,4)
Leisure activity	9.9 (2.7)	9.5 (3.3)	8.5 (1.9)	7.7 (2.5)	(1,2 > 4)
Major life stress	5.8 (2.6)	5.8 (2.8)	7.0 (2.2)	7.3 (3.1)	(1,2 < 4)
Subjective health status	2.7 (1.0)	2.4 (1.0)	1.9 (0.9)	1.8 (1.0)	(1 > 2,3,4/2 > 3,4)
Social support	3.6 (0.7)	3.5 (0.9)	3.0 (1.2)	2.9 (1.3)	(1,2 > 3/1,2 > 4)

Note: Significance refers to IV level mean difference results from Tukey HSD tests.

based on their PES and depression status (no PES/no depression, PES/no depression, no PES/depression, PES/depression). Again, analyses were performed by gender because of the unbalanced nature of the group sizes ($n = 384$, 61% of participants were women), and to allow for assessment of gender differences across the predictors. See Table 4 for within-group correlations and standardised canonical coefficients of six predictor variables for the significant discriminant function by gender.

For the 251 men (39%), the overall Wilks' Lambda was statistically significant, $\Lambda = 0.65$, χ^2 (18, $n = 251$) = 107.39, $p < .001$. Results indicated that the predictors significantly differentiated stress and depression status. Only the first discriminant function was significant and accounted for 91% of the between-groups variability (see Table 4 for the within-group correlations of the predictors and the standardised weights). Using canonical discriminant functions and a 0.30 cut-off (Tabachnick and Fidell, 2001), social support (0.60), life stress (−0.52), and age (0.45) demonstrate the strongest relationship with the significant discriminant function. The means on the discriminant function are consistent with the interpretation. The no PES/no depression group had the highest mean on the discriminant function (0.46) while the PES/depression group had the lowest (−1.38) (see Table 3 for means and standard deviations for the predictors by the four groups). For the no PES/no depression group, social support ($M = 3.5$) was the highest, life stress ($M = 4.8$) the lowest, and

TABLE 4. Within-group Correlations and Standardised Canonical Coefficients of Six Predictor Variables for the Significant Discriminant Function by Gender

Predictors	Men (n = 251) Function 1		Woman (n = 384) Function 1	
	Within-group correlation coefficient	Standardised canonical coefficient	Within-group correlation coefficient	Standardised canonical coefficient
Age	0.33	−0.45	0.49	0.58
Diagnosed physical problems	−0.31	−0.10	−0.51	−0.35
Leisure activity	−0.26	−0.08	0.43	0.29
Major life stress	−0.57	−0.52	−0.34	−0.19
Subjective health status	−0.47	−0.25	0.54	0.34
Social support	−0.65	−0.60	0.44	0.37

mean age (70 years) the highest. Conversely, for the PES/depression group, social support ($M = 2.1$) was the lowest, life stress the highest ($M = 7.3$), and this group displayed the lowest mean age (65 years). Results indicated that the prediction of the PES/depression group status for men was 43%. The κ-coefficient was 0.22, indicating a slight to moderate better than chance for prediction. Finally, to assess how well the classification procedure would predict a new sample, the 'leave one out' procedure was used and indicated that 38% of cases would be correctly classified.

For the 384 women (61%), the overall Wilks' Lambda was statistically significant ($\Lambda = 0.67$, χ^2 (18, $n = 384$) = 154.40, $p < .001$). Results indicated that the predictors were significantly differentiated by PES/depression status. Only the first discriminant function was significant and accounted for 94% of the between-groups variability (see Table 4 for the within-group correlations for the predictors and the standardised weights). Canonical discriminant functions above 0.30 were age (0.58), social support (0.37), diagnosed physical health problems (−0.35), and subjective health (0.34), and had the strongest relationship with the significant discriminant function. The means on the discriminant function are consistent with the interpretation. The no PES/no depression group showed the highest mean on the discriminant function (0.62) while the PES/depression group had the lowest (−1.15) (see Table 3 for means and standard deviations for the predictors by the four groups). For the no PES/no depression group, the mean age ($M = 73$ years) was the highest, social support ($M = 3.6$) the highest, diagnosed physical health problems the lowest ($M = 3.1$), and subjective health rating ($M = 2.7$) the highest. Conversely, for the PES/depression group, mean age (65 years) was the lowest, social support ($M = 2.9$) the lowest, diagnosed physical health problems the second highest ($M = 4.9$), and subjective health status the lowest ($M = 1.8$). Results indicate that the prediction of PES/depression group status for women was 46%. The κ-coefficient was 0.26, indicating a slight to moderate better than chance for prediction. Finally, to assess how well the classification procedure would predict a new sample, the 'leave one out' procedure was used and indicated that 42% of cases would be correctly classified.

DISCUSSION

The goal of this chapter was to explore QoL defined by the combination of PES and mental health status (depression). Our first hypothesis used ANOVAs to test the justification and validation that the four PES/depression groups could be statistically differentiated. All ANOVAs were statistically significant and follow-up Tukey tests indicated that there were significant mean group differences *between each* of the four independent variable levels for at least one of the DVs for each gender. In addition, each of the four groups was significantly different from each of the other groups at least once, supporting the differential potential of the four PES/depression groups for helping to describe QoL.

Results for the second hypothesis showed that group membership could be predicted using a set of indicators related to QoL, in this case, defined by both PES and depression statuses. For men, social support, major life stressors, and age best predicted group membership. Men who showed the lowest QoL, those in the PES/depression group, had the lowest social support scores and highest life stress scores, and were the youngest of the four groups. They also had the lowest level of involvement in leisure activities. For women, similar results were found. The best predictors for group membership were age, social support, diagnosed physical health problems, and subjective health. Women who showed the lowest QoL, those in the PES/depression group, were the youngest, had the lowest social support scores, second-highest number of physical health problems, and lowest subjective health scores. For both men and women, the no PES/depression and the PES/depression groups had the lowest QoL.

It is clear that participants in the lowest QoL groups have the strongest scores for the negative predictors. As noted by Evans (2003) and Krause (1996), the relationships between the negative factors and QoL are often contextually based and very complex. Given this complex nature of relationships, the major findings are partially explained using an adaptation of Evans' (2003) framework: QoL is contextually based on the interrelationship between variables related to perceived control, social support, and building structure.

First, lack of personal control over the environment constitutes a plausible explanation for those experiencing increased PES and/or depression. Limited opportunities for social engagement may result from concerns about the safety of the environment, and the potential for negative interactions among residents (Schensul *et al.*, 2003; Ward *et al.*, 2004).

Second, social support was highest for those in the no PES/depression group regardless of gender. This supports previous findings showing that social support is a buffer for depression. Residents with more positive social supports experienced higher QoL. Negative social interaction (or social stress) including demands by others is directly associated with depression in older urban adults, especially African Americans and Latinos (Felton and Berry, 2000). Residents in this study derive much of their primary social support from their building networks (Schensul *et al.*, 2003), and if these social support networks are compromised, mental health problems and negative perception of the building environment increase.

Third, the structural characteristics of the buildings (safety, security, accessibility, and aesthetics) affect QoL. Several factors are found in this study including ill-repaired buildings, non-existent or poorly functioning security systems, little or no control over who enters the building, high crime areas, overcrowding, disharmony among residents, and excessive noise. Many of these buildings are located in high crime neighbourhoods (Schensul *et al.*, 2003; Ward *et al.*, 2004), and for some of the residents, the presence of drug- and sex-related activity in and around many of the buildings (drug dealing, commercial sex work) can be linked to a loss of perceived control. The consequences can be linked to decreased physical and mental health status, as indicated by the results: the most physical health problems and poorest subjective health was found in those belonging to the low QoL groups.

There were limitations to this study. As Evans (2003) and Krause (1996) point out, it is difficult to explain the complex relationships between environmental stressors, depression and mental health, physical health, and QoL. The identification of underlying effects of mediators and moderators in complex environments is difficult, and the challenge is trying to isolate the primary causal and explanatory factors (Evans, 2003). For example, even though we know that each of the predictor variables (age, diagnosed physical health problems, leisure activity, major life stress, subjective health status, and social support) can be used to classify QoL status as determined by the PES/depression status relationship, it is difficult to assess which variables are contextually most or least important. We are continuing our work in these buildings knowing that more research is needed to determine the strength and contextual salience of each of the variables using a better-defined QoL model.

As previously mentioned, the relationships between PES, depression, physical and mental health status, and QoL are complex and highly contextual (Evans, 2003; Krause, 1996). We have focused here primarily on the negative QoL findings from the study. However, it also needs to be mentioned that many of our participants, despite living in environments that are highly predictive of diminished QoL, did not fall into the low QoL groups. In fact, 76% of the men ($n = 190$) and 74% of the women ($n = 281$) were classified into the groups with higher QoL (no PES/no depression, PES/no depression). We know that, despite the environmentally stressful contexts in which these participants live, many of them possess strong social support systems, participate in leisure and other activities when possible, and despite physical age-related problems, tend to have rather realistic views concerning their physical health. As we continue to work in many of the buildings of this study, we recognise the importance of QoL-related issues specific to older, low-income, ethnically and culturally diverse older adults in senior housing.

This population needs environmental intervention and continued support from the private and public sectors if overall QoL is to be improved. In the current national environment in the USA, however, policies designed to improve life in low-income housing are not likely to be approved and financially supported. Community members and advocacy agencies must work with building residents, staff and administration, and policymakers to form partnerships that continue to challenge policy limitations, in order to improve QoL in these residential settings.

Acknowledgements This research was funded by a grant from The Patrick and Catherine Weldon Donaghue Medical Research Foundation. The authors would like to express their appreciation to members of the research team who contributed greatly to this project including Evelyn Baez, Karen Blank, Kenneth Brockman, Nuria Ciofalo, Kelly Desmarais, Gretchen Diefenbach, Leslie Escobales, Sonia Gaztambide, Eugene Hickey, Stephanie Kneip, Gustavo Lopez, Dawn McKinley, the Hartford Housing Authority, and members of the tenants associations and residents groups from the 13 buildings involved in this study. We would also like to thank members of the Clinical Advisory Committee who have supported the work of the project including mental health services at St. Francis Hospital, Capital Region Mental Health Center, Charter Oak Rice Heights Health Clinic, Hartford Behavioral Health, Institute for the Hispanic Family, and Institute of Living/Hartford Hospital Geriatric Program.

REFERENCES

Andrews, G. and Peters, L. (1998) 'The Psychometric Properties of the Composite International Diagnostic Interview', *Social Psychiatry and Psychiatric Epidemiology*, 33, 80–88.

Bazargan, M. and Hamm-Baugh, V.P. (1995) 'The Relationship Between Chronic Illness and Depression in a Community of Urban Black Elderly Persons', *Journal of Gerontology*, 50(2), S119–S127.

Bojrab, S.L., Sipes, G.P., Weinberger, M., Hendrie, H.C., Hayes, J.R., Darnell, J.C., and Martz, B.L. (1988) 'A Model for Predicting Depression in Elderly Tenants of Public Housing', *Hospital and Community Psychiatry*, 39(3), 304–309.

Bullinger, M. (1989) 'Psychological Effects of Air Pollution on Healthy Residents: A Time Series Approach', *Journal of Environmental Psychology*, 9(2), 103–118.

Cornoni-Huntly, J., Brock, D., Ostfeld, A., Taylor, J., and Wallace, R. (1986) *Established Population for Epidemiologic Studies of the Elderly: Resource Data Book*, Rockville, MD: US DHHS, National Institute on Aging.

Drewnowski, A. and Evans, W.J. (2001) 'Nutrition, Physical Activity and Quality of Life in Older Adults: Summary', *Journal of Gerontology*, 56A, (11), 89–94.

Duggleby, W. (2002) 'Helping Each Other: Social Engagement of Elderly Participants in Qualitative Research', *Nursing and Health Sciences*, 4(S3), A3.

Eriksson, S. (1999) 'Social and Environmental Contributants to Delirium in the Elderly', *Dementia and Geriatric Cognitive Disorders*, 10(5), 350–352.

Evans, G.W. (2003) 'The Built Environment and Mental Health', *Journal of Urban Health*, 80(4), 536–555.

Feldman, P.J. and Steptoe, A. (2004) 'How Neighborhoods and Physical Functioning Are Related: The Roles of Neighborhood Socioeconomic Status, Perceived Neighborhood Strain and Individual Health Risk Factors', *Annals of Behavioral Medicine*, 27(2), 91–99.

Felton, B.J. and Berry, C.A. (2000) 'Do the Sources of the Urban Elderly's Social Support Determining Its Psychological Consequences?' *Psychology and Aging*, 7, 89–97.

Fitzpatrick, K.M. and LaGory, M.E. (2000) *Unhealthy Places: The Ecology of Risk in Urban Landscape*, New York, Routledge.

Garland, J. (1990) 'Environment and Behavior: A Clinical Perspective' in J. Bond and P. Coleman, *Aging and Society: An Introduction to Social Gerontology*, Thousand Oaks, CA, Sage, pp.123–143.

Guarnaccia, P.J., DeLaCancela, V., and Carillo, E. (1989) 'The Multiple Meanings of Ataques de nervios in the Latino Community', *Medical Anthropology*, 11, 47–62.

(GSA) Gerontological Society of America (2002) *The State of Aging and Health in America*, Washington, DC, GSA.

Hays, J.C., Steffens, D.C., and Flint, E.P. (2001) 'Does Social Support Buffer Functional Decline in Elderly Patients with Unipolar Depression?' *American Journal of Psychiatry*, 158(11), 1850–1855.

Herskind, A.M., McGue, M., Holm, N.V., Sroensen, T.I., Harvald, B., and Vaupel, J.W. (1996) 'The Heritability of Human Longevity: A Population-based Study of 2872 Danish twin pairs born 1870–1900', *Human Genetics*, 97, 319–323.

Herzog, A.R., Ofstedal, M.B., and Wheeler, L.M. (2002) 'Social Engagement and Its Relationship to Health', *Clinics in Geriatric Medicine*, 18(3), 593–609.

Heurtin-Roberts, S., Snowden, L., and Miller, L. (1997) 'Expressions of Anxiety in African Americans: Ethnography and the Epidemiological Catchment Area Studies', *Culture, Medicine and Psychiatry*, 21(3), 337–363.

Holmes, T.H. and Rahe, R.H. (1967) 'The Social Readjustment Rating Scale', *Journal of Psychosomatic Research*, 11(2), 213–218.

Kessler, R.C., Andrews, G., Mroczek, D., Ustun, B., and Wittchen, H.-U. (1998) 'The World Health Organization Composite International Diagnostic Interview Short-term (CIDI-SF)', *International Journal of Methods in Psychiatric Research*, 7, 171–185.

Kleinman, A., Das, V., and Lock, M. (eds) (1997) *Social Suffering*, Berkeley, CA, University of California Press.

Krause, N. (1996) 'Neighborhood Deterioration and Self-related Health in Later Life', *Psychology and Aging*, 11(2), 342–352.

Levitt, M.J., Clark, M.C., and Rotton, J. (1987) 'Social Support, Perceived Control and Well-being: A Study of an Environmentally Stressed Population', *International Journal of Aging and Human Development*, 25(4), 247–258.

Manton, K.G. (1982) 'Changing Concepts of Morbidity and Mortality in the Elderly Population', *Milbank Memorial Fund Quarterly*, 60(2), 183–191.

Moody, H.R. (2002) *Aging: Concepts and Controversies*, 4th edn, Thousand Oaks, CA, Sage.

Morgan, L. and Kunkel, S. (2001) *Aging: The Social Context*, 2nd edn, Thousand Oaks, CA, Pine Forge Press.

Mutran, E.J., Reed, P.S., and Sudha, S. (2001) 'Social Support: Clarifying the Construct with Applications for Minority Populations', *Journal of Mental Health and Aging*, 7, 67–78.

NAMI (National Alliance for the Mentally Ill) (2003) Housing: NAMI's Position. Retrieved 14 October 2003 from http://web.nami.org/update/unitedhousing.html

Norquist, J.O. (2003) *How the Government Killed Affordable Housing*. Small Property Owners of America. Retrieved 14 October 2003 from http://www.spoa.com/pages/govkill.html

Powell, L.M., Winter, L., Kleban, M.H., and Ruckdeschel, K. (1999) 'Affect and Quality of Life: Objective and Subjective', *Journal of Aging and Health*, 11(2), 169–198.

Guarnaccia, P.J., DeLaCancela, V., and Carillo, E. (1989). The Multiple Meanings of ataques de nervios in the Latino Community. *Medical Anthropology*, 11, 47–62.

Schensul, J.J., Levy, J., and Disch, W.B. (2003) 'Individual, Contextual and Social Network Factors Affecting Exposure to HIV/AIDS Risk Among Older Adults in Low-Income Senior Housing Complexes', *JAIDS: Journal of Acquired Immune Deficiency Syndrome*, 33(S2), S138–S152.

Shipp, K.M. and Branch, L.G. (1999) 'The Physical Environment as a Determinant of the Health Status of Older Populations', *Canadian Journal on Aging*, 18(3), 313–327.

Svanborg, A. and Djurfeldt, H. (1987) 'Social Isolation and Inactivity in a Population of 70-Year-Olds' in L. Levi (ed.), *Society, Stress and Disease*, Vol. 5, London, Oxford University Press, pp.118–120.

Tabachnick, B.G. and Fidell, L.S. (2001) *Using Multivariate Statistics*, 4th edn, Needham Heights, MA, Allyn & Bacon.

Tran, T.V., Fitzpatrick, T., Berg, W.R., and Wright, Jr., R. (1996) 'Acculturation, Health, Stress and Psychological Distress among Elderly Hispanics', *Journal of Cross-Cultural Gerontology*, 11, 149–165.

HUD (US Department of Housing and Urban Development) (2003a) *History of Designated Housing*. Retrieved 14 October 2003 from http://www.hud.gov:80/offices/pih/centers/sac/designated/history.cfm

HUD (US Department of Housing and Urban Development) (2003b) *Fair Housing Laws and Presidential Executive Orders*. Retrieved 14 October 14 2003 from http://www.hud.gov:80/offices/fheo/fhlaws/index.cfm

Ward, E., Disch, W.B., Levy, J., and Schensul, J.J. (2004) 'Perception of HIV/AIDS Risk among Urban, Low-income Senior Housing Residents', *AIDS Education and Prevention*, 16(6), 571–588.

Ware, J.E., Kosinski, M., Bayliss, M.S., *et al.* (1995) 'Comparison of Methods for the Scoring and Statistical Analysis of SF-36 Health Profile and Summary Measures: Summary of Results from the Medical Outcomes Study', *Medical Care*, 33(S4), 264–279.

Weinberger, M., Darnell, J.C., Martz, B.L., and Hiner, S.L. (1988) 'Differences between Elderly Public Housing Tenants and Community Residents: A Case-control Study', *Journal of Applied Gerontology*, 7, 73–84.

Wittchen, H.-U. (1994) 'Reliability and Validity Studies of the WHO-Composite International Diagnostic Interview (CIDI): A Critical Review', *Journal of Psychiatric Research*, 28(1), 57–84.

WHO (World Health Organization) (1990) *Composite International Diagnostic Interview, Version 1.0*, Geneva, World Health Organization.

10. AGEING AND QUALITY OF LIFE IN ASIA AND EUROPE

A comparative sociological appraisal

INTRODUCTION

Age, being a characteristic that every society uses to move people into and out of statuses, roles, rights, and obligations, is perceived differently in various societies. The process of creating social categories based on age is known as age grading and ageing, and varies from culture to culture, and from one historical period to another. We will see how changes in the proportion of people in a population at each age level have important social consequences in different societies. One of our objectives in this chapter is to outline the connotation of such changes in Asia and Europe. Population ageing or graying (Conception, 1996), due to increased longevity and a declining birth rate, is more prevalent in the industrial world (Europe) rather than the developing world (Asia). The chapter explains how, due to changes in population structure, ageing will alter trends in the decades ahead with special reference to Asia.

Population ageing as an unprecedented phenomenon in human history is increasingly observed in the developed and the developing world – entailing social, economic, health, and other problems. Currently, increases in the proportion of older people (60 years and over), accompanied by declines in the proportion of the young age groups (under 15), are potentially responsible for challenges in various policy dimensions. According to projections, by 2050, the number of older people in the world will exceed the number of young for the first time in history (UN Population Division, 2001). Such a scenario will lead to new challenges in human life. However, by 1998, this historic reversal in relative proportions of the young and the old had already taken place in the more developed regions.

The phenomenon of ageing is affecting each and every one of us in every society irrespective of age and sex. It has a direct bearing on the intergenerational equity and solidarity which are the very foundations of the societies. Hence, quality of life (QoL) is widely affected due to this current change. Likewise, the consequences and implications of ageing are reflected in all facets of life and affecting the QoL in all areas. For example, in the economic area, population ageing will have impacts on economic growth, saving, investment, consumption, labour market, pensions, taxation, and so on. Also, in the social sphere, ageing affects health and health care, family composition, living arrangements, housing, etc. All these factors and more inevitably affect various dimensions of QoL.

However, the trend towards ageing is largely irreversible in the decades to come simply as a result of demographic transition taking place in the world in which both fertility and mortality have decreased in an unprecedented manner.

167

H. Mollenkopf and A. Walker (eds.), Quality of Life in Old Age, 167–177.
© 2007 *Springer.*

According to UN estimates, the world added approximately 600 million older people to its population at the turn of the century, i.e. almost three times the number it had in the middle of the 20th century. However, by the middle of the 21st century, the world's older population will again triple – reaching two billion. Such a great change in population structure needs more attention, more relevant resources, and more appropriate planning.

Although the developed regions have experienced ageing earlier, the less developed ones including Asia are following the same path. In the more developed world, in particular in Western Europe, almost one-fifth of the population was estimated to be aged 60 years and over in 2000. By 2050, this proportion is projected to reach one-third. On the other hand, while only about 8% of the population in Asia is currently over the age of 60, this proportion will increase to 20% by the middle of the 21st century. Such a dramatic change will need relevant and appropriate infrastructures to be able to handle the Asian ageing population, and to be adequately responsive to the QoL needs of the emerging elderly.

As the speed of population ageing is much faster in Asia compared with Europe, and the whole developed world, Asia has much more to do to adjust to the consequences of such population ageing. Likewise, population ageing in Asia is taking place at much lower levels of socio-economic development than was the case in Europe in the middle of the 20th century.

Demographically speaking, in 2000, the median age for the world was 26 years. The country with youngest population is Yemen, with a median age of 15 years, and the oldest is Japan, with a median age of 41 years. By 2050, the world median age is projected to increase by about 10 years to 36 years. The country with the youngest population at that time is predicted to be Niger in Africa, with a median age of 20 years, and the oldest is expected to be Spain, with a median age of 55 years by that year (UN Population Division, 2001). Such a change will give a different perspective to the ageing population so far as their QoL is concerned.

A new phenomenon of the 'elderly ageing' is also growing, and it is estimated that those aged 80 years and over are currently increasing at the rate of 3.8% per annum and they comprise more than one-tenth of the total number of older people. Under such conditions one-fifth of older people will be 80 years and over by the middle of the 21st century. Such a scenario indicates that the potential dependency burden on working-age groups (15–46) could be remarkable and heavy.

While the majority of the older population are women, due to the fact that female life expectancy is higher than male, as estimated in 2000, there were 36 million more women than men aged 60 years and over. Also, as the ratio widens at the age of 80 and above, when there are almost two men for every five women, more specific plans should be implemented so as to protect the QoL of such potentially vulnerable people.

So far as income is concerned, countries with a high per capita income tend to have lower rates of labour force participation and, in contrast, older people participate to a greater extent in labour markets in the less developed regions including Asia largely due to the limited coverage of retirement schemes and the low incomes when they are provided. Therefore, many have to work even at ages unsuitable for their physical capacity, which eventually leads to their poor QoL.

Another factor responsible for low QoL among older people is illiteracy. Although a lot of effort has been made to eradicate illiteracy, it is still common especially among Asian older people. According to estimates, almost half of those aged 60 years and over in the less developed regions including Asia were declared illiterate by 2000. Only about one-third of older women and three-fifths of older men could read and write at a basic level whereas, in Europe, literacy has almost approached full coverage except in some countries.

In the study of older people in modern society, growing attention has been focused on their life satisfaction and QoL (Tinker, 1983; Hughes, 1990). Life satisfaction is related to the degree to which people feel they achieve their aspirations, morale, and happiness. But precisely how QoL is measured is difficult to decide. In a nutshell, ways of measuring QoL of older people could include: their individual characteristics, their physical and mental health, their dependency, their housing, their social environment, their comforts and security. However, to develop a system of health care and security for older people, paying special attention to the needs of women is highly recommended with a view to enhancing the ability of families to take care of their older relatives.

AN OLD AGE CRISIS

While the age of retirement is lowering in many parts of the developing world due to the large number of young people waiting to get into jobs, it is in contrast, increasing in the Western world, especially in Europe, due to an increase in the number of older people and lack of young labour market entrants. The emerging problem is being solved by some European countries by attracting guest workers from the developing countries.

Systems of financial support for older people are in trouble worldwide. To ensure that these systems continue to protect pensioners and promote economic growth, countries need to consider comprehensive pension reforms. Based on estimates, over the next 25 years, the proportion of the world's population over 60 will nearly double, i.e. from 9% to 16%. However, populations are ageing much faster in the developing countries than they did in industrial countries. As today's young workers near retirement around the year 2030, 80% of the world's older people will live in what today are developing countries (mainly Asian). More than half will live in Asia, and more than a quarter in China alone (Finance and Development, 1995). These countries need to develop their old-age security systems quickly and make them sufficiently resilient to withstand rapid demographic change. Under the conditions where the extended family system and village support networks on which two-thirds of the world's older people depend, tend to break down due to pressures of urbanisation, industrialisation, and rapid sociocultural mobility, older people come to be at loss. As a result of all these factors some old-age systems are in serious financial trouble. However, the situation happens to be more acute in Asia and more moderate in Western Europe.

Challenges to Societies

In traditional communities work and organisational structure of the family were interconnected. Relationships and contacts within age groups were close and there was mutual dependence between the young and older groups. Such close connections and exchange of functions between generations within the family network ensured the survival of older people where there were no other forms of guaranteed social support in old age. This type of network allowed older people to have enough authority and participate in family functions based on a family division of labour. However, industrialisation and the process of social change in both Asia and Europe, have led to social differentiation of age groups with reference to economic functions, official retirement, and other such conditions.

Currently, due to the modernisation of societies in different educational, scientific, and technical respects, the younger generations are capable of providing for themselves. Therefore, the older generations are sometimes left isolated and dependent on pensions and other kinds of social help. This process eventually promotes the relative independence of generations from each other, diminishes the necessity for cooperation, and results in the destruction of family solidarity and mutual dependence. Therefore, in modern societies, responsibility for older people is more and more becoming formal and depersonalised. Under such systems older people do not play their former roles. They depart from the family, for example not carrying out the role of grandparents, and the younger generations tend to require less support from the older ones (Aleksandrova, 1974).

Socio-economic Effects of Ageing

The inevitable harmful social and economic effects of ageing are becoming obvious more than ever before particularly in Asia (Population Council, 2001). Most prominent among the concerns that are being voiced with respect to ageing is how to fund social security programmes in the face of increasing numbers of retired people, and how to pay for rising health care costs generated by older people (Mullan, 2000). These concerns have at times led to the conclusion that population ageing is bound to be more a catastrophic drain on economic resources. While the Western (European) countries are and will continue to be rather well equipped to handle the present and projected increase in the older population, the whole scenario is more problematic for Asian countries where there are shortages of necessary infrastructures and societies that are rapidly changing to new cultural forms. Thus, the Asian older population is much more socially and economically insecure in different dimensions.

Living in a demographically diverse world has also led to unprecedented ageing too. While the global population increased by two billion during the last quarter of the 20th century reaching six billion in 2000, resources have not increased sufficiently to respond the increasing number of older people particularly in Asia. As projected, the population will increase by another two billion during the first decades of the 21st century and, as nearly all of the increase has been and will be in the developing countries including Asia, ageing problems will emerge more than ever before in the south.

As we live in a world of unprecedented demographic diversity, we should be more cautious and planning-minded. As the traditional demographic groupings of countries are breaking down, more socio-economic problems relating to ageing populations are emerging. Over the next 25 years, increases in population in South Asia and the Middle East are expected to be larger than the last 25 years. In contrast, in European countries, and in East Asia, population growth has slowed or stopped and rapid population ageing has become a serious concern (Population Council, 2002). Increasing levels of ageing accompanied by increasing mobility and urbanisation are affecting the economic and social outlooks of many countries.

The challenges found due to such diversities require adequate responses. The most urgent of these occur where rapid population growth, high levels of poverty, and low levels of economic growth coincide. Under such conditions older people face various problems.

THE VULNERABILITY OF OLDER PEOPLE

Deteriorating environmental conditions and extreme events do not affect all countries and populations in the same way. Hence, many factors contribute to the vulnerability of older people including poverty, poor health, low levels of education, gender inequality, lack of access to resources and services, and unfavourable geographical location. All of these affect older people more in Asia than in Europe. Under conditions where populations in general are socially disadvantaged or lack a political voice, older people in particular are at the greatest risk. Vulnerable ageing populations include the poorest, the least empowered segments and especially the women. These vulnerable older people have limited capacity to protect themselves from current and future environmental and social hazards, such as polluted air and water, disasters, and the adverse consequences of large-scale environmental change such as biodiversity loss and climate change.

So far as older women are concerned, they particularly face greater risk of physical and psychological abuse due to discriminatory societal attitudes and the non-realisation of the human rights of women. Women's poverty is directly related to the absence of economic opportunities and autonomy, lack of access to economic resources including credit, land ownership and inheritance, lack of access to education and support services, and minimal participation in the decision-making process. Poverty can also force women into a situation in which they are vulnerable to sexual exploitation (United Nations, 2002).

Older Widows

The aged members, especially older women, face a serious situation in today's family structure. The demographic scenario of ageing indicates a rise in the longevity of women (Desai and Thakkar, 2003). As the proportion of older people increases, that of widows and widowers is very likely to rise. Comparing the proportion of widows with the widowers, the number of the former is higher due to the fact that women marry earlier than men and also tend to outlive men. Similarly, after the age of 60, women have the chance of longer life. The chance of remarriage for men in later life keeps the proportion of widowers lower than widows almost everywhere. However, the consequences of widowhood leading to isolation and loneliness are faced more by women than men.

Research shows that widowhood appears as an effect of marital dissolution world-wide. Apart from divorce, in most cases it happens as a natural event due to the death of a spouse. In both cases, women tend to suffer longer-term of negative social and economic consequences compared with men (Neubeck, 1996, p.478).

In spite of recognising the problems faced by older widows in many parts of Asia, governments appear unready to take more responsibility and, instead, want the individual family to help its members in crisis situations such as widowhood. The challenges faced by widows towards the end of the 20th century have increased even today among large numbers of widows. To solve and improve the problem, assistance, cooperation, and contributions of different institutions are required.

THE THEORETICAL CONTEXT OF AGEING

Ageing as a *transition in the life course* is fundamentally different from other ascribed statuses, such as race and gender. Being black or white, male or female, and the like, is a lifelong status, except in rare cases. Age, in contrast, is a transitional status because people periodically move from one age category to another. As people age, they face different sets of expectations and responsibilities, enjoy different rights and opportunities, and possess different amounts of power and control. Consequently, transitions from one age status to another are societally important (Keller *et al.*, 1994). They are often marked by rites of passage and public cere-monies that are full of ritual symbolism that record the transition being made. Weddings, retirement dinners, funerals, and so on are all examples of rites of passage in an industrial society.

To better understand the ageing process, five key sociological concepts will be helpful as we explore further the ideas of age, ageing, and age structure with Asian and European connotations. Age structure is a specific element of the *social struc-ture* of all human societies. Also, historical or cultural differences in age structure create different contexts for *social action* by individuals and groups. Changes in age structure also bring about problems of *functional integration*. Different proportions of age groups in a population affect *power* too, such as the age for voting. Discussion of the meanings of age in different societies is different from one *culture* to another. Generally speaking, age shapes the flow of people into, and out of, social roles and statuses (social networks), and the rights and responsibilities that go with them, which is different between one society and another. Age also organises the distribu-tion of valued resources in a society such as money, power, and prestige (O'Rand and Krecker, 1990).

From the point of view of *conflict theory*, older people became a social problem when those in power in the industrial world found it advantageous to push them aside. As the industrial revolution spread more than 100 years ago, managers of big businesses found older people a nuisance. At that time, they drew more wages than young workers who wanted the jobs of the older workers. As older workers were pushed out of their jobs the percentage of those over 65 who worked declined steadily. As older people lost out to younger groups with new technical and institutional resources, the meaning of 'to retire' changed from 'to withdraw from public notice' to 'to be no longer qualified for active service' (Achenbaum, 1978). To be old came to mean to be castaway; that is, to have almost nothing, and to be dependent on whatever someone might give you (in those days).

Conflict theory also explains how older people reacted to the social changes that brought them poverty and deprivation. They consolidated into a powerful lobbying force for social security. Therefore, the social security benefits currently available for the ageing population are the result of direct conflict between competing interest groups. The old banded together to push their interests and concerns and that was a starting point for the old age benefits in the West.

The conflict perspective emphasises that power, privilege, and other resources are limited, and that they are distributed unequally among the various groups in the society. As it pursues its own interests and values, each group comes in conflict with the others. Thus say conflict theorists, whenever you examine a social problem, you should look at the distribution of power and privilege, for social problems centre around the conflicting interests and values of a society's groups (Henslin and Light, 1983). Conflict in society, then, is both natural and inevitable. Although it always exists, it played a vital role in providing older people with retirement security, especially in the West in the early 20th century. Therefore, the poor and neglected older population could achieve their rights with the framework of conflict theory.

THE ASIAN OUTLOOK ON AGEING

In Asia ageing has become an issue of concern for the different sectors of governments dealing with the socio-economic needs of older people. Although older people are expected to be respected, many societies are witnessing a new trend. Because of rural–urban migration, industrialisation and shifting employment patterns among the younger adult population, older people are facing increased social isolation and many other challenges in many Asian countries, particularly in rural areas. From a socio-psychological point of view these isolated people in a community such as a large city feel alienated (Experts, 2000, p.161).

However, in some countries such as Thailand, older people are valued for their contribution to society and are encouraged to remain active (UNFPA, 2002). In addition to many other roles played by the seniors, most of the older people in Thailand play a leading role in religious observances by supervising and providing information concerning religious activities to younger members of the family and community. They also transmit their traditions and culture to the younger generations.

Although the developing Asian countries have been experiencing rapid social, cultural, and economic changes, the conditions of older people have not improved. So far as older women are concerned, they are in an even worse situation. They are identified as subordinate to men throughout their lives and, when ultimately left alone when they get old, they are often deeply poor and destitute. While in the developed countries retirement is expected to be the period to enjoy personal and leisure activities, in the developing Asian countries, older people are still preoccupied with their basic livelihood. As social welfare and health insurance in many developing Asian countries have limited funds, individual financial resources play a significant role in enhancing and improving one's QoL (IFA, 2001). Under such a scenario, older people in developing Asia remain financially dependent on others. To illustrate this more thoroughly, they psychologically and physically depend more on others rather than the state, or the relevant agencies. Very few Asian countries have infrastructures sufficient to help their disabled and older people. One of the countries functioning well in this respect is Singapore' wherein substantial financial resources have helped the ageing population, and thereby enhanced their QoL in different ways.

Modernisation in many parts of Asia has greatly influenced the lives of older people due to increasing change in the family structure and ties, more mobility among the families and more employment by the women. All of these have caused the families to be more segmented and, consequently, not to have enough time to invest in older people. Also, with the increasing decline in fertility and mortality rates, population ageing is appearing more than ever before: generating significant demands for long-term care (IFA, 2001). Hence, the demographic trends are dramatically changing the face of many nations in Asia, or will soon do so. One way of measuring the speed of these shifts is through a measure of 'population ageing'. Although the phenomenon is very recent in Asia, it is rapidly spreading in many parts of the continent.

However, as explored, the majority of older people still wish to live with their adult children. There is clear evidence showing the familism and family-feeling among older people in most parts of Asia. As observed, modernisation is seen as a paradoxical phenomenon in Asia since it is eroding the traditional support system.

Today, older people have come onto the agenda of many Asian countries as happened in the West previously. Asia too needs to develop more literature on this topic. It is becoming the region where the majority of older people are concentrated. That is to say, the majority (52%) of the world's senior citizens (people aged 60 and over) live in Asia; 4 in every 15 are concentrated in Eastern Asia including China, and 1 in 6 inhabit South-central Asia including India (ESCAP, 1996). Similarly, about 1 in 15 live in South-east Asia including Indonesia, and Western Asia includes the rest.

Such population development is largely due to economic success in the region and a result of success in population control since early 1980s. Increased life expectancy which also resulted in, or is a consequence of, improved health care and living standards, has led to increasing old age in all societies, but more so in the Western world. However, while until around the 1970s many countries especially in South-east Asia were still considered to have young populations, since the 1980s the older age categories have increased, making it necessary to examine the conditions of these growing numbers of older people.

Since the 1950s life expectancy of men has increased by 20 years or more in Indonesia, Republic of Korea, and Thailand, and by 15 years in Japan. While the number of women has increased even more dramatically (*Human Development Report*, 1997), these developments have eventually resulted in an accelerated increase in the proportion of the older population in almost all parts of Asia, but with some fluctuations.

Older people's conditions are not the same all over Asia. For example, in South-east Asia, the proportion of those aged 65 and over is not yet as high as in Japan. There is growing concern in this regard since the necessary institutional arrangements for taking care of older people outside the family are not yet in place. Therefore, much has to be done to ensure adequate responses. To ease and solve the problems of older people, especially in Asia, more interdisciplinary research and education addressing the above topics is necessary at all levels. The different disciplines should also conduct their studies in ways that make the result mutually accessible to older people.

A EUROPEAN OUTLOOK ON AGEING

The establishment of individual and universal mandatory pension rights has come to be seen as an efficient way to eradicate poverty in old age among both women and men. Health promotion and well-being of older people in Europe are among the issues which have been of priority and well attended in Europe, compared with Asia, in the course of the 20th century.

Modernisation, which first occurred in Europe, is a multidimensional concept (Gruyter, 1993). It is divided into four distinct elements:
• Economic modernisation (industrialisation)
• Political modernisation (democratisation)
• Societal modernisation (realisation of freedom and equality)
• Cultural modernisation (the move towards rationalism)
All four of these dimensions affected older people's lives in one way or another. The process of modernisation is still advancing and is changing the lives of older people in almost all European countries, namely, changing their QoL.

Progress in general QoL has caused the major social risks such as illness, accidents and impecunious old age to be protected in Europe on a larger scale as compared with Asia (Mire, 1997). At the same time, while poverty is lower among older people in Europe compared with Asia social exclusion is appearing in the continent as a new concept. Poverty and social exclusion being central issues of social policy, so far as older people are concerned, they have been well addressed in Europe.

Since new forms of administration occurred in Europe much earlier than Asia due to the emergence of industrialisation, problems affecting older people, and methods

TABLE 1. Comparative Ageing Indicators of Asia and Europe for Selected Countries in Two Periods (%)
(From World Population Data Sheet(s) 1995 and 2005)

Asia			Europe		
	Per cent 65+			Per cent 65+	
(Region) Country	1995	2005	(Region) Country	1995	2005
Asia	5	6	Europe	13	16
Armenia	7	11	Denmark	15	15
Azerbaijan	5	7	Finland	14	16
Iraq	3	3	Ireland	11	11
Jordan	3	3	Norway	16	15
Lebanon	5	6	Sweden	18	17
Saudi Arabia	2	3	Britain	16	16
Turkey	4	6	Austria	15	15
India	3	2	Belgium	16	17
Iran	4	4	France	15	16
Nepal	3	4	Germany	15	18
Pakistan	3	4	Luxembourg	13	14
Srilanka	4	6	Holland	13	14
Indonesia	4	5	Switzerland	15	16
Singapore	7	8	Czech Rep.	10	14
Thailand	4	7	Hungary	14	16
Viet Nam	5	7	Poland	11	13
China	6	8	Romania	11	14
Japan	14	20	Russia	11	13
North Korea	4	8	Italy	16	19
South Korea	5	9	Portugal	15	17
Taiwan	7	9	Spain	14	17

of eliminating them started earlier in that continent, especially in the Western part, compared with Asia, and that is why QoL there started to be enhanced earlier too. Comparative sociological research indicates that there are major differences between the QoL of older people in Asia and Europe. The main causes of difference between the two stems from lack of resources, lack of capital and underdevelopment of administration.

CONCLUSION

There is a clear need for research on the type and magnitude of the conditions and problems of older people in relation to gender, age groups, physical and mental health status, socio-economic status, and ability to continue to be productive. Research is also necessary on the enabling environment and the resources available in the family, community, society, and the state to care for older people in a way that is conducive to making them independent, self-reliant, and productive.

It is quite evident that the unprecedented demographic, social, and economic changes which had their origins in the 19th and 20th centuries, and are continuing into the 21st century, are transforming the world. The declines in fertility reinforced by increasing longevity have produced and will continue to produce unprecedented changes in the structure of all societies, notably the historic reversal in the proportion of the young and the old in Europe, and in some cases in Asia. Many parts of Asia are still in their infancy with respect to the development of formal services. Hence, despite rapid social change, family care-giving for older people is still the dominant type of care-giving in Asia. Likewise, the profound, pervasive, and enduring consequences of population ageing present enormous opportunities as well as enormous challenges for all societies. That is a scenario which needs research, development, planning, and investment.

REFERENCES

Achenbaum, W.A. (1978) *Old age in the New Land: The American Experience Since 1970*, Baltimore, Johns Hopkins University Press.

Aleksandrova, M.D. (1974) *Problems of Social and Psychological Gerontology*, Leningrad, University of Leningrad Press.

Conception M.B. (1996) 'The Greying of Asia: Demographic Dimensions' in Asian Population Studies Series, No.141, *Added Years of Life in Asia, Current Situation and Future Challenges*, Bangkok, ESCAP.

Desai, N. and Thakkar, U. (2003) *Women in Indian Society*, New Delhi, National Book Trust, 85pp.

ESCAP (1996) Data Sheet, Bangkok, UN Economic and Social Commission for Asia and the Pacific.

Experts (2000) *Advanced Learner's Dictionary of Sociology*, New Delhi, Anmol Publications.

Finance and Development (1995) Washington DC, IMF Publications.

Gruyter, W. (1993) *European Sociology*, New York, ISA Publications.

Henslin, J.S. and Light, D.W. (1983) *Social Problems*, London, McGraw-Hill.

Hughes, B. (1990) 'Quality of Life' in S. Peace (ed.) *Researching Social Gerontology*, London, Sage, pp.46–58.

Human Development Report (1997) New York, Oxford University Press.

IFA (International Federation on Ageing) (2001) *Montreal Conference: Selected Papers,* Montreal.

Keller, S., Calhoun, D., Light, D., and Harper, D. (1994) *Sociology*, London, McGraw-Hill.

MIRE (1997) *Comparing Social Welfare Systems in Southern Europe*, Vol. 3, Florence Conference.

Mullan, P. (2000) *The Imaginary Time Bomb: Why Ageing Problem is Social Problem?* New York, I.B. Tauris.

Neubeck, K.J. (1996) *Sociology*, New York, McGraw-Hill.

O'Rand, A. and Krecker, M. (1990) 'Concepts of the Life Cycle', *Annual Review of Sociology*, 16, 241–262.

UN Population Division (2001) *Population Newsletter*, 27, December, New York, Department of Economic and Social Affairs.

Population Council (2002) *Population and Development Review*, 28, 2, June, New York.

Population Council (2001) *Population and Development Review*, 27, 1, March, New York.

Tinker, H. (1983) *Improving the Quality of Life and Promoting Independence of Elderly People*, London, HMSO.

UNFPA (2002) *Population Ageing and Development*, New York.

United Nations (2002) *Madrid International Plan of Action on Ageing 2002*, International Association of Gerontology, Co-sponsored by UNFPA.

World Population Data Sheet(s) 1995 and 2005, Washington, DC, Population Reference Bureau.

11. ETHNICITY AND QUALITY OF LIFE

INTRODUCTION

As Gabriel and Bowling (2004) note, the concept of quality of life (QoL) is used to refer to both macro (societal, objective) and micro (individual, subjective) definitions. Bowling (2004) distinguished eight different models of QoL within these two broader categories: objective standard of living; health and longevity; satisfaction of human needs; life satisfaction and psychological well-being; social capital; ecological and neighbourhood resources; health and functioning; cognitive competence, autonomy, self-efficacy; etc.; and values, interpretations, and perceptions. Within the research on ethnicity and QoL in old age this same diversity is evident and there is less research available. Recently, Moriarty and Butt (2004) noted that very little of the work on QoL has focused on people from minority ethnic groups. Furthermore, the literature on ethnicity includes a large number of different ethnic and subcultural groups within many different host nations. These two factors, the relatively few studies focusing in the area, together with the fact that those that do often deal with different ethnic groups in different countries, mean there is insufficient research to draw international conclusions on ethnic groups generally or on particular ethnic groups.

With these caveats there is nevertheless a body of research that focuses on ethnicity and on particular ethnic groups within Western societies. This chapter reflects on that research in terms of QoL in old age and raises questions for future research. The chapter begins by narrowing the focus to QoL studies that deal with subjective social indicators of life satisfaction and psychological well-being. It then discusses findings relating to ethnic groups generally, i.e. similarities and differences between majorities and minorities, with the latter being viewed as a single group. This is followed by an examination of correlates of QoL among particular ethnic groups, with the older Chinese people living in Canada as an illustrative example. Finally, issues encountered in ethnic studies that make achieving valid data especially difficult (such as the general confounding of ethnicity with socio-economic status in the empirical world) are described.

NARROWING THE FOCUS

Narrowing research on QoL to subjective social indicators such as life satisfaction and psychological well-being does not remove the problem of multiple terms with variations in meaning and a whole host of different indicators. Subjective social indicators include, for example, studies on morale, self-esteem, individual fulfilment and happiness, subjective well-being, and overall well-being (Walker, 2005;

179

H. Mollenkopf and A. Walker (eds.), Quality of Life in Old Age, 179–194.
© 2007 *Springer.*

Veenhoven, 1999). Much of the gerontological literature has focused largely on the cognitive component of life satisfaction and its many domains (including but not exclusive to: family, social relationships, finances, leisure, spirituality, and health). The popularity of life satisfaction has been noted by Westerhof *et al.* (2001). When used as a measure of successful ageing, two-thirds of the studies in a meta analysis of the correlates of subjective well-being in old age chose life satisfaction (Pinquart and Sorenson, 2000). This gerontological focus has a long-standing history, particularly in the USA, which contrasts somewhat with a European focus within QoL on decline and disability, and therefore health (Gabriel and Bowling, 2004).

Subjective well-being refers to a person's evaluative reactions to his or her life – either in terms of life satisfaction (cognitive evaluations) or affectivity (ongoing emotional reactions). The domains of QoL are relatively consistent across a variety of measures and include family and other relationships, emotional well-being, social activities, personal health and health of others, finances and standard of living, independence, religion and spirituality (Brown *et al.*, 2004), as well as satisfaction with life as a whole (Diener and Diener, 1995; Argyle *et al.*, 1989). Not surprisingly, health and functioning are more important for older people than for younger adults. Gabriel and Bowling (2004) derive similar domains from in-depth interviews with older people and open-ended survey questions to older people, namely social relationships, finances, and activities.

Subjective well-being is important to study both because of its inherent value (people's experiences, where they live their lives) and because of what Walker (2005) refers to as the apparent paradox between positive subjective evaluations expressed by so many older people while living under objectively adverse conditions including poverty and poor housing. In Canada, for example, more than three-quarters of all older people living at home report being diagnosed with at least one chronic condition (the most common of which is arthritis or rheumatism), and approximately one-third suffer from some functional disabilities; diseases that impair cognition, specifically dementia, increase with age. However, subjective perceptions of physical health do *not* decline with age (more than three-quarters of older people rate their health as good, very good, or excellent). Those aged 75 and over are three times more likely than 18–19-year-olds to score high on a measure of sense of coherence (a view that the world is meaningful, events are comprehensible, and challenges are manageable) (Chappell *et al.*, 2003). Gitelman (1976) reports that, in a study of black and Jewish aged poor in the USA, Jewish respondents had objectively better lives but were less satisfied than older black people. As is evident from the following section, this same disparity between objective living conditions and subjective experiences of the quality of their life is evident among older people of ethnic groups as well.

RACE/ETHNICITY

Race, ethnicity, and minority refer, within gerontology, to individuals who live in a country where the majority of the population is from a different race, or ethnic or cultural background, than that of the individual. Conceptually, researchers often combine individuals from different ethnic groups in their analyses on the assumption that all individuals who are living in a country in which the majority does not share their background are similar in important respects to one another. This is the rationale for ethnic studies as a distinct area. Studies confirm that ethnic groups within a Western country often share disadvantages. Decades ago, the US Human Resources Planning Institute (1972) reported that minority older adults had lower perceived health and economic status than older white people. Blau *et al.* (1979) documented racial differences as greater than age variations that tended to remain when socio-economic status was controlled. Similarly Sutton and Persaud (1989) reported discrimination and unequal opportunity as pervasive for minority older people. The consequences of a historic lack of education and employment opportunities included shorter life expectancies, earlier loss of spouse, and greater declines in health. In the 1990s, the continuing inequities in income and assets, educational background, and societal factors for minority older people in the USA were still being documented (Miranda, 1992; Rubensteen and Kramer, 1994). The same inequities have been reported outside of the USA. Blakemore (1985) reports severe disadvantages in terms of income and health for minority Asian and West Indian ethnic groups in both Great Britain and the USA.

These inequities continue into the present. Takamura (2002) notes a marked disparity in the health and insurance status of older minority Americans and the majority white population. In Britain, Bajekal *et al.* (2004) report that on conventional indicators of social inequalities such as material circumstances, health, participation in formal social networks, and quality of the physical environment, Pakistani, Indian, and Caribbean groups all rank low. In Canada, similar findings are evident (McDonald *et al.*, 2001).

It is this enduring inequality that has led to a major concern with ethnicity and QoL. Indeed, Blau *et al.* (1979) have argued that race/ethnicity is the most fundamental division within American society. To argue that ethnicity is a significant factor shaping evaluation of life areas is not to deny that there are some whites who also live in such disadvantage. However, the probability of someone from an ethnic minority living in disadvantage is greater than it is for whites. Ethnic minorities also often have the added difficulty of living with discrimination and prejudice from the host society.

These inequalities that plague minority groups within Western societies reflect the broader economic, political, and social structures that result in disadvantage. A political economy perspective draws attention to these broader structural forces in society that impact ethnic groups, positing that the foremost character of the mode of production, currently the move from competitive monopoly to global capitalism, embeds an inherent division between those who own and control the means of production and those who do not. The nature of ethnic, racial, and minority group consequences of structural inequalities are less well researched than are class or gender-based inequalities (Chappell and Penning, 2005; Coburn, 2001; Calasanti and Zajicek, 1993). However, as evident in the preceding section, ethnic-based inequalities are similar to class and gender-based inequalities in that ethnic groups are more likely to report worse health and socio-economic circumstances. Even among recent immigrants who tend to have better health than native Canadians due to a selectivity factor, over time this healthy immigrant effect appears to dissipate (Chen and Wilkins, 1996). Furthermore, older people have often experienced a history of life-long disadvantage. The accumulative effect of this disadvantage by the time these individuals reach old age is often much worse than experienced during the younger years (Nazroo *et al.*, 2004).

Interestingly though, these structured inequalities, while without question causing hardship, are not necessarily correlated with social psychological measures of QoL, i.e. with one's satisfaction with and happiness in life. Bajekal *et al.* (2004) note that Pakistani, Indian, and Caribbean groups in the UK ranked higher than whites on their perception of their residential neighbourhood and frequency of family contact, despite reporting objectively worse conditions (also see Blakemore, 1985). Moriarty and Butt (2004) find that social support is a major component of QoL for ethnic groups in Great Britain and that, despite poor health and low incomes, minority groups consistently report social support to be high.

Social support, including interactions with family, is a popular explanation for the disjuncture between the objective indicators of QoL and the psychosocial indicators of QoL. Our lives, while heavily influenced by larger structural forces such as the organisation of capitalism within society, are experienced much closer to home. The key to these experiences is our relationships with those close to us, particularly with our families. Family values and behaviour derive from the cultural context within which we live our lives. However, acculturation of immigrant groups into host cultures is not well understood. While assimilationists implicitly assume that the dominant culture is superior, there is evidence to suggest that a model of bi-culturalism or a hybridising of the donor culture and the receiving culture is appropriate to identify the experience of many immigrant groups (Chen, 1997; Van Ziegert, 2002). The creation of new transnational identities consists of a blending of the two cultures (the donor and the receiving).

Much of the gerontological research on ethnicity and the family suggests that ethnic subcultures have a much greater value of familialism and provide support and care for one another when the health of older members declines. Indeed, in the 1980s it was commonly assumed that members of ethnic cultures had much more extensive social interactions and caregiving for their older members, obviating the necessity for formal care services in old age. In the late 1980s Driedger and Chappell (1987) reported that empirical research at the time revealed that some ethnic groups demonstrated larger, and in some cases stronger, informal social support than did other ethnic groups, and that some but not all had larger social networks than was true for whites. As they note, however, the cultural interpretation of 'friend' is variable with some groups, such as aboriginals and blacks in North America often interpreting what whites would consider 'acquaintances' as friends whereas whites have a much narrower definition of 'friend' – which has strong implications for interpreting data. Culturally insensitive measures may contribute to the widely held belief in larger and stronger social networks among ethnic minorities.

They also raise the possibility that because many ethnic groups are disadvantaged economically, stronger social networks may reflect social arrangements based on need and may or may not reflect preferred modes of caregiving if other options were available.

It is still unknown whether the ways ethnic groups care for their older people is largely because of poverty (they cannot afford, or cannot access, appropriate services due to cultural differences such as language barriers) or whether it is a part of their cultural norm to provide care for the elderly members more often than do whites. Research does suggest that there are barriers of access for ethnic groups to utilise health services, including the lack of health care workers who speak their language or understand their culture (Koehn, 2005). So many of the ethnic groups are also disadvantaged economically that it is difficult to empirically disentangle the effects of poverty from that of culture.

In addition, the extent to which ethnic minorities differ from one another and/or from the host society in terms of their familial in-group embeddedness is not clear. Almost all people, whether part of the majority or minority, have network ties and almost all of those who are elderly today are, or have been, married and have children and live close to at least one of them. Almost all of us have regular contact with our families and friends. The family, in other words, is the first line of support in meeting the needs of ageing members in all societies (Kivvet, 1993).

A related area of interest is whether greater or lesser assimilation within one's own ethnic culture while living in the host society is beneficial or harmful. Existing attachments may be pulled apart by migration but, on the other hand, members of a subcultural group in the host country may provide support to newcomers. Being embedded within the minority culture can buffer the traumas of change and of discrimination from the host society and it can also prevent individuals from becoming more integrated into the host culture. Moriarty and Butt (2004) report that, for the immigrant groups they studied in the UK (black, South Asian, Chinese), the men relied upon the workplace for their social contacts noting that the vast majority of them had emigrated for work, and in retirement maintained some of these contacts. Women were more centred on their families.

There were, however, differences among ethnic groups so that, for example, among Indians, Pakistanis, and Bangladeshis, 'chain migration' was common – they lived in close proximity to other ethnic group members and tended to have many members of their own families close by. Among the Chinese, blacks, and black-Caribbeans, kin were more likely to live further away and oftentimes outside of the UK. These authors argue cautiously, given their small numbers, that there seems to be a trend for greater geographical differences between parents and children among immigrants with higher levels of education and employment in professional jobs; they are more similar to the white communities. This is similar to Chinese immigrants in Canada, where older generations migrated to ethnic enclaves (Chinatowns) but current immigrants are younger, well educated, and come to well-paying jobs within Canada. This current group of immigrants is not settling in ethnic enclaves but rather among members of the host society (Li, 1998).

The previous section has examined available research which takes into account several, or at least more than one, ethnic groups. The research demonstrates the generally disadvantaged position of ethnic and subcultural groups within Western nations. The disadvantage appears to be indisputable at the more objective level using indicators such as income and health. Whether examining these aspects as QoL or as part of the broader economic, political, and social structure in which these individuals are embedded, the conclusion is the same – ethnic subcultural groups have a disadvantaged position relative to whites. This is generalisable in the West almost irrespective of the group that one is examining.

However, examining the psychosocial indicators of QoL shows a very different picture, with subcultural groups often revealing equal or better subjective QoL than whites. While there are differences between groups (see below) the general conclusion that the greater socio-economic disadvantage does not necessarily translate into psychosocial QoL disadvantage is fairly conclusive. The explanation seems to lie in cultural norms of familial support. It is unknown, though, whether social support and caregiving among subcultural groups is driven primarily out of economic necessity and lack of access to alternatives or it is primarily driven by cultural preference, irrespective of economic circumstance. A questioning of the extent of differences in this area between subcultural groups and the host society was raised, given the research on the care of older people provided by whites in developed countries, despite a rhetoric of individualism which suggests that such caring does not take place. Empirical research demonstrates the contrary.

There is also research available on individual ethnic groups, i.e. studies on one ethnic group that do not include data on other groups. There is a large amount of research available on blacks and Hispanics in the USA (which is not the focus of this chapter). Otherwise, there tends to be limited research on any specific ethnic group, particularly if one wishes to make cross-national comparisons. The next section turns to some of this literature.

SEPARATE ETHNIC GROUPS

The issue of whether or not involvement in one's ethnic subculture is advantageous or disadvantageous is open to debate and would appear, as noted above, that it can be either or both depending on the individuals involved and their situation. Some studies report that greater acculturation to the host society is more beneficial. Among diasporic Chinese, for example, Casado and Leung (2001) and Lam et al. (1997) report lower depression scores for immigrant Chinese women with greater English usage and acculturation in the USA. Miller and Chandler (2002) find that, among Russian immigrants to the USA, those with greater English usage report lower depression. Among Mexican-Americans, Gonzalez et al. (2001) find that the least acculturated are more likely to be at high risk of depression. Lee et al. (1996) also report that older Korean immigrants to the USA assess their lives as lower if they have retained more ethnic identity.

Contrary evidence is also readily available. For example, Silveira and Alleback (2001) find that, among Somali men in the UK, reliance on Somali peers and religious practices buffer against depression (see Hao and Johnstone, 2000 for a discussion on the importance of religious participation among immigrants). Angel and Angel (1992) discuss the benefits for older Hispanic residents who live within ethnic enclaves where their culture of origin is largely duplicated. Remennick (2003) studied Russian immigrants who do not view their poor knowledge of the native language as an obstacle to integration because their worlds are primarily within the immigrant communities. Similarly, Kim (1999) reports stronger ethnic attachment related to lower levels of loneliness and more satisfaction with supports among Korean immigrants. Tsang *et al.* (2004) note that strong ties to the ethnic community correlate with a good QoL among older Chinese in Australia. There are fewer studies that report no differences between acculturation and subjective QoL. Lee *et al.* (1996) and Tran *et al.* (1996) are two that report no such relationship among Korean-Americans and Hispanic immigrants respectively.

One message is clear – all ethnic subcultural groups are not similar to one another. They differ in terms of both objective circumstance and subjective life experiences. Nazroo *et al.* (2004) report that in the UK Bangladeshi, Pakistani, Caribbean, and Indian older people are disadvantaged objectively compared with whites, but among these groups, the Bangladeshis and Pakistanis are worse off than are Caribbeans and Indians. They also report differences in social support depending on the group examined, with South Asians more likely to reveal intense routine support systems. Among black-Caribbeans and mixed white- and black-Caribbeans, a pattern of older people living close to one child with others more dispersed is more prevalent. Morton *et al.* (1992) report that older Vietnamese residents of San Diego, California, show fewer associations between acculturation and level of impairment than is true for older Chinese residents in the same city. Janson and Mueler (1983) report a positive relationship between age and well-being in the USA, i.e. with increasing age there is increasing well-being, but that this relationship increased more among both blacks and whites than it did among Mexican-Americans.

Furthermore, a discrepancy between subjective psychosocial measures is also reported between ethnic minorities and those individuals from the donor country, i.e. those who remain in their home country where they constitute the majority. For example, Chappell *et al.* (2000) and Chappell and Lai (1999) report that older Chinese people living in Canada reveal higher life satisfaction than do older Chinese people living in Suzhou, China. Similarly, older Chinese people living in Vancouver show substantially higher life satisfaction than older people living in either Hong Kong or Shanghai. When examining the domains of life satisfaction (friendships, transportation, finances, recreation, housing, family members, family responsibilities, everyday food, health, and self), older Chinese people living in Canada had higher satisfaction in each of these separate domains and overall than was true for older Chinese people living in China. Often, minorities in Western countries, while disadvantaged compared to whites in the host country, are advantaged on objective indicators compared with those remaining in their original homeland. Whether these differences in subjective well-being reflect those circumstances awaits further research.

The study of separate ethnic groups is intriguing and reveals the distinctiveness of the subcultures of each of the groups. Any of these studies demonstrates the importance of culture and the ethnocentricity that is often evident within Western research. It is, however, a body of research that is very much in its infancy from which general conclusions are difficult to draw. One of the reasons is because of the tremendous difficulties of disentangling confounding effects, not to mention added difficulties of conducting research that is culturally congruent. Nevertheless, in-depth research into particular subcultural groups is a tremendous learning experience for Western researchers. We now turn attention to such an example, drawing upon research among older Chinese people living away from their homeland in Canada.

If one examines the gerontological literature on life satisfaction, one finds relatively little that focuses on subcultural groups of older people such as the Chinese despite the popularity of this topic especially in the USA. We do know that the predictors of life satisfaction among the Chinese in China tend to be similar to those among whites in Western countries, i.e. health, socio-economic status, and social support (Ho *et al.*, 1995). These same predictors also emerge as the significant factors related to life satisfaction among Chinese minorities living abroad (Chappell, 2005a).

This is not to say that the specific indicators or measures of either the predictors or the outcomes are the same from study to study. Health, for example, is sometimes measured as number of chronic conditions, any number of specific diseases, functional disability, perceived health, and/or utilisation of particular health care services such as physician visits. Social support is even more diverse in its measurement and can include contact with family members, specifically with children, marital status, qualitative evaluations of the interaction, involvement within the community, and having a confidante in one's life. Socio-economic status might be measured in terms of personal income, household income, assets, and/or expenditures. Outcomes are measured in terms of one of a number of life satisfaction scales, happiness, morale, subjective well-being, affect, depression, etc. Nevertheless, at a higher order of abstraction, the variables that emerge consistently as correlates are the three noted.

In North America, Lai and McDonald (1995) find that psychological health, sense of personal control, and social support are strong correlates of life satisfaction among older Chinese people in Canada. Also among Chinese Canadians, Gee (1998) reports living arrangements, health status, and place of residence as significant correlates of subjective well-being. Chappell and Lai (1999) and Chappell *et al.* (2000), using the same data, report social support, health, and socio-economic status as significant correlates of life satisfaction. In the USA, Mui (1996) reports perceived health, living with others, and satisfaction with help from family members as correlated with depression, and Hao and Johnstone (2000) find economic and human capital factors important for emotional well-being.

Within a seven-city study of older Chinese nationals in Canada, Chappell (2005a) examined the role of involvement in traditional Chinese culture as a predictor of life satisfaction while taking the traditional predictors (health, social support, and socio-economic status) into account. Each of the known traditional factors (socio-econimic status, health, and social support) emerged as significant; their involvement in the traditional culture was also significant. In particular, a scale measuring involvement in the local Chinese community, such as frequenting senior centres, attending Chinese social functions organised by the Chinese community, and maintaining close ties with the Chinese community in Canada, was significant overall and was correlated with the domains of social relationships, community participation, attitude towards life, and spirituality. Visits to their homeland and family reunion immigration were also significantly correlated. Actual involvement with other Chinese within their subculture emerged as an important predictor but another scale measuring what might be described as ethnocentric values (e.g. the importance to them of their children marrying a Chinese, of their children speaking Chinese, etc.) was not related either positively or negatively to their QoL.

The importance for the Chinese of involvement in their traditional culture in Canada might be enhanced because of their relative inability with the English language. All of 43.2% in Chappell and Lai's (1999) British Columbia sample spoke no English and only 21.3% considered their English-speaking ability to be good. In the seven-city study (Chappell, 2005a) only 3.0% chose to be interviewed in English. Cheng et al. (1978) report a similar situation in San Diego.

As discussed earlier, the value of involvement in traditional culture for QoL of ethnic older people is unresolved. Involvement in traditional culture is often seen as an indicator of adherence to subcultural values among people who are exposed to different cultural values and in the case of Chinese immigrants to vastly different value systems (Stewart, 2000), one reflecting a mainstream white society that restricted and excluded their participation in the majority society. The prejudice directed towards Chinese people was at one time legally and socially sanctioned. For example, for a period of time, the Chinese in Canada were forbidden from holding a Canadian citizenship even if they were born in Canada, and they were also denied the right to vote although it might be noted that it was legal to conscript them for war (Li, 1998). In North America, at least, the environment for Chinese immigrants has been characterised as one of democracy, capitalism, and individualism representing the majority society together with anti-Chinese discrimination and prejudice against others (Chen, 1997). The result, both Chen and Van Ziegert (2002) argue, is a blending of Chinese and Western cultures that results in a new transnational identity. This biculturalism represents neither exclusionism nor assimilation. It is, rather, a hybridising of two cultures that incorporates a continuing influence of traditional culture as well as the majority culture. Chan (2001) argues that the involvement of Chinese immigrants in traditional festivals sustains and demands their cultural negotiation. Lin (2000) similarly discusses ethnic festivals as sites for enquiring into one's past and creating cultural identity.

It could be argued that involvement in traditional culture is advantageous for elderly Chinese immigrants because, unlike in the West where age does not automatically confer prestige, traditional Chinese culture embraces a collectivist perspective valuing older people. Filial piety has deep historical roots where providing care for older people is seen as a duty and responsibility, and a cultural virtue (Aranda, 2002). However, the traditional patriarchal system values sons over daughters and indeed has viewed daughters as 'shibun' (goods of lost value) so that sons, particularly the eldest son, has the major responsibility for caring for his parents. Daughters become members of their husband's families and provide assistance to the son in caring for his parents, i.e. her parents-in-law (Liu and Kendig, 2000).

There is, however, evidence that the practice of filial piety is changing both within China and among Chinese communities overseas (although many Chinese researchers argue that the value and ideal of filial piety is retained and that while there is no single version in any Asian nation there does appear to be a shared core of understandings; Ikels, 2004; Whyte, 2004). In both China and Canada the role for daughters-in-law is lessening; daughters now provide much more assistance to their own parents than to their parents-in-law (Lee et al., 2000; Ng et al., 2002; Chappell, 2005b). In addition, spouses are becoming more dominant in providing care and assistance than one would expect based on the traditional notion (Chappell, 2003). This research on filial piety demonstrates the continuing role of women as the care providers in Chinese society, whether in China, or among diasporic Chinese. The gendered differentiation of social support and of caregiving would appear to be universal, applicable equally to majorities and minorities. The form that gendering takes, however, can differ.

The modern face of filial piety is transforming into what Whyte (2004) refers to as the 'networked family', where older people are less likely to live with their own children. The similarity to care provided in the West through 'intimacy at a distance' is striking. One does, however, still see a greater involvement of sons in providing care among Chinese-Canadian families than is evident within the majority culture (Chappell, 2005a). Similarly, Pang et al. (2003) report a shift from traditional expectations of filial piety to more dependence on neighbours and friends among Chinese-American immigrants aged 60 and over. These researchers also see the changes as an adaptability that combines Eastern and Western perspectives. Is this hybridising of eastern and western culture better than adherence to one alone? The answer remains elusive.

Part of Chinese culture is not to recognise mental health problems, not to seek help, and to value 'saving face'. This, combined with significant language barriers, not to mention lack of cultural sensitivity on the part of service providers, suggests that the older Chinese people may find much comfort within their own subcultural community. Chen and Kazanjian (2005) report preliminary results suggesting that Chinese immigrants use less overall health care than others, and this is particularly true with regard to mental health problems and visiting psychiatrists and psychiatric hospitals. However, Chappell and Lai (1998) find among older British Columbians that both their health and the utilisation of physician services and homecare are about the same as one finds in the host society. Mental health services were not examined separately in that study.

This discussion of diasporic Chinese highlights the complexities of a single group and the necessity for research delving into each group in its own right. It also demonstrates the cultural sensitivity required to conduct this research, a topic that is expanded in the next section.

THE RESEARCH QUAGMIRE

Research with subcultural groups requires an extra amount of care and due diligence, particularly when the researchers are not members of that subcultural group. Even when they are within group members, if they have been raised primarily within the majority culture, they may not be sensitive to the differences in meaning within the subculture. Western research, dominated as it is by middle-class white researchers, requires cultural sensitivity. The issue refers to the validity of data. Two examples are provided here from research with diasporic Chinese within Western nations.

Among the research on Chinese immigrants, particularly in North America, it has been demonstrated again and again that older Chinese people are less likely to live alone than is true of older whites, and are more likely to live with their children even when they are married and living with their spouse (Burr and Mutschler, 1993; Kamo and Zhou, 1994; Chappell, 2003). This same tendency not to live alone and to live with children even when married has been reported in mainland China as well. In the West, this has been interpreted as evidence of closer family bonds. However, as Ikels (1990) has noted, in China these living arrangements usually reflect a child living in the parents' home, not the parents living in their child's home, and is dictated by scarcity of accommodation and resources.

We do not know the extent to which it is a preference. Among Canadian studies, Chappell (2003, 2005a) has reported that the majority of older people among the Chinese population do not speak or understand English well. The vast majority also lives in poverty and has immigrated to the country because of family reunion. In other words, their options for living alone, or even living with their spouse (who is typically in the same situation as themselves) are severely restricted. It could be lack of alternatives due to both poverty and an inability to navigate the majority culture because of language difficulties that dictate their living situation rather than cultural preference or that cultural preference is dictated by necessity. Yet, so often, the research literature is interpreted as indicative of Chinese in North America displaying Chinese cultural norms concerning the value of older people. We do not know the extent to which Chinese families would choose this living arrangement if older people were economically independent and were fluent in the host language. This is not to deny that there may be advantages and benefits for all generations if they live together. The point here is the interpretation by Western researchers that this is indicative of high value placed on elderly members of the family when living arrangements may reflect resource constraints.

The second example comes from the area of measuring instruments. Lu et al. (2001) have noted how cross-cultural studies tend to adopt measures developed in the West within Western theoretical frameworks and for use in Western countries. However, their uncritical adoption for studies in other cultures, including subcultural groups, can be severely problematic. The example they use is of the word 'happiness', which did not appear in the Chinese language until recently, and when it did the term 'fu' or 'fu-qi' is vague, referring to anything positive and good in life. Because of the Confucian emphasis on collective well-being, of the good of society over that of the individual, older people tend not to complain irrespective of their difficulty. This is a cultural-specific coping strategy; it reinforces a Westerner's interpretation that the older Chinese people have few problems and are happy (Lam, 1994). Chinese culture suggests that individuals should not brag about themselves and that includes their psychosocial state of being if, for example, they are feeling very good and very happy. This example highlights the particular sensitivity that is required within the area of subjective or psychosocial QoL.

CONCLUSION

While QoL has been used very broadly, spanning conceptualisations at the societal and objective level to those referring to the individual and subjective levels, this chapter has targeted research on QoL and ethnicity in old age focused specifically on subjective social indicators such as subjective well-being, life satisfaction, and psychological well-being. Reflecting on the research, even in this more restricted area of QoL, is challenging. It is plagued by the same problems as is all of the QoL research (a diversity of different concepts to measure subjective QoL and a plethora of operationalisations of even the same concept of QoL together with the multitude of concepts and measures of outcomes related to QoL). This is confounded with a lack of longitudinal research for examining cause and effect together with the added complexities of measuring ethnicity.

Ethnic research as a body of knowledge lacks sufficient studies on the same ethnic group either within one country or across countries to draw generalisations. There tends to be relatively little research conducted on any particular group. Even where there is a relative wealth of research (e.g. on blacks or Hispanics in the USA), it is concentrated in one country, making cross-national comparisons difficult.

Despite these challenges, there is much interest in the study of ethnicity because of the unfortunate relationship between disadvantage and subcultural group status within developed countries. This shared commonality across subcultural groups justifies the interest in 'ethnic studies', which combine subcultural groups within a particular developed nation under one umbrella. The inequities that are shared across subcultural groups tend to refer to the objective indicators that are often used as a measure of QoL, specifically to economic hardship, which is related to employment, health, and other disadvantages. Such macro-structural factors are also often examined when the focus is on a more sociopsychological perspective as, not indicators of QoL *per se*, but in the context of the broader economic, political, and social structures that influence our subjective QoL.

Whatever the nomenclature, the fact remains that there is a disjuncture among ethnic groups in terms of the objective indicators of QoL (or alternatively stated, the social, economic, and political structures within which they live their lives) and sociopsychological indicators of QoL. Despite the real structural disadvantage that they experience, their overall subjective well-being is often higher than for host populations, which do not share this same disadvantage. This is explained using the concept of social support wherein subcultural groups experience greater social support than whites. Without denying or diminishing the profound impact that larger societal forces have on these individuals, their local day-to-day lives including the relationships with those close to them, particularly their families, appear to greatly enhance their QoL. It is within the local context of day-to-day living that individuals can maintain agency despite structural disadvantage (Wray, 2003).

We do not know the extent to which much of the in-group supportiveness arises out of necessity (lack of resources in order to purchase help, lack of adequate access to culturally appropriate services, etc.), and there is some indication that as these groups gain in economic advantage, the historic supportiveness may diminish. Much more research is required before conclusions can be drawn. Relatedly, acculturation of immigrant groups into host societies is not well understood. Nevertheless, research suggests that older notions of assimilation vs pluralism are inaccurate and that subcultural groups instead create new identities consisting of a blending of donor and receiving cultures. Whether and how greater or less acculturation is beneficial or harmful for subcultural groups seems to depend on the circumstance. This is particularly evident in studies of separate ethnic groups within different countries. Ethnic subcultural groups are not all the same.

Finally, the chapter ends with a couple of examples intended to stress the difficulties of studying subcultural groups and the absolute need to insure cultural validity in this research. Without it, not only are our findings invalid, but incorrect results could lead to false recommendations, especially in cases where applied research is informing government decision-making.

REFERENCES

Angel, J. and Angel, R.J. (1992) 'Age at Migration, Social Connections and Well-Being among Elderly Hispanics', *Journal of Aging and Health*, 4(4), 480–499.

Aranda, O.O.P. (2002) 'Chinese Immigrants' Caregiving Experiences with Their Elderly Parents', MA thesis, Long Beach, CA, California State University.

Argyle, M., Martin, M., and Crossland, J. (1989) 'Happiness as a Function of Personality and Social Encounters' in J.P. Forgas and J.M. Innes (eds) *Recent Advances in Social Psychology: An International Perspective*, New York, Elsevier/North-Holland, pp.189–203.

Bajekal, M., Bland, D., Grewal, I., Karlsen, S., and Nazroo, J. (2004) 'Ethnic Differences in Influences on Quality of Life at Older Ages: A Quantitative Analysis', *Ageing and Society*, 24(5), 709–728.

Blakemore, K. (1985) 'Ethnic Inequalities in Old Age: Some Comparisons Between Britain and the United States', *Journal of Applied Gerontology*, 4(1), 86–101.

Blau, Z.S., Oser, G.T., and Stephens, R.C. (1979) 'Aging, Social Class and Ethnicity: A Comparison of Anglo, Black and Mexican-American Texans', *Pacific Sociological Review*, 22(4), 501–525.

Bowling, A. (2004) 'Loneliness in Later Life', in A. Walker and C. Hagan Hennessy (eds) *Growing Older: Quality of Life in Old Age*, Maidenhead, UK, Open University Press, pp.107–126.

Brown, J., Bowling, A., and Flynn, T. (2004) 'Models of Quality of Life: A Taxonomy and Systematic Review of the Literature', Sheffield, FORUM Project, University of Sheffield (http://www.shef.ac.uk/ageing-research).

Burr, J.A. and Mutschler, J.E. (1993) 'Nativity, Acculturation and Economic Status: Explanations of Asian American Living Arrangement in Later Life', *The Journal of Gerontology*, 48(2), S55–S63.

Casado, B.L. and Leung, P. (2001) 'Migratory Grief and Depression among Elderly Chinese American Immigrants', *Journal of Gerontological Social Work*, 36(1–2), 5–26.

Calasanti, T.M. and Zajicek, A.M. (1993) 'A Socialist-Feminist Approach to Aging: Embracing Diversity', *Journal of Aging Studies*, 7, 117–131.

Chan, M.R.W.W. (2001) 'Chinese-Canadian Festivals: Where Memory and Imagination Converge for Diasporic Chinese Communities in Toronto (Ontario)', PhD dissertation, Canada, York University.

Chappell, N.L. (2005a) 'Perceived Change in Quality of Life among Chinese Canadian Seniors: The Role of Involvement in Chinese Culture', *Journal of Happiness Studies*, 6(1), 69–91.

Chappell, N.L. (2005b) 'Filial Piety among Diasporic Chinese', public lecture, Hong Kong, University of Hong Kong.

Chappell, N.L. and Penning, M.J. (2005) 'Family Caregivers in the Context of Health Reform', in M. Johnson (ed. in chief) *The Cambridge Handbook of Age and Ageing*, Cambridge, Cambridge University Press.

Chappell, N.L. (2003) 'Correcting Cross-cultural Stereotypes: Aging in Shanghai and Canada', *Journal of Cross-Cultural Gerontology*, 18(2), 127–147.

Chappell, N.L., Gee, E., MacDonald, L., and Stones, M. (2003) *Aging in Contemporary Canada*, Toronto, Canada, Prentice Hall.

Chappell, N.L., Lai, D., Lin, G., Chi, I., and Gui, S. (2000) 'International Differences in Life Satisfaction among Urban-Living Elders: Chinese and Canadian Comparisons', *Hallym International Journal of Aging*, 2(2), 105–118.

Chappell, N.L. and Lai, D.C. (1999) 'Chinese Elders: China and North America Comparisons', in I. Chi (ed.) *Researches on the Problems of Elderly Chinese*, Hong Kong, Hanchuan Chinese Professional Management Centre, pp.23–46.

Chappell, N.L. and Lai, D. (1998) 'Health-Care Service Use among Chinese Seniors in British Columbia, Canada', *Journal of Cross-Cultural Studies*, 13, 21–37.

Chen, S. (1997) 'Being Chinese, Becoming Chinese-American: The Transformation of Chinese Identity in the United States 1910–1928', PhD dissertation, Salt Lake City, UT, University of Utah.

Chen, A.W. and Kazanjian, A. (2005) 'Rate of Mental Health Service Utilisation by Chinese Immigrants in British Columbia', *Canadian Journal of Public Health*, 96(1) 49–51.

Chen, J., Ng, E., and Wilkins, R. (1996) 'The Health of Canada's Immigrants in 1994–1995' *Health Reports*, 74, 33–45.

Cheng, E., Horn, G., and Hong, M.J. (1978) 'A Cross-Cultural Study of Minority Elders in San Diego', San Diego, CA, San Diego University, Centre on Aging.

Coburn, D. (2001) 'Health, Health Care and Neo-Liberalism', in P. Armstrong, H. Armstrong, and D. Coburn (eds) *Unhealthy Times: Political Economy Perspectives on Health and Care in Canada*, Don Mills, Ontario, Oxford University Press.

Diener, E. and Diener, M. (1995) 'Cross-Cultural Correlates of Life Satisfaction and Self-Esteem', *Journal of Personality and Social Psychology*, 68, 653–663.

Driedger, L. and Chappell, N.L. (1987) *Aging and Ethnicity: Toward an Interface*, Butterworths Perspectives on Individual Population Aging series, Toronto, Ontario, and Vancouver, British Columbia, Butterworths.

Gabriel, Z. and Bowling, A. (2004) 'Quality of Life in Old Age from the Perspective of Older People' in A. Walker and C. Hagan Hennessy(eds) *Growing Older: Quality of life in Old Age,* Maidenhead, UK, Open University Press, pp.14–34.

Gee, E.M. (1998) 'Well-being among Chinese Canadian Elders', International Sociological Association Paper.

Gitelman, P.J. (1976) 'Morale, Self-Concept and Social Integration: A Comparative Study of Black and Jewish Aged, Urban Poor', Doctoral dissertation, New Brunswick, NJ, Rutgers University.

Gonzalez, H.M., Haan, M.N., and Hinton, L. (2001) 'Acculturation and the Prevalence of Depression in Older Mexican-Americans: Baseline Results of the Sacramento Area Latino Study on Aging', *Journal of the American Geriatrics Society*, 49(7), 948–953.

Hao, L. and Johnson, R.W. (2000) Economic, Cultural and Scial Origins of Emotional Well-Being: Comparisons of Immigrants and Natives at Midlife, *Research on Aging*, 22(6), 599–629.

Ho, S-C., Woo, J., Lau, U., Chan, S-G., Yuen, Y-K., Chan, Y-K., and Chi, I. (1995) 'Life Satisfaction and Associated Factors in Older Hong Kong Chinese', *Journal of the American Geriatrics Society*, 43, 252–255.

Ikels, C. (2004) 'Introduction', in C. Ikels (ed.) *Filial Piety, Practice and Discourse in Contemporary East Asia*, Stanford, Stanford University Press, pp.1–15.

Janson, P. and Mueler, K.F. (1983) 'Age, Ethnicity, and Well-Being: A Comparative Study of Anglos, Blacks and Mexican Americans', *Research on Aging*, 5, 353–367.

Kamo, Y. and Zhou, M. (1994) 'Living Arrangements of Elderly Chinese and Japanese in the United States', *Journal of Marriage and the Family*, 56, 544–558.

Kim, O. (1999) 'Mediation Effect of Social Support Between Ethnic Attachment and Loneliness in Older Korean Immigrants', *Research in Nursing and Health*, 22, 169–175.

Kivvet, V.R. (1993) Informal Supports among Older Rural Minorities in N.C. Bull (ed.) *Aging in Rural America*, Newbury Park, CA, Sage Publications, pp.204–215.

Koehn, S. (2005) *Community-based Research Seeks to Address Barriers to Access to Care for Ethnic Minority Seniors*, Vancouver, British Columbia, Gerontology Research Centre, Simon Fraser University.

Lai, D.W.L. and MacDonald, J.R. (1995) 'Life Satisfaction of Chinese Elderly Immigrants in Calgary, *Canadian Journal on Aging*, 14, 536–552.

Lam, L. (1994) 'Self-assessment of Health Status of Aged Chinese-Canadians', *Journal of Asian and African Studies*, 29(1–2), 77–90.

Lam, R.E., Pacala, J.T., and Smith, S.L. (1997) 'Factors Related to Depressive symptoms in an Elderly Chinese American Sample', *Clinical Gerontologist*, 17(4), 57–70.

Lee, R.P.L., Yu, E., Sun, S., and Liu, W.T. (2000) 'Living Arrangements and Elderly Care: The Case of Hong Kong', in W.T. Liu and H. Kendig (eds) *Who Should Care for the Elderly: An East-West Value Divide*, Singapore, Singapore University Press/National University of Singapore and World Scientific Publishing Co. Pte. Ltd, pp.269–296.

Lee, M.S., Crittenden, K.S., and Yu, E. (1996) 'Social Support and Depression among Elderly Korean Immigrants in the United States', *International Journal of Aging and Human Development*, 42(4), 313–327.

Li, P.S. (1998) *The Chinese in Canada*, 2nd edn, Toronto, Ontario, Oxford University Press.

Lin, P.Y-W. (2000) 'Cultural Identity and Ethnic Representation in Arts Education: Case Studies of Taiwanese Festivals in Canada', PhD dissertation, Canada, The University of British Columbia.

Liu, W.T. and Kendig, H. (2000) 'Critical Issues of Caregiving: East–West Dialogue' in W.T. Liu and H. Kendig (eds) *Who Should Care for the Elderly: An East–West Value Divide*, Singapore, Singapore University Press/National University of Singapore and World Scientific Publishing Co. Pte. Ltd, pp.1–23.

Lu, L., Gilmour, R., and Kao, S.F. (2001) 'Cultural Values and Happiness: An East–West Dialogue', *The Journal of Social Psychology*, ISSN: 0022–4545, 141(4), 477–493.

McDonald, L., George, U., Daciuk, J., Yan, M., and Rowna, H. (2001) 'A Study on the Settlement Related Needs of Newly Arrived Immigrant Seniors in Ontario', Centre for Applied Social Research,Toronto, University of Toronto.

Miller, A.M. and Chandler, P. (2002) 'Acculturation, Resilience, and Depression in Midlife Women from the Former Soviet Union', *Nursing Research*, 51(1), 26–32.

Miranda, M.R. (1992) 'Quality of Life in Later Years: Why We Must Challenge the Continuing Disparity between Whites and Minorities', *Perspectives on Aging*, 21(1), 4–10.

Moriarty, J. and Butt, J. (2004) 'Social Support and Ethnicity in Old Age' in A.Waker and C. Hagan Hennessy (eds) *Quality of Life in Old Age*, Maidenhead, UK, Open University Press.

Morton, D.J., Stanford, E.P., Happersett, C.J., and Molgaard, C.A. (1992) 'Acculturation and Functional Impairment among Older Chinese and Vietnamese in San Diego County, California', *Journal of Cross-Cultural Gerontology*, 7(2), 151–176.

Mui, A.C. (1996) 'Depression among Elderly Chinese Immigrants: An Exploratory Study', *Social Work*, 41, 633–645.

Nazroo, J., Bajekal, M., Blane, D., and Grewal, I. (2004) 'Ethnic Inequalities', in A. Walker and C. Hagan Hennessy (eds) *Growing Older: Quality of life in Old Age,* Maidenhead, UK, Open University Press, pp.35–59.

Ng, A.C.Y., Phillips, D.R., and Lee, W.K. (2002) 'Persistence and Challenges to Filial Piety and Informal Support of Older Persons in a Modern Chinese Society: A Case Study', *Journal of Aging Studies,* 16, 135–153.

Pang, E.C., Jordon, M.M., Silverstein, M., and Cody, M. (2003) 'Health-seeking Behaviors of Elderly Chinese Americans: Shifts in Expectations', *The Gerontologist,* 43(6), 864–874.

Pinquart, M. and S. Sorensen (2000) Influences of Socioeconomic Status, Social Network and Competence on Subjective Well-Being in Later-Life, *Psychology and Aging,* 15(2), 187–224.

Remennick, L. (2003) 'Retired and Making a Fresh Start: Older Russian Immigrants Discuss their Adjustment in Israel', *International Migration,* 41(5), 153–176.

Rubensteen, L.Z. and Kramer, J.B. (1994) 'Health Problems in Old Age', *Gerontology and Geriatrics Education,* 15(1), 23–43.

Silveira, E. and P. Alleback (2001) 'Migration, Ageing and Mental Health: An Ethnographic Study on Perceptions of Life Satisfaction, Anxiety and Depression in Older Somali Men in East London', *International Journal of Social Welfare,* 10(4), 309–320.

Stewart, C.C. (2000) 'Chinese American Families from Taiwan: A Transnational Study of Filial Connection', *Dissertation Abstracts International, A: The Humanities and Social Sciences* 61(1), 384-A, University of Minnesota.

Sutton, D.L. and Persaud, D. (1989) 'Minority Elderly New Yorkers: The Social and Economic Status of Rapidly Growing Population', New York State Office for the Aging, Division of Program Development and Evaluation, Albany, New York, Research and Analysis Unit.

Takamura, J. (2002) 'Social Policy Issues and Concerns in a Diverse Aging Society: Implications of Increasing Diversity', *Generations* 26(3), 33–38.

Tran, T.V., Fitzpatrick, T., Berg, W.R., and Wright Jr., R. (1996) 'Acculturation, Health, Stress and Psychological Distress among Elderly Hispanics', *Journal of Cross-Cultural Gerontology,* 11(2), 149–165.

Tsang, E.Y.L, Liamputtong, F., and Pierson, J. (2004) 'The View of Older Chinese People in Melbourne about Their Quality of Life', *Aging and Society,* 24, 51–74.

United States the Human Resources Planning Institute (1972) *Social Indicators Survey of the Aging,* Seattle, Washington, Office on Aging: Human Resource Planning Institute.

Van Ziegert, S. (2002) 'Global Spaces of Chinese Culture: A Transnational Comparison of Diasporic Chinese Communities in the United States and Germany', PhD dissertation, Houston, TX, Rice University.

Veenhoven, R. (1999) 'Quality-of-Life in Individualistic Society', *Social Indicators Research,* 48, 157–186.

Walker, A. (2005) 'A European Perspective on Quality of Life in Old Age' in H-W. Wahl and D. Deeg (eds) *European Journal on Ageing: Social, Behavioural and Health Perspectives,* 2(1), March. Heidelberg, Germany, Springer-Verlag, pp.2–12.

Westerhof, G. (2001) Beyond Life Satisfaction: Lay Conceptions of Well-being among Middle-Aged and Elderly Adults, *Social Indicators Research,* 56(2), 179–204.

Whyte, M.K. (2004) 'Filial Obligations in Chinese Families: Paradoxes of Modernization', in C. Ikels (ed.) *Filial Piety, Practice and Discourse in Contemporary East Asia,* Stanford, Stanford University Press, pp.106–127.

Wray, S. (2003) 'Women Growing Older: Agency, Ethnicity, and Culture', *Sociology,* 37(3), 511–527.

12. HEALTH AND QUALITY OF LIFE

INTRODUCTION

A host of research from the past decades has accumulated evidence that health is a consistent determinant of quality of life (QoL). However, both health and QoL are poorly defined concepts. Moreover, the association between health and QoL may not be equally strong across the life course. The salience of health in the lives of older people may change with increasing age. This chapter sets as its task to show distinctions between aspects of health, conceptualisations of health-related QoL, and furthermore, to critically examine the notion of a relation between health and QoL in older age. In particular, the salience of health in healthy and ill older people and changes with ageing in this salience are examined. A review of the literature will be followed by empirical evidence from both quantitative and qualitative research.

HEALTH

Health is a concept usually taken for granted, but there is no commonly agreed operational definition. The concept is approached from both objective and subjective perspectives, in part depending on the discipline of the researcher. This section discusses aspects of health, keeping in mind that disease measures alone do not adequately describe health, and that functional ability and self-rated health become more important with increasing age.

Although body and mind cannot be conceptually separated, to facilitate description, a first distinction can be made between physical and mental health. The development of physical health problems is often modelled as the so-called disablement process (Verbrugge and Jette, 1994). The stages of this process range from morbidity to disability. In addition to these health concepts, individuals' perception of their health (or self-rated health) is considered to be a comprehensive measure of all aspects of health (Idler and Benyamini, 1997). Below, these concepts are described and prevalences in the older population are given.

Morbidity

Morbidity refers to the prevalence of chronic conditions. Among the most prevalent self-reported chronic conditions, musculoskeletal diseases, including arthritis and osteoporosis, rank on top (Minicuci *et al.*, 2003). In the age group 65–84 years, their combined prevalence ranges from almost 40% to more than 70%, depending on the country and the sampling frame. Heart diseases, including myocardial infarction, angina pectoris, heart failure, and peripheral artery disease, rank second. Also for this

195

H. Mollenkopf and A. Walker (eds.), Quality of Life in Old Age, 195–213.
© 2007 *Springer.*

category of diseases, prevalence differs widely across countries. In particular, heart diseases show a north-south gradient in Europe. It is highest in Finland (46% in the age group of 75–84 years) and lowest in Italy (18%). Respiratory diseases, including chronic bronchitis, emphysema, and asthma, rank next, with particularly high prevalences in Spain and Italy (about 34%), and lower prevalences in the Netherlands and Sweden (about 15%). Diabetes ranks fourth, again with different prevalences across countries. In contrast to the distribution of heart disease, diabetes is more prevalent in Mediterranean countries (15–20%), but less prevalent in the north of Europe (about 11%). Stroke and cancer share the fifth rank in prevalence. Their prevalence ranges from 5% to 11%, again with significant differences across Europe. Like that of diabetes, the prevalence of stroke is relatively high in the Mediterranean countries, and lower in the Scandinavian countries. A comparison of prevalences of self-reported chronic conditions between the UK and the USA showed consistently higher prevalence rates for the USA (Banks *et al.*, 2006).

Despite their different prevalences, each of the chronic conditions discussed has similar associations with mortality across countries (Noale *et al.*, 2005). Cancer has the highest risk of mortality, followed by diabetes, heart diseases, and respiratory diseases. Musculoskeletal diseases are not associated with increased mortality risk.

Disability

Disability is the inability to perform usual daily activities. More complex activities such as shopping, preparing a meal, and doing housework are termed instrumental activities of daily living (IADLs). More basic, and at the same time more obligatory activities, such as bathing, dressing, and toileting, are termed activities of daily living (ADLs). The former are more likely to be affected by socio-cultural factors than the latter. Nevertheless, a surprising variety is observed in prevalence of ADL disability across European countries, ranging from 6–35%. Although available studies show different patterns, ADL disability turns out to be highest in Italy and Spain, and lowest in Finland (Pluijm *et al.*, 2005; Van den Brink *et al.*, 2003). A comparison between the Netherlands and the USA showed higher prevalence of walking disability in the USA (Melzer *et al.*, 2004). In Europe a north-south gradient is observed, which may be explained by socio-economic as well as cultural differences. There is ample evidence for socio-economic differences in morbidity and disability across Europe (Huisman *et al.*, 2003). Among older people in Mediterranean countries, only a small minority has completed elementary education, whereas this is true for the majority of northern countries. Differences in ADL prevalence may also stem from differences in the meaning of dependence and the availability of family help. Self-reported disability may be greater when help is available within the family, which is the norm in the Mediterranean countries, as opposed to northern countries (Seeman *et al.*, 1996; Minicuci *et al.*, 2004).

As was the case with chronic conditions, despite the diversity in prevalence across countries, the predictive ability of ADL disability for mortality is very similar and higher than for each of the five most prevalent single chronic diseases (Noale *et al.*,

2005). Also, the association between disability and a more objective performance based disability score is similar across countries (Van den Brink *et al.*, 2003).

IADL disability shows a different pattern from ADL disability, with Italy and Finland ranking among the lowest prevalences (about 25% in the age group 75–84 years) and Spain, the Netherlands, and Sweden among the highest (almost 50% in the age group 75–84 years). IADLs are particularly sensitive to cultural biases, and this is shown when these activities are compared across countries for older men and women separately. The prevalences appear to vary with norms about the gender-division of household work. Across Europe, women are more independent in cooking, men in shopping. Housework is least affected by gender biases. Cultural biases related to family norms and living arrangements are demonstrated in preparing meals, in which older women show more disability in countries where households are usually shared with children, and less, where older unmarried women generally live alone (Nikula *et al.*, 2003).

Disability is differentially associated with chronic conditions. Stroke, diabetes, and arthritis are generally the most disabling conditions (Kriegsman *et al.*, 1997; Verbrugge and Patrick, 1995). Heart disease has become less disabling over time, probably due to advances in medical treatment and in disease management (Deeg *et al.*, 1994).

Self-rated Health

Older people's own evaluation of their health is considered to be a good summary of overall health, but is also known to be sensitive to socio-cultural factors (Jylhä *et al.*, 1998). It is thus not surprising that self-rated health differs across European older populations. However, unlike for morbidity and disability, there is not a clear north-south gradient (Bardage *et al.*, 2005). Instead, there is an east-west gradient: older people in Eastern Europe rate their health as worse than those in Western Europe (Appels *et al.*, 1996; Carlson, 1998).

In most countries, but not all, women rate their health as somewhat worse than men do (Bardage *et al.*, 2005; Carlson, 1998). There are, moreover, clear gender-differences in the predictive ability of self-rated health for mortality. In most countries, but again not in all, older men's self-perceptions of health are more predictive of mortality than older women's self-perceptions (Idler, 2003). The gender differences, to a large extent, can be explained by differences in health profiles and in lifestyle. In addition, differences in standards of health may play a role. Women generally know more about their health, and are more willing to act on perceived health declines than men do (Idler, 2003). These behavioural differences may bring older women more years of life, albeit often in less-than-ideal health.

Mental Health

Common mental health conditions in older age are depression, anxiety, and dementia. In all three conditions, the psychiatric diagnosis can be distinguished from the milder, but still clinically relevant syndromal or 'subthreshold' condition. For the affective disorders, the psychiatric diagnosis does not increase – or even decreases

– with age. The diagnosis of depression has a prevalence of about 2% (Beekman
et al., 1999). The diagnosis of dementia is not very prevalent at age 65 (1%), but
increases steeply with age, especially after age 80. At age 90 and over, 30% of the
population has dementia (Hofman *et al.*, 1991). Among the subthreshold conditions
in people aged 65 and over, depression has the greatest prevalence.

A widely reported phenomenon in depression is that women have higher levels
than men do. The gender differential is consistent across countries, but varies in size.
The higher symptom load of women can partly be attributed to a greater amount of
physical health problems, partly to more disadvantaged social and economic conditions (Sonnenberg *et al.*, 2000). However, these factors explain only up to 50% of
the gender differential, so that other yet undiscovered pathways exist.

An association between subthreshold depression and disability is observed at
older ages (Penninx *et al.*, 1998; Penninx *et al.*, 2000). Also subthreshold depression
is associated with increased mortality, but this association is significant only in men.
However, diagnosed depressive disorder is associated with mortality in both men and
women (Penninx *et al.*, 1999; Schoevers *et al.*, 2000).

QoL AND HEALTH-RELATED QoL

There are two main conceptual approaches to QoL assessment (Browne *et al.*, 1997):
the standard-needs approach where QoL is seen as the extent to which certain universal needs are met, and the psychological processes approach, where QoL is considered to be constructed from individual evaluations of personally salient aspects of
life. The vast majority of studies on the QoL in relation to health use the standard-
needs approach, in which health is considered as a basic need, and measurement
instruments are intended to capture the extent to which patients consider this need to
be met. Instruments measuring 'health-related QoL' can be generic or disease-
specific (Hickey *et al.*, 2005). Disease-specific instruments are often used in treat-
ment-effect studies, e.g. in patients with visual impairment (McKee *et al.*, 2005).
Other studies monitor QoL in patients with diseases that cannot be treated but the
severity of which can be managed with appropriate interventions, e.g. heart failure
patients (Gott *et al.*, 2006). In specific applications, health-related QoL measurement
is used to identify the burden of diseases, and as a tool towards cost-effectiveness
analyses (Sprangers *et al.*, 2000; Sullivan *et al.*, 2005). The applications add a quality-of-life perspective to the more traditional remaining life-expectancy perspective.
Ultimately, such calculations might be used for resource allocation, either to cost-
effective treatments, or to patient groups with seriously reduced QoL (Fayers and
Bjordal, 2001).

Generic instruments on health-related QoL assess QoL by specific domains of
health status, such as functional ability or pain (for a recent review, see Haywood
et al., 2005). Among the many instruments, only few have been found to be recom-
mendable in terms of reliability, measure of detail and sensitivity. In particular, the
Short Form-36 (SF-36) is recommended where a detailed and broad ranging assessment of health is required, particularly in community-dwelling older people with

limited morbidity; the EuroQol-5D (EQ-5D) is recommended where a more succinct assessment is required, particularly where a substantial change in health is expected (Haywood *et al.*, 2005). Despite their good quality, the question arises how much these instruments differ from measures of health status that abstract from specific diseases (Covinsky *et al.*, 1999).

To circumvent this issue of conceptual overlap, the psychological processes approach seems promising (Browne *et al.*, 1997). Here, instead of using a predetermined model of QoL, the domains of QoL to be considered are based on individual evaluations of their importance. The relevant domains are weighted to an individual, overall rating of QoL.

THE IMPORTANCE OF HEALTH IN QoL

The following sections start from the notion that an individual's perception of their QoL is determined by the perceived balance of various aspects of life in a given situation. Because old age is characterised by frequent changes in living conditions, the majority of which are losses (Baltes, 1987), it can be expected that the importance of aspects of life is liable to change. If an aspect of life that is lost is continued to be perceived as important, this may create 'cognitive dissonance' (Festinger, 1957) or 'mental incongruence' (Dykstra and De Jong Gierveld, 1993). Thus, a change in the perceived importance of an aspect of life that is lost may be part of the cognitive adjustment to a new situation that cannot be changed. As this adjustment is expected to reduce 'mental incongruence', it is expected to be beneficial in those older people who attach less importance to the domain of life in which a loss is experienced (Bowling and Gabriel, 2004).

AN EMPIRICAL ILLUSTRATION

In this section, longitudinal changes in QoL in older age are examined in an older cohort in the Netherlands. Two concepts of QoL are selected: satisfaction with life and positive affect. This section continues to examine the relative importance of nine aspects of life in the older population, thereby distinguishing between healthy and unhealthy older people. In particular, it is examined how the importance of good health changes with increasing chronic health problems. Finally, the question is addressed of how a change in the priority of good health affects QoL in those who experienced the onset of chronic conditions.

Data

The Longitudinal Aging Study Amsterdam (LASA) is an ongoing, interdisciplinary study on changes in physical, emotional, cognitive, and social functioning, interrelations between these domains of functioning, as well as potential determinants and consequences of changes in functioning. Its main goal is to supply insights that enhance the autonomy, QoL, and social integration of older people (Deeg *et al.*, 1993; Deeg and Westendorp-de Serière, 1994).

The LASA cohort started with a stratified random sample of 3,107 participants aged 55–85 in 1992–1993. These participants were born in 1908–1937. The sample was taken from municipal registries in 11 cities and towns, based in three socio-culturally distinct geographical areas. The north-east of the Netherlands is traditionally protestant, the south is traditionally roman catholic, and the west is the most secularised area. In each area, a larger-size city and several smaller towns were included. Thus, the sample can be considered to reflect the national distribution of population density. The sample consists of age and sex strata that are weighted according to expected mortality, so that five years after baseline each 5-year age group would include equal numbers of men and women. For the initial sample, the upper age limit of 85 was chosen because beyond this age a high non-response was expected, whereas it was considered easier to keep participants in the study once they had participated, and so with time, the sample would include sufficient numbers of the oldest-old. The lower age limit of 55 was chosen because it was considered important to have data on the changes in functioning that precede the strict definition of older age of 65 and over. Moreover, a sizeable proportion of the Dutch population aged 55–64 shares characteristics with those 65 and over because of early retirement.

All LASA participants are followed up every three years. Each data collection cycle includes identical interviews and measurements of physical, cognitive, emotional, and social functioning as well as possible determinants and consequences of changes in functioning. In addition to a face-to-face interview, participants were asked to fill in a self-administered questionnaire, including questions about life satisfaction and importance of domains of life. These data provide the opportunity to examine various changes in living conditions, and to relate these changes to the development of QoL.

Study sample. At the origin of the LASA in 1992–1993 there were 3,107 participants, 1,506 (48.5%) men and 1,601 (51.5%) women. Longitudinal data are used up to 1998–1999. At the end of this study period of six years, the participants' ages were 64–94 years ($n = 2,076$). The self-administered questionnaire was returned by 74.1% of participants at baseline, and by 82.4% of participants at 6-year follow-up. From those with complete data at baseline ($n = 2,302$), across the six years, complete data are available for 1,268 people (55.1%). From those without complete data, 761 (24.5%) had died, 270 (8.7%) had dropped out for other reasons, and 808 (26.0%) had not filled in the self-administered questionnaire or had item non-response at any one of the three measurement cycles. As compared to the subjects with incomplete data, the subjects with complete data were younger, better educated, more often had a partner, were more often based in the north-east, had better health and higher positive affect, but did not differ significantly in terms of gender and life satisfaction.

Measures Quality of life. Two aspects of QoL are selected, cognitive appraisal and positive affect or happiness (Diener *et al.*, 1999). The first aspect is measured using two items on satisfaction with current and with past life, each with five response categories. These were summed to a scale ranging from 2 (low) to 10 (high). The second aspect is measured using four items from the Center for Epidemiologic Studies Depression scale, which constitute a positive affect factor (Radloff, 1976; Beekman *et al.*, 1994). Each measures an aspect of positive affect (happy, hopeful, enjoy, feel self-worth) with four response categories. After summing, the resulting scale ranges from 0 (low) to 12 (high).

Importance of domains of life. Participants were asked to indicate which three of a list of nine aspects of life they considered important in their current lives. The list included the following domains: physical health, mental health, housing, income, marriage, family, friends, time spending, and religion. Although these domains were presented to the respondents in random order, they are listed here in an order that parallels the need hierarchy conceived by Maslow (1954), with the following stages: autonomy, competence, relatedness to others, and spirituality. This questionnaire is regularly used by the Netherlands Social and Cultural Planning Office (Gijsberts, 1993), and adapted for use in an older population (Deeg and Braam, 1997). In particular, the aspect 'good health' is split up in physical and mental health, the aspect 'pleasant work' is substituted with 'meaningful spending of time', and the aspect 'much leisure time' is replaced by 'good housing'.

Health. Health is defined as the absence or presence of one or more of the following chronic conditions: heart disease, peripheral artery disease, stroke, diabetes, chronic lung disease, cancer, and arthritis (Kriegsman *et al.*, 1996).

Socio-demographic characteristics. These included age, gender, level of education, and living with a partner.

Findings

Overall development of quality of life. As can be seen in Figure 1, positive affect shows more variation over time than life satisfaction. Whereas life satisfaction remains virtually stable on average, positive affect shows a decrease with ageing. Men show better QoL than women on both life satisfaction and positive affect, but both genders show similar change and stability over time. The same is true when two other groups are distinguished: those who remain in relatively good health (maximum of one disease) and those who experience two or more diseases. Both positive affect and life satisfaction are higher in the former than in the latter group. Multivariate analysis of variance shows that in both groups, positive affect decreases significantly over time whereas life satisfaction remains stable (Figure 2). Again, the size of the decline in positive affect does not differ between the two groups.

Other authors have reported cross-sectional gender and health differences in QoL (Hilleras *et al.*, 2001; Westerhof, 2003), as well as negligible change in life satisfaction over time in older age (Bowling *et al.*, 1993; Mroczek and Spiro, 2005). These authors also reported that neither gender nor health conditions substantially affect

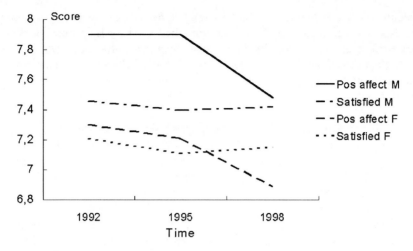

Figure 1. Development of QoL across 6 years by gender. (From Longitudinal Aging Study Amsterdam, 1992–1998.) Positive affect and life satisfaction, both rescaled to range: 0–10; M = males; F = females.

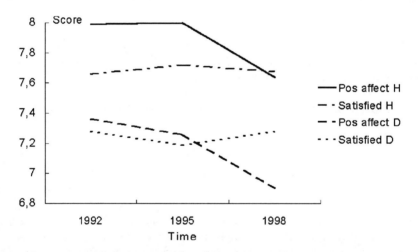

Figure 2. Development of QoL across 6 years by health status. (From Longitudinal Aging Study Amsterdam, 1992–1998.) Positive affect and life satisfaction, both rescaled to range: 0–10; H = maximum one disease at all measurement cycles; D = two or more diseases at any measurement cycle.

life satisfaction over time (Bowling *et al.*, 1993; Bowling *et al.*, 1996). Little evidence on changes in positive affect in older age is available so far.

Importance of aspects of life. The large majority of older men and women (76%) attach great importance to good physical health. More than half of the older people consider a good marriage among the three most important aspects of life. Here is a significant difference between men and women: fewer women than men consider a good marriage as important, corresponding to the higher prevalence of widowhood among women than among men. Good mental health ranks in the third place, indicated as important by 37% of both men and women. Next is a happy family (30%) and good housing (29%). The remaining aspects of life were selected among the three most important aspects of life by less than one quarter of the respondents. Among these aspects, a good income is considered important by more men than women, whereas a strong faith, a meaningful spending of time, and many friends are considered important by more women than men.

The aspects of life included in this study have been found by other authors as important to QoL (Bowling *et al.*, 2003; Bowling and Gabriel, 2004; Bryant *et al.*, 2001; Gabriel and Bowling, 2004; Leung *et al.*, 2004; Nilsson *et al.*, 2005; Westerhof, 2003). However, the predominant importance of good physical health to older people has not emerged as clearly from previous research (Bowling, 1995; Farquhar, 1995; Wilhelmson *et al.*, 2005).

When distinguishing between older people with no chronic diseases and those with one or more chronic diseases, our data show that the former consider good physical health more often (84%) as important than the latter (73%) – a significant difference. Similarly, good mental health is considered important more often by older adults with good cognitive function (38%) than by those with cognitive limitations (29%). Clearly, not all ill people give up good health as a priority, and this corresponds to the low correlation between health status and perceived importance of health found by others (Browne *et al.*, 1994). Nevertheless, our findings fit in with the concept of an adjustment mechanism to reduce 'mental incongruence'.

Aspects of life that appear as significantly more salient among subjects with chronic diseases as compared with subjects without chronic diseases are good housing and a strong faith (Figure 3). With respect to good housing, it can be assumed that having a chronic disease makes good housing more critical to continued functioning in daily life (see also Chapter 7). A strong faith, furthermore, may facilitate reconciliation with the limitations imposed by the disease and perhaps with the awareness of the finitude of life that is associated with diseases with a poor prognosis (Braam *et al.*, 2004). In studies in patient groups, good physical and cognitive functioning, economic security, good relationships with family and friends, and religious beliefs were found to be important factors to QoL (Berman *et al.*, 2004; Widar *et al.*, 2004).

Six-year changes in importance of aspects of life. So far, we have described cross-sectional data on differences in priorities between groups in different situations, in particular between healthy and ill people. However, it cannot be assumed that priorities remain unchanged over time (Browne *et al.*, 1994). Let us therefore now turn to changes in priorities over time. Because we intend to show the variability in the development during the 6-year period, instead of just comparing the importance of an aspect at baseline with 6-year follow-up, we performed a cluster analysis of the responses at baseline, 3-year and 6-year follow-up for each aspect of life (Singer and Ryff, 2001; Deeg, 2005). For most aspects of life, a three-cluster solution fitted best

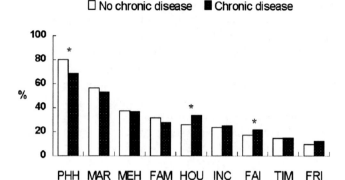

Figure 3. Importance of aspects of life (% indicating each aspect as among the three most important aspects) in older persons without and with chronic diseases. (From Longitudinal Aging Study Amsterdam, 1992–1993.) PHH = good physical health; MAR = good marriage; MEH = good mental health; FAM = happy family; HOU = good housing; INC = good income; FAI = strong faith; TIM = meaningful spending of time; FRI = many friends. * Difference between subjects without and with chronic diseases significant at $p < 0.05$

to the data, including three out of the following four clusters or trajectories: equally important, more important, less important, and equally unimportant.

Good physical health remained of uncontested importance (Table 1): in as many as 71% this remained the priority across 6 years. In almost half of the cohort, the importance of a good marriage remained high. Interestingly, the importance of many aspects of life increased, i.e. they were selected among the three most important aspects in the course of 6 years, although they were not selected as such at baseline. These aspects included good mental health, good housing, meaningful spending of time, a nice family, many friends, and a strong faith. For these aspects, no trajectory 'less important' was distinguished. The importance of a good income varied the most. Whereas as many as 25% of the cohort considered this aspect more important over the years, 11% considered it less important, and no trajectory 'equally important' was distinguished.

TABLE 1. Six-Year Changes in Importance of Aspects of Life (%)

	Equally important	More important	Less important	Equally unimportant
PHH	71	–	15	14
MAR	46	–	13	41
MEH	30	20	–	50
FAM	28	11	–	61
INC	–	25	11	64
HOU	18	16	–	66
FAI	18	5	–	77
TIM	6	14	–	80
FRI	7	8	–	85

Good physical health and a good marriage were the only aspects of which the importance did not increase. Once older persons had selected these aspects at baseline, they did not often change their views about their importance. However, a substantial minority considered them as not important after 6 years. Again, these findings correspond to the concept of 'mental incongruence'. Good health and a spouse, when lost over the years, are not likely to be replaced. In light of the new situation, it is then adaptive to consider these aspects of life as no longer important.

As the subjects were asked to report the three areas most important to them, when good physical health was no longer important to them, they must select other aspects of life (Figure 4). Substantial frequencies of the trajectories 'more important' among those who no longer reported good physical health as their priority, were found for good mental health, suggesting that the loss of physical health may be considered bearable, as long as the mind stays in good shape. Other aspects with increased importance among those who reported good physical health as less important over time were material aspects such as good housing and a good income, corresponding to the cross-sectional results reported above. Also, a substantial number of subjects reported a meaningful spending of time as more important. The importance of meaningful activities to frail older people has been noted by other authors (Gabriel and Bowling, 2004; Wilhelmson *et al.*, 2005). A smaller minority attached increased importance to a nice family, many friends, and a strong faith. They appeared to shift priorities either to social relationships or to a relationship with God, perhaps in seeking support for declining health.

Change in importance of physical health and quality of life. The ultimate question is whether adaptation to deteriorating health by considering health as less important is indeed beneficial to QoL. Comparing those who considered good physical health as equally important throughout the 6 years with those who found this aspect of life less important, positive affect appears to be lower for the latter group (Figure 5). A decline in positive affect can be seen in both groups, but the rate of decline was

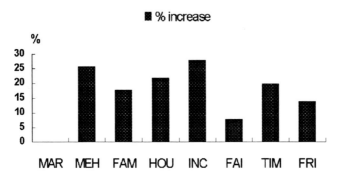

Figure 4. Increased importance of aspects of life in older persons who report decreased importance of good physical health. (From Longitudinal Aging Study Amsterdam, 1992–1998.) MAR = good marriage; MEH = good mental health; FAM = happy family; HOU = good housing; INC = good income; FAI = strong faith; TIM = meaningful spending of time; FRI = many friends.

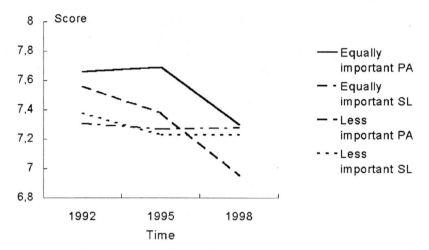

Figure 5. Development of QoL across 6 years by change in importance of good physical health. (From Longitudinal Aging Study Amsterdam, 1992–1998.)
Range: 0–10; PA = positive affect; SL = life satisfaction.

not statistically significant when tested using multivariate analysis of variance adjusting for age. Notably, satisfaction with life did not differ between the two groups throughout the study period. This suggests that the QoL of those who changed their priorities in life away from health changed in a way similar to those who did not change their priority regarding health. Keeping in mind that those who changed their priority were in poorer health than those who did not, this finding supports the beneficial effect of solving mental incongruence.

If changing priorities indeed is a useful mechanism, we could go one step further and ask if starting to consider one aspect of life as important is better than starting to consider another as important. To address this question, we compared the QoL of those who changed their priorities from good physical health to another aspect of life with those who considered good physical health equally important throughout the 6-year period using multivariate analysis of variance adjusting for age. For most changes in priorities, there were no significant differences in the course of either positive affect or life satisfaction. There were two exceptions: those who shifted their priority to good housing fared worse in terms of positive affect, and those who shift-ed their priority to a good income fared worse in terms of life satisfaction. Both these aspects are concrete and relate to safety and autonomy. In contrast to aspects of life such as health, which were reported as important by subject who were in good health, good housing and income were reported as important more often by older subjects who did not have these aspects (no home owner and low income, respec-tively). Tentative explanations are that trying to adapt one's housing and trying to get by on insufficient income takes more effort than one has to spend, or that worrying about one's housing or one's income affects one's QoL more because they pertain to more basic needs (Maslow, 1954).

STANDARD MEASUREMENT VERSUS PEOPLE'S OWN PERCEPTION OF QoL

It has been repeatedly asserted that standard measurement instruments of QoL are limited in their ability to capture QoL of individuals (Farquhar, 1995; Walker, 2005). We have seen that the concept of QoL includes different aspects to different people, and different aspects to the same people over time. When it comes to research, the concept of QoL measured using standard instruments is in fact the concept of the researcher, not the one of the subject. Several authors have compared people's own perceptions of QoL with outcomes of standard measurement instruments (Bowling and Gabriel, 2004; Strawbridge *et al.*, 2002), but a comparison of the overall rating of QoL based on these own perceptions with ratings based on standard measurement instruments has not been reported so far.

In the context of the LASA, qualitative interviews were conducted in a subsample of 14 non-frail and 11 frail older people (Puts *et al.*, 2005; Puts *et al.*, 2007). After making an inventory of what constituted QoL for a subject, the subject was asked to give an overall rating of their QoL taking into account everything that had been brought up during the inventory. This rating was summarised as 'good', 'sufficient', or 'insufficient'. At the end of the interview, the subjects were asked to fill in a questionnaire including two widely used QoL instruments. The first was the Short Form-12 (SF-12) developed in the Medical Outcomes Study (Ware *et al.*, 1996), which includes a physical and a mental component summary (PCS and MCS, respectively). The second was the EQ-5D (Brooks *et al.*, 1996). This instrument has five dimensions: mobility, self-care, usual activities, pain or discomfort, and anxiety or depression. By eliciting a valuation (utility) for each dimension, it claims to provide a multidimensional description of QoL. Each instrument is scaled from 0 to 100, where 100 indicates optimal QoL. For this study, each instrument was divided into three categories: good (> 75th percentile), sufficient (25th–75th percentile), and insufficient (< 25th percentile).

Only 3 out of the 25 subjects perceived their QoL as insufficient according to their own definitions. Another six perceived their QoL as sufficient, and the remaining majority of 16 rated their QoL as good (Table 2). In contrast, the number of subjects with 'good' QoL using standard measurement instruments was only six or seven, depending on the instrument. Furthermore, QoL was classified as 'sufficient' among 11–13 subjects, and as 'insufficient' among 6–7 subjects. The latter subjects included those who perceived their QoL as insufficient according to their own standards. Most interesting are the 2–3 subjects who rated their QoL as 'good' according to their own standards but were classified as having 'insufficient' QoL using standard instruments. Vice versa, no subjects classified as having 'good' QoL using standard instruments perceived their QoL as 'insufficient' using their own standards.

These data, although based on a small sample, demonstrate considerable discrepancy between QoL according to standards of the individual and ratings using standard instruments. The general conclusion is that standard instruments underrate the QoL of older people. It should be noted that the standard instruments centre around physical health, and to a lesser extent, around mental health. Using their own

TABLE 2. Comparison Between Perceived Quality of Life Based on Qualitative Interviews and as Measured by Standard Instruments

| | Perceived QoL[a] | | |
	Insufficient	Sufficient	Good
EQ-5D[b]			
Insufficient	3	1	2
Sufficient	0	5	8
Good	0	0	6
SF-12 (PCS)[c]			
Insufficient	3	2	2
Sufficient	0	4	8
Good	0	0	6
SF-12 (MCS)[d]			
Insufficient	2	2	3
Sufficient	1	2	8
Good	0	2	5

[a] Overall quality of life according to standards of subject
[b] EuroQol based on utilities in five dimensions
[c] 12-item Short-Form, physical component summary
[d] 12-item Short-Form, mental component summary

standards, older individuals still perceive sufficient quality in their lives, although they may lack in physical or mental health. As one author has formulated: standard instruments 'discriminate' against older people because they are phrased predominantly in relation to physical function (Hickey *et al.*, 2005).

CONCLUSION

With the aim to address the relation between health and QoL, this chapter started with a review of various aspects of health, their conceptual interrelation, and their prevalence in the older population. It was shown that each of the concepts morbidity, disability, self-rated health, and mental health, are clearly interrelated, and that each of them hold varying mortality risks. Furthermore, although the prevalence may differ widely across national contexts, the risk of mortality is comparable cross-nationally for each concept.

In the following part, commonly used definitions of QoL in relation to health were briefly reviewed, and the uses of QoL measures in health policy and practice were indicated. It was observed that the distinction between measures of health status and health-related QoL measures is not very clear. It seemed helpful to distinguish between the standard needs approach where QoL is seen as the extent to which certain universal needs are met, and the psychological processes approach, where QoL is considered to be constructed from individual evaluations of personally salient aspects of life. The majority of health-related QoL instruments use the first approach, assuming that health remains equally salient among people with increasing diseases and disabilities. The second approach, in contrast, does not assume that health or any other aspect of life is more important than others. This approach first investigates what aspects of life are important for an individual, and then asks them to rate their QoL weighting all aspects. This approach also acknowledges that the salience of aspects of life may change for an individual. The psychological process according to which changes may take place was suggested to be cognitive dissonance or mental incongruence. This process is an adaptive strategy to losses of important aspects of life that cannot be regained.

In our empirical illustration, it was found using non-domain-specific measures of QoL, that life satisfaction remains rather stable over time, but positive affect shows a decline. Both life satisfaction and positive affect are higher among healthy than among ill older people. However, the change over time in these measures does not show differences. Moreover, while health is considered as important by most older people, this is less so for ill than for healthy people. Aspects of life that become more important for ill older adults are mental health, housing, family and friends, meaningful spending of time and religion. Those who consider the non-material aspects as more important were shown to keep their QoL at a level that is comparable to those for whom health is of undiminished importance. For these individuals, it was concluded that the adaptation of their priorities was successful. Those, however, who considered material aspects such as housing and income more important, showed a greater decline in QoL than the compared group.

Finally, a further empirical illustration showed that QoL as rated by older individuals after making an inventory of aspects of life important to them, was higher than their score on a standard health-related QoL measurement instrument. How can we reconcile our earlier conclusion that health is considered the most important aspect of life by the majority of older people, with the conclusion that the personal standard based QoL is higher than the QoL measured by a standard instrument which stresses health? First, QoL is multidimensional, including material, social, and spiritual aspects. Even though health-related QoL instruments are presented as multidimensional, they only include various dimensions of health, and not other aspects of life. As we have seen, ill older people consider health as less important over time, while they adopt other aspects of life as more important. Second, another psychological process that may play a role is 'social comparison', i.e. people tend to compare their situation with that of others as important (Beaumont and Kenealy, 2004). It then depends whom they choose for this comparison: those who are worse off or those who are better off. There is evidence that people who compare themselves with those who are worse off, 'downward comparison', perceive their QoL as better (Frieswijk et al., 2004). Most standard instruments do not provide the possibility to compare oneself with others, so that this psychological process remains unobserved.

To summarise, this chapter provides an overview of health and QoL, and in particular provides evidence that the relation between the two may not be as strong in older age as has been assumed. More research should be directed towards other aspects of life that may be important to ill older people, and towards psychological processes that play a role in evaluations of QoL. Ultimately, it is important to have an adequate picture of QoL in older people experiencing decreasing health, as it can be used to identify and prioritise problems, to facilitate communication between health professionals and patients, to screen for hidden problems, to facilitate shared clinical decision making and to monitor reactions to treatment (Higginson and Carr, 2001). This would serve the ultimate objective of caring for ill or frail older people: to maintain or if possible to improve QoL.

REFERENCES

Appels, A., Bosma, H., Grabuskas, V., Gostautas, A., and Sturmans, F. (1996) 'Self-rated Health and Mortality in a Lithuanian and a Dutch Population', *Social Science and Medicine*, 42, 681–689.
Baltes, P.B. (1987) 'Theoretical Propositions of Life-Span Developmental Psychology: On the Dynamics Between Growth and Decline', *Developmental Psychology*, 23, 611–626.
Banks, J., Marmot, M., Oldfield, Z., and Smith, J.P. (2006) 'Disease and Disadvantage in the United States and in England', *Journal of the American Medical Association*, 295, 2037–2045.
Bardage, C., Pluijm, S.M.F., Pedersen, N.L., Deeg, D.J.H., Jylhä, M., Noale, M., Blumstein, T., and Otero, A., for the CLESA working group (2005) 'Self-Rated Health among the Elderly: An International Comparison', *European Journal of Ageing*, 2, 149–158.
Beaumont, J.G. and Kenealy, P.M. (2004) 'Quality of Life Perceptions and Social Comparisons in Healthy Old Age', *Ageing and Society*, 24, 755–769.
Beekman, A.T.F., Copeland J.R., and Prince, M.J. (1999) 'Review of Community Prevalence of Depression in Later Life', *British Journal of Psychiatry*, 174, 307–311.
Beekman, A.T.F., Van Limbeek, J., Deeg, D.J.H., Van Tilburg, W., and Wouters, L. (1994) 'Een screeningsinstrument voor depressie bij Nederlandse ouderen in de algemene bevolking: De bruikbaarheid van de Center for Epidemiologic Studies Depression Scale (CES-D)', [Screening for Depression in the Elderly in the Community: Using the Center for Epidemiologic Studies Depression Scale (CES-D) in the Netherlands]. *Tijdschrift voor Gerontologie en Geriatrie*, 25, 95–103.
Berman, E., Merz, J.F., Rudnick, M., Snyder, R.W., Rogers, K.K., Lee, J., Johnson, D., Mosenkis, A., Isrant, A., Wilpe, P.R., and Lipschutz, J.H. (2004) 'Religiosity in a Hemodialysis Population and its Relationship to Satisfaction with Medical Care, Satisfaction with Life, and Adherence', *American Journal of Kidney Diseases*, 33, 488–497.
Bowling, A. (1995), The Most Important Things in Life. Comparisons between Older and Younger Population Groups by Age and Gender', *International Journal of Health Sciences*, 6, 169–175.
Bowling, A., Farquhar, M., Grundy, W., and Formby, J. (1993) 'Changes in Life Satisfaction over a Two and a Half Year Period among Very Elderly People Living in London', *Social Science and Medicine*, 36, 641–655.
Bowling, A., Farquhar, M., and Grundy, E. (1996) 'Associations With Changes in Life Satisfaction among Three Samples of Elderly People Living at Home', *International Journal of Geriatric Psychiatry*, 11, 12, 1077–1087.
Bowling, A. and Gabriel, Z. (2004), An Integrational Model of Quality of Life in Older Age. Results from the ESRC/MRC HSRC Quality of Life Survey in Britain', *Social Indicators Research*, 69, 1–36.
Bowling, A., Gabriel, Z., Dykes, J., Dowding, L.M., Evans, O., Fleissig, A., Banister, D., and Sutton, S. (2003) 'Let's Ask Them: A National Survey of Definitions of Quality of Life and Its Enhancement among People Aged 65 and over', *International Journal of Aging and Human Development*, 56, 269–306.
Braam, A.W., Hein, E., Deeg, D.J.H., Beekman, A.T.F., and Van Tilburg, W. (2004) 'Religious Involvement and Course of Depressive Symptoms in Older Dutch Citizens: Results from the Longitudinal Aging Study Amsterdam', *Journal of Aging and Health*, 16, 467–489.
Brooks, R. with the EuroQol Group (1996) 'EuroQol: The Current State of Play', *Health Policy*, 37, 53–72.
Browne, J.P., O'Boyle, C.A., McGee, H.M., Joyce, C.R., McDonald, N.J., O'Malley, K., and Hiltbrunner, K. (1994) 'Individual Quality of Life in the Healthy Elderly', *Quality of Life Research*, 3, 235–244.

Browne, J.P., OBoyle, C.A., McGee, H.M., McDonald, M.J., and Joyce, C.R.B. (1997) 'Development of a Direct Weighting Procedure for Quality of Life Domains', *Quality of Life Research*, 6, 4, 301–309.

Bryant, L.L., Corbett, K.K., and Kutner, J.S. (2001) 'In Their Own Words: A Model of Healthy Aging', *Social Science and Medicine*, 53, 927–941.

Carlson, P. (1998) 'Self-Perceived Health in East and West Europe: Another European health divide', *Social Science and Medicine*, 46, 1355–1366.

Covinsky, K.E., Wu, A.W., Landefeld, C.S., Connors, A.F., Phillips, R.S., Tsevat, J., Dawson, N.V., Lynn, J., and Fortinsky, R.H. (1999) 'Health Status Versus Quality of Life in Older Patients: Does the Distinction Matter?' *American Journal of Medicine*, 106, 435–440.

Deeg, D.J.H. (2005) 'Longitudinal Characterization of Course Types of Functional Limitations', *Disability and Rehabilitation*, 27, 253–262.

Deeg, D.J.H. and Braam, A.W. (1997) 'De betekenis van kwaliteit van leven voor ouderen zelf: een kwantitatieve benadering' [What is Important to Older Persons and how Does it Affect Their Quality of Life? A quantitative approach], *Medische Antropologie*, 9, 136–149.

Deeg, D.J.H., Knipscheer, C.P.M., and Van Tilburg, W. (1993) *Autonomy and Well-Being in the Aging Population. Concepts and Design of the Longitudinal Aging Study Amsterdam* (LASA). NIG-Trendstudies no. 7. Bunnik, The Netherlands, Netherlands Institute of Gerontology.

Deeg, D.J.H., Kriegsman, D.M.W., and Van Zonneveld, R.J. (1994) 'Trends in Fatal Chronic Diseases and Disability in the Netherlands 1956–1993 and Projections of Active Life Expectancy 1993–1998' in C.D. Mathers, J. McCallum, and J.M. Robine (eds) *Advances in Health Expectancy. Proceedings of the 7th International Workshop of the Network on Health Expectancy REVES*, Canberra, Australia, February 1994. Canberra, Australian Institute of Health and Welfare, pp.80–95.

Deeg, D.J.H. and Westendorp-de Serière, M. (1994) (eds) *Autonomy and Well-Being in the Ageing Population I: Results from the Longitudinal Aging Study Amsterdam (LASA) 1992–1993*, Amsterdam, VU University Press.

Diener, E., Suh, E.M., Lucas, R.E., and Smith, H.L. (1999) 'Subjective Well-being: Three Decades of Progress', *Psychological Bulletin*, 125, 276–302.

Dykstra, P.A. and De Jong Gierveld, J. (1993) 'The Theory of Mental Incongruity, with a Specific Application to Loneliness among Widowed Men and Women', in R. Gilmour (ed.) *Theoretical Frameworks in Personal Relationships*, Hillsdale, NJ, Erlbaum.

Farquhar, M. (1995) 'Elderly People's Definitions of Quality of Life', *Social Science and Medicine*, 41, 1439–1446.

Fayers, P. and Bjordal, K. (2001) 'Should Quality-of-Life Needs Influence Resource Allocation?' *Lancet*, 357 (March 31), 978.

Festinger, L.A. (1957) *Theory of Cognitive Dissonance*, Evanston, Ill, Row, Peterson and Company.

Frieswijk, N., Buunk, B.P., Steverink, N., and Slaets, J.P. (2004) 'The Interpretation of Social Comparison and Its Relation to Life Satisfaction among Elderly People: Does Frailty make a Difference?' *Journal of Gerontolology B: Social Sciences*, 59, 250–257.

Gabriel, Z. and Bowling, A. (2004) 'Quality of Life from the Perspectives of Older People', *Ageing and Society*, 24, 675–691.

Gijsberts, M. (1993) *Culturele veranderingen: Het wegen waard? [Cultural Changes: Worth Measuring?]* Rijswijk, the Netherlands: Social and Cultural Planning Office.

Gott, M., Barnes, S., Parker, C., Payne, S., Seamark, D., Gariballa, S., and Small, N. (2006) 'Predictors of the Quality of Life of Older People with Heart Failure Recruited from Primary Care', *Age and Ageing*, 35, 172–177.

Haywood, K.L., Garratt, A.M., and Fitzpatrick, R. (2005) 'Quality of Life in Older People: A Structured Review of Generic Self-assessed Health Instruments', *Quality of Life Research*, 14, 7, 1651–1668.

Hickey, A., Barker, M., McGee, H., and O'Boyle, C. (2005) 'Measuring Health-related Quality of Life in Older Patient Populations – A Review of Current Approaches', *Pharmacoeconomics*, 23, 971–993.

Higginson, I.J. and Carr, A.J. (2001) 'Measuring Quality of Life: Using Quality of Life Measures in the Clinical Setting', *British Medical Journal*, 322, 1297–1300.

Hilleras, P.K., Aguero-Torres, H., and Winblad, B. (2001) 'Factors Influencing Well-Being in the Elderly', *Current Opinion in Psychiatry*, 14, 4, 361–365.

Hofman, A., Rocca, W.A., Brayne, C., Breteler, M.M., Clarke, M., Cooper, B., Copeland, J.R.M., Dartigues, J.F., Da Silva Droux, A., Hagnell, O., Heeren, T.J., Engedal, K., Jonker, C., Lindesay, J., Lobo, A., Mann, A.H., Mölsä, P.K., Morgan, K., O'Connor, D.W., Sulkava, R., Kay, D.W.K., and Amaducci, L. (1991) 'The Prevalence of Dementia in Europe: A Collaborative Study of 1980–1990 Findings', *International Journal of Epidemiology*, 20, 736–748.

Huisman, M., Kunst, A.E., and Mackenbach, J.P. (2003) 'Socioeconomic Inequalities in Morbidity among the Elderly; a European Overview', *Social Science and Medicine*, 57, 861–873.

Idler, E.L. (2003) 'Discussion: Gender Differences in Self-rated Health in Mortality, and in the Relationship Between the Two', *The Gerontologist*, 43, 372–375.

Idler, E.L. and Benyamini, Y. (1997) 'Self-rated Health and Mortality: A Review of Twenty-seven Community Studies', *Journal of Health and Social Behavior*, 38, 21–37.

Jylhä, M., Guralnik, J.M., Ferrucci, L., Jokela, J., and Heikkinen, E. (1998) 'Is Self-rated Health Comparable across Cultrues and Genders?' *Journals of Gerontology: Social Sciences*, 53B, 144–152.

Kriegsman, D.M.W., Deeg, D.J.H., Van Eijk, J.T.M., Penninx, B.W.J.H., and Boeke, A.J. (1997) 'Do Disease Specific Characteristics Add to the Explanation of Mobility Limitations in Patients with Different Chronic Diseases? A Study in The Netherlands', *Journal of Epidemiology and Community Health*, 51, 676–685.

Kriegsman, D.M.W., Penninx, B.W.J.H., Van Eijk, J.T.M., Boeke, A.J.P., and Deeg, D.J.H. (1996) 'Self-reports and General Practitioner Information on the Presence of Chronic Diseases in Community-dwelling Elderly. A Study on the Accuracy of Patients' Self-reports and on Determinants of Inaccuracy', *Journal of Clinical Epidemiology*, 49, 1407–1417.

Leung, K.K., Wu, E.C., Lue, B.H., and Tang, L.Y. (2004) 'The Use of Focus Groups in Evaluation Quality of Life Components among Elderly Chinese people', *Quality of Life Research*, 13, 1, 179–190.

Maslow, A.H. (1954) *Motivation and Personality*, New York, Harper.

McKee, M., Whatling, J.M., Wilson, J.L., and Vallance-Owen, A. (2005) 'Comparing Outcomes of Cataract Surgery: Challenges and Opportunities', *Journal of Public Health*, 27, 348–352.

Melzer, D., Lan, T.-Y., Tom, B.D.M., Deeg, D.J.H., and Guralnik, J.M. (2004) 'Variation in Thresholds for Reporting Mobility Disability Between National Population Sub-groups and Studies', *Journal of Gerontology: Medical Sciences*, 59, 1295–1303.

Nilsson, J., Grafstrom, M., Zaman, S., and Kabir, Z.N. (2005) 'Role and Function: Aspects of Quality of Life of Older People in Rural Bangladesh', *Journal of Aging Studies*, 19, 363–374.

Minicuci, N., Noale, M., Bardage, C., Blumstein, T., Deeg, D.J.H., Gindin, J., Jylhä, M., Nikula, S., Otero, A., Pedersen, N.L., Pluijm, S.M.F., Zunzunegui, M.V., and Maggi, S. for the CLESA Working Group (2003) 'Cross-national Determinants of Quality of Life from Six Longitudinal Studies on Aging: The CLESA Project', *Aging Clinical and Experimental Research*, 15, 187–202.

Minicuci, N., Noale, M., Pluijm, S.M.F., Zunzunegui, M.V., Blumstein, T., Deeg, D.J.H., Bardage, C., and Jylhä, M., for the CLESA working group (2004) 'Disability-free Life Expectancy: A Cross-national Comparison of Six Longitudinal Studies on Ageing. The CLESA Project', *European Journal of Ageing*, 1, 37–44.

Mroczek, D.K. and Spiro, A. III (2005) 'Change in Life Satisfaction during Adulthood: Findings from the Veterans Affairs Normative Aging Study', *Journal of Personality and Social Psychology*, 88, 189–202.

Nikula, S., Jylhä, M., Bardage, C., Deeg, D.J.H., Gindin, J., Minicuci, N., Pluijm, S.M.F., and Rodriguez-Laso A., for the CLESA Working Group (2003) 'Are IADLs Comparable Across Countries? Sociodemographic Associates of Harmonized IADL Measures', *Aging Clinical and Experimental Research*, 15, 451–459.

Noale, M., Minicuci, N., Maggi, S., Bardage, C., Gindin, J., Nikula, S., Pluijm, S., and Rodríguez-Laso, A. for the CLESA working group (2005) Predictors of Mortality: An International Comparison of Socio-Demographic and Health Characteristics from Six Longitudinal Studies on Aging: The CLESA Project, *Experimental Gerontology*, 40, 1–2, 89–99.

Penninx, B.W.J.H., Deeg, D.J.H., Van Eijk, J.T.M., Beekman, A.T.F., and Guralnik, J.M. (2000) 'Changes in Depression and Physical Decline in Older Adults: A Longitudinal Perspective', *Journal of Affective Disorders*, 61, 1–12.

Penninx, B.W.J.H., Geerlings, S.W., Deeg, D.J.H., Beekman, A.T.F., and Van Tilburg, W. (1999) 'Minor and Major Depression and the Risk of Death in Older Persons', *Archives of General Psychiatry*, 56, 889–896.

Penninx, B.W.J.H., Guralnik, J.M., Ferrucci, L., Simonsick, E.M., Deeg, D.J.H., and Wallace, R.B. (1998) 'Depressive Symptoms and Physical Decline in Community-Dwelling Older Persons', *Journal of the American Medical Association*, 279, 1720–1726.

Pluijm, S.M.F., Bardage, C., Nikula, S., Blumstein, T., Jylhä, M., Minicuci, N., Zunzunegui, M.V., Pedersen, N.L., and Deeg, D.J.H. for the CLESA Study Working Group (2005) 'A Harmonized Measure of Activities of Daily Living was a Reliable and Valid Instrument for Comparing Disability in Older People across Countries', *Journal of Clinical Epidemiology*, 58, 1018–1023.

Puts, M.T.E., Lips, P., and Deeg, D.J.H. (2005) 'Sex Differences in the Risk of Frailty for Mortality Independent of Disability and Chronic Diseases', *Journal of the American Geriatrics Society*, 53, 40–47.

Puts, M.T.E., Shekary, N., Widdershoven, G., Heldens, J., Lips, P., and Deeg, D.J.H. (2007) What Does Quality of Life Mean to Older Frail and Non-frail Community-Dwelling Adults in the Netherlands? Quality of Life Research, 16, 263–277.

Radloff, L.S. (1976) 'The CES-D Scale: A Self-report Depression Scale for Research in the General Population', *Applied Psychological Measurement*, 1, 385–401.

Seeman, T.E., Bruce, M.L., and McAvay, G.J. (1996) 'Social Network Characteristics and Onset of ADL Disability: MacArthur Studies of Successful Aging', *Journals of Gerontology: Social Sciences*, 51B, S191–S200.

Schoevers, R.A., Geerlings, M.I., Beekman, A.T.F., Penninx, B.W.J.H., Deeg, D.J.H., Jonker, C., and Van Tilburg, W. (2000) 'Association of Depression and Gender with Mortality in Old Age. Results from the Amsterdam Study of the Elderly (AMSTEL)', *British Journal of Psychiatry*, 177, 336–342.

Singer, B. and Ryff, C.D. (2001) 'Person-centered Methods for Understanding Aging: The integration of numbers and narratives', in R.H. Binstock, and L.K. George (eds), *Handbook of Aging and the Social Sciences*, San Diego, CA, Academic Press, pp.44–65.

Sonnenberg, C.M., Beekman, A.T.F., Deeg, D.J.H., and Van Tilburg, W. (2000) 'Sex-differences in Late-life Depression' *Acta Psychiatrica Scandinavica*, 101, 2000, 286–292.

Sprangers, M., De Regt, E.B., Andries, F., Van Agt, H.M.E., Bijl, R.V., De Boer, J.B., Foets, M., Hoeymans, N., Jacobs, A.E., Kempen, G.I.J.M., Miedema, H.S., Tijhuis, M.A.R., and De Haes, H.C.J.M. (2000) 'Which Chronic Conditions are Associated With Better or Poorer Quality of Life?' *Journal of Clinical Epidemiology*, 53, 895–907.

Strawbridge, W.J., Wallhagen, M.I., and, Cohen, R.D. (2002) 'Successful Aging and Well-being: Self-rated Compared with Rowe and Kahn', *The Gerontologist*, 42, 727–733.

Sullivan, P.W., Lawrence, W.F., and Ghushchyan, V. (2005) 'A National Catalog of Preference-based Scores for Chronic Conditions in the United States', *Medical Care*, 43, 736–749.

Van den Brink, C.L., Tijhuis, M., Kalmijn, S., Klazinga, N.S., Nissinen, A., Giampaoli, S., Kivinen, P., Kromhout, D., and Van den Bos, G. (2003) 'Self-reported Disability and Its Associates with Performance-based Limitation in Elderly Men: A Comparison of Three European Countries', *Journal of the American Geriatrics Society*, 51, 782–8.

Verbrugge, L.M. and Jette, A. (1994) 'The disablement process', *Social Science and Medicine*, 38, 1–14.

Verbrugge, L.M. and Patrick, D.L. (1995) 'Seven Chronic Conditions: Their Impact On US adults' Activity Levels and Use of Medical Services', *American Journal of Public Health*, 85, 173–182.

Walker, A. (2005) 'A European Perspective on Quality of Life in Old Age', *European Journal of Ageing*, 2, 2–10.

Ware, J. Jr., Kosinski, M., and Keller, S.D. (1996) 'A 12-item Short-form Health Survey: Construction of Scales and Preliminary Tests of Reliability and Validity', *Medical Care*, 34, 220–233.

Westerhof, G.J. (2003) 'The Personal Experience of Aging: Multidimensionality and Multidirectionality in Relation to Successful Aging and Well-being', *Tijdschrift voor Gerontologie en Geriatrie*, 34, 96–103. (In Dutch.)

Widar, M., Ahlstrom, G., and Ek, A.C. (2004) 'Health-related Quality of Life in Persons with Long-term Pain after a Stroke', *Journal of Clinical Nursing*, 13, 497–505.

Wilhelmson, K., Andersson, C., Waern, M., and Allebeck, P. (2005) 'Elderly People's Perspectives on Quality of Life', *Ageing and Society*, 25, 585–600.

MARJA VAARAMA, RICHARD PIEPER, AND ANDREW SIXSMITH

13. CARE-RELATED QUALITY OF LIFE

Conceptual and empirical exploration

INTRODUCTION

In this chapter the concept of care-related quality of life (crQoL) is discussed as a basis for research within social gerontology and as a framework for evaluation of quality and performance within health and social care services for older people. The motivation for this, both theoretically and practically oriented work, lies in an increasing awareness that issues of quality of life (QoL) are particularly relevant in the study of older people who are vulnerable, frail or disabled. The changes in personal capacities, abilities, and circumstances that often accompany old age may fundamentally challenge the basis of a person's well-being and may undermine their ability to cope with everyday life (Sixsmith, 1994; Hughes, 1990). For those people who rely on daily support from health and social care services this is likely to have a major impact on their QoL. Enhancing QoL should be a major component in how we assess the value and impact of the services.

Considerable attention has been given to issues of health-related QoL (Bowling, 1995, 2004), e.g. in respect to particular illnesses or conditions. Attention has been given to QoL for people, especially older people, who are suffering from chronic, long-term conditions, such as congestive heart failure, stroke, and arthritis. Rather less attention has been given to older people who are described as 'frail', or who experience multiple low-level conditions that have impact on their abilities to cope with everyday life (Birren *et al.*, 1991). Many of these people are dependent on the care and support they receive from formal (e.g. health and social care) and informal (e.g. family and neighbours) sources and their well-being is inevitably bound up in these care relationships. If care is fundamental to the well-being of frail older people, then a framework that specifically incorporates the role of care in the production of well-being is needed, rather than a more general concept of well-being. From an applied perspective, organisations involved in the monitoring, commissioning, and delivery of care services are specifically interested in evaluating the impact of care services.

The work reported in this chapter has been carried out as part of the Care Keys – a project funded under the European Union's Quality of Life Research and Development programme.[1] Care Keys is a multidisciplinary project that aims to develop a conceptual model of crQoL, and a 'tool kit' for the evaluation and management of the quality of long-term care of older people, with emphasis on client voice and outcomes. This study was performed at the initial stage of the project to find our fitting outcome measures, and to test the connection between care and well-being for justifying the basic Care Keys approach and whether it provides a promising avenue for the project working.

H. Mollenkopf and A. Walker (eds.), Quality of Life in Old Age, 215–232.
© 2007 *Springer.*

TOWARDS A CONCEPTUAL MODEL OF crQoL

While the term QoL is commonly used and an increasing body of literature since around 1970 has been the topic, there is still no established definition and conceptual framework. Quite often a definition is avoided and a pragmatic position taken using QoL as an 'umbrella term' (Brown and Brown, 2003). There is an underlying idea that QoL is about the 'good life' and the evaluation of a person's life with reference to standards of 'goodness', but there is also the fact that the concept is used rather indiscriminately with different authors 'filling in' different things depending on their theoretical or practical approaches. As Diener (1994, p.105) reminds us, 'the most useful definition of subjective well-being will be based on a compelling theory'. This theory is missing, Diener observes, although he makes some important steps toward integrating psychological approaches to subjective well-being as part of a broader concept of QoL. We need conceptual models to ensure that all relevant components of the problem area are included and to orient our thinking both in research and practical application.

One common approach to reach a more systematic concept is to list relevant aspects of human life which would constitute a 'good' QoL. In the Care Keys project Vaarama et al. (2004) outline a number of broad components of QoL relevant to frail older people (see also Hughes, 1990; Cummins, 1997):

- Socio-economic factors (income, household structure, and ethnicity)
- Individual characteristics (age, health, cognitive, and emotional capacities)
- Social factors (family, social networks, and social participation)
- Life changes (traumatic or disruptive events, or lack of change)
- Environmental factors (housing, facilities, amenities, and neighbourhood)
- Social and health care services (including expectations, preferences and amount, and type of support)
- Personal autonomy factors (ability to make choices, and control)
- Activities (leisure, sports, productive activities, and work)
- Psychological health (psychological well-being, morale, loneliness, and happiness)
- Subjective life satisfaction (evaluation of own QoL in relevant life domains)

Starting with such a list, we may distinguish four approaches to the development of a systematic concept or model. In a *first* approach the list is reduced to a more aggregated and systematic taxonomy of relevant life domains such as physical well-being, material well-being, social well-being, emotional well-being, and productive well-being (see Felce and Perry, 1997; Cummins, 1997). The typical perspective of these taxonomies is on measurement of QoL in relevant *domains*. Diverse aspects of a person's life and the environment may be included without theoretical specification of the relationship they have to (the theory of) the person. The taxonomy is guided by expert judgement and usually supported by statistical analyses. An example is the QoL assessment developed by the WHOQoL group (Skevington et al., 2004).

A *second* approach – also resulting in a taxonomy – starts with a theoretical framework and develops a model of the 'good life' or the 'successful life'. The approach will start from an integrated or 'holistic' model of the human being or person and specify theoretical *dimensions* which may be more structural or developmental depending on the theory. The framework might be quite general, e.g. drawing on a theory of system development (Freund *et al.*, 1999; Veenhoven, 2000) or more specific like the model of QoL of older people' with frailty or dementia by Lawton (1991) employing environmental psychology, or the model for older people in transition from home to institution by Tester *et al.* (2003) using a gerontological framework.

A *third* approach organises components in a causal process model specifying the *variables* or indicators as conditions, causes and effects of QoL. Typically, the perspective is explanatory or involves testing hypotheses about certain condi-tions of QoL. Thus, the selection of variables is usually guided and restricted by empirical research. There is a wealth of research from this perspective (Renwick *et al.*, 1996; Schalock and Siperstein, 1996; Brandstädter and Renner, 1999). This approach tends to adopt a narrower concept of 'final outcomes' limited to subjective well-being (Diener, 1994). This has the advantage that objective conditions and subjective outcomes can be measured independently. But the distinction objective vs. subjective QoL can create a lot of conceptual confusion, as argued in the following section.

A *fourth* approach – often practice oriented – looks at QoL in the context of social and health intervention or the production of welfare specifying *factors* of input or resources, process (interventions) and outcome, and combining them in strategies. Intervention and production models structure the field in conditions of production and in factors which can be manipulated to produce an outcome or product. QoL appears as a complex product which will be analysed with reference to goals and interventions and, indeed, other goals or products may be intended, e.g. benefits for informal carers, equity of distribution, or other collective social benefits in the production of welfare (Davies and Knapp, 1981; Knapp, 1984, 1995; Brown and Brown, 2003; Vaarama and Pieper, 2005). Domain specific concepts often have this perspective, like health-related QoL or, in the present case, crQoL. This approach also corresponds to a care management or a social policy and planning perspective.

These approaches are not mutually exclusive. This is demonstrated by a useful starting point of any review of QoL of frail older people and for a conceptualisation of QoL – the work of Lawton. His concept of person-environment fit (Lawton and Nahemow, 1973) is based on the idea that increasing frailty in old age causes significant loss of competence, affecting the ability to perform activities of daily living (ADLs). People with low or reduced personal capacities are more vulnerable to the demands of the environment compared with people with high capacity, and environmental support or opportunities become very important in terms of their everyday tasks of living and their QoL. Lawton (1991, p.6) extended this basic concept further by including subjective and objective evaluations and offers the following definition:

Quality of life is the multidimensional evaluation, by both intrapersonal and socio-normative criteria, of the person–environment system of an individual in time past, current, and anticipated.

He describes QoL in terms of four overlapping subdimensions or 'sectors':

Objective 'person-environment fit'
• *Behavioural competence*, or the capacity of the person to deal with the demands of everyday life.
• *Environment* or the demands and opportunities of the physical and social circumstances within which the person lives.
Subjective evaluation
• *Life satisfaction*, or the person's subjective evaluation of their objective life circumstances in different life domains.
• *Psychological well-being*, or the subjective or experiential well-being including happiness, loneliness, etc.

The concept is explicitly multidimensional proposing a 'four-dimensional plotting of how the person stands' (Lawton, 1991, p.12), and rejects one-dimensional concepts. It also includes reference to time, as life satisfaction summarises the past, psychological well-being reflects the present, and the behavioural competence refers to mastery of the future. It also argues for the combination of subjective and objective measures in QoL concepts.

Unfortunately, Lawton (1997) is not very clear about the theoretical status of the four subdimensions in the model and uses slightly different ways of describing them. To clarify their meaning, we would like to draw on the subjective QoL concept of Diener (1994), a different conceptual model by Veenhoven (2000) and on an interpretation of Lawton's model as a transactional model by Davies and Knapp (1981, p.126) for a few comments (for a more detailed discussion see Vaarama, Pieper, and Sixsmith, forthcoming).

In his definition, Lawton refers to 'both intrapersonal and social-normative criteria'. While there are good arguments to consider both types of criteria, they should not be combined in one model of QoL. Individual subjective standards are unavoidably implied in statements on 'life satisfaction'. Certainly, social services and planning need some more objective evaluation, especially since the person may report himself or herself to be 'happy' despite poor life circumstances from an objective or external perspective (see Lawton, 1991; Cummins, 1997). We should clearly distinguish between evaluative standards employed in social planning and social politics for *their* purposes and the QoL evaluations of the clients themselves – if only to make the systematic analysis of a misfit between the two value standards possible. Lawton moves here to the fourth approach towards QoL as distinguished above.

He suggests combining subjective and objective factors in the general QoL model. But it should be clear that the terms 'subjective' and 'objective' are somewhat misleading in this context. They refer actually to the subjective or objective *methods* to measure QoL factors (see also Diener, 1994). Especially when describing the model for clients with dementia, it becomes clear that the subjective self-reports are not always feasible for the measuring of, say, life quality and psychological well-being.

In this case, the observations by carers may take their place (see Lawton, 1997). Following the general approach of the person-environment fit, ideally, we should make use of a triangulation of methods with both subjective and objective measures for all dimensions of QoL.

There is a discussion in the literature on whether we should develop group-specific models of QoL for, for instance, people with diseases, impairments, or frailty. However, a specific focus on frail older people is rare. Tester *et al.* (2003) comment that 'where frail older people are concerned, the results of such work remain unsatisfactory, both theoretically and methodologically'. They go on to say: 'it is doubtful that a generic definition of QoL will be useful for all research purposes. Instead, QoL models specific to particular groups of older people are being developed, for example, dementia-specific QoL models'. While this approach is meaningful to specify the person–environment fit, it does not call for dismissal of Lawton's generic model and the adoption of a different theoretical model for each group. In fact, the gist of the argument in Lawton's model is precisely that one needs to understand the *relationships* between the person and his/her environment *in each case*, because it is not the environment as such which has an influence, but the environmental features relevant for a certain person's way of life (see also Chapter 7). The specification of group specific environments is, certainly, a meaningful strategy to avoid the practical and empirical problems of analysing the relationships for each person.

Veenhoven (2000) presents a fourfold taxonomy of QoL as four 'qualities of life': (1) liveability of the environment, i.e. the external conditions within which the person lives; (2) the life-ability of the person, i.e. the competence of a person to cope with the problems of life or to exploit its potential; (3) utility of life, i.e. the broader value of the person's life or the meaning that a person's life has for others within society; (4) appreciation of life, i.e. the inner outcomes of life, including subjective well-being, life satisfaction, and happiness. While Veenhoven's framework does not address frail older people specifically, it has clear similarities with Lawton's work. The only subdimension which does not really correspond is the 'utility of life'. But here we would follow Lawton and insist that it is the person's evaluation of his/her own life which is relevant and not some external evaluation of the utility for others. Keeping this in mind, the Veenhoven model helps to interpret the subdimensions of Lawton in a fourfold table of 'four qualities of life' (Table 1) and combine them with an interpretation by Davies and Knapp (1981, p.126) also incorporating a well-established distinction in gerontology between coping processes of assimilation, accommodation, adaptation, and affective regulation. In the Care Keys project, Pieper and Vaarama developed the following preliminary model for crQoL (Vaarama *et al.*, forthcoming).

To emphasise the role of care we place it in the scheme to see how it interconnects with the components of QoL. While care certainly should aim to enhance all aspects of the QoL of older people, it is primarily an essential feature of the supportive environment of the person providing a better 'fit' by social and material resources. In the model it would belong to the 'outer relations'. The subjective qualities

TABLE 1. A Four-Dimensional crQoL Model (Adapted from Lawton, 1991 and Veenhoven, 2000)

Dimensions	Person-environment fit (Veenhoven: potentials)	Subjective evaluation (Veenhoven: results)
Inner relations (Veenhoven)	Behavioural competence *Assimilation*	Psychological well-being *Affective regulation*
Outer relations (Veenhoven)	Environmental demands and (care) support *Adaptation*	Life satisfaction in different life domains *Accomodation*

of care would reflect the quality of the relation which the person has to his or her environment, and indicate that a person receives and can make use of the support he or she needs for sustaining QoL. Not only is the objective satisfaction of assessed needs by care important for crQoL but, also, the degree to which preferences and expectations are met, thus, measuring also the relevance of a service for the client. This has been an area that has had little attention within the QoL literature for older people (Bowling, 1997, p.7).

Finally, there is a substantial literature on the role of psychological resources in determining the ability to cope with problems and situations. For example, ideas of 'self-control' (Abeles, 1991), Rotter's concept of locus of control (1966) and Staudinger's concept of resilience (Staudinger *et al.* 1999) have been used to explain why some people appear better able to cope and adapt to everyday life changes. As indicated above, general coping processes can be included in the model. But the model also has a place for the autonomy and control of the person or care client in the subdimension of behavioural competence. Recent ideas about good practice emphasise the need to involve clients in order to make those decisions responsive to perceived needs and to engender a sense of personal involvement, commitment, and control over one's own life.

THE STUDY DESIGN

More or less complex and dynamic causal models can be conceived on the basis of the generic model of Lawton. Our aim is to specify a preliminary model of crQoL that differentiates between the factors important to the QoL for care-dependent older people living at home, and explores the relations between QoL outcomes and determinants. Our overall assumption is that care is crucial for the QoL of frail older people, but the connection may be mediated by other factors (Figure 1).

We explored our model with data from a previous study with face-to-face interviews with randomly sampled people aged over 75 living at home and using community social and health services in 1998 in Finland. The sample size was 331 complete cases: female 79%, mean age 84 years, married 11%, higher occupational education 35%, and living alone 85%. The instruments used in the interview were mostly nationally or internationally validated single-item questions on objective living conditions, subjective health, happiness and life satisfaction. To measure

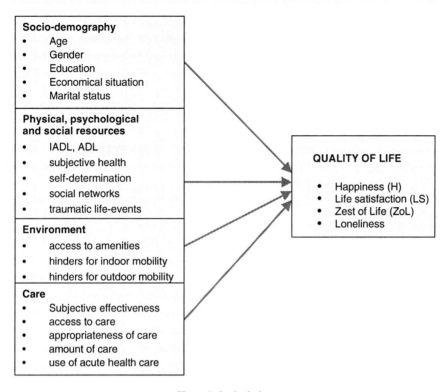

Figure 1. Study design

subjective QoL, we used an application of Philadelphia Geriatric Centre Morale Scale (PGCMS) (Lawton, 1975). The PGCMS combines the aspects of agitation, attitude towards own ageing and lonely dissatisfaction. In the original scale, the 17-scale items are answered only with 'yes' or 'no' and from all the answers an index is calculated. In this study, PGCMS was applied using a modified 3-scale version (0 = no, 1 = yes, 2 = cannot say). The reliability of the applied scale was high (Cronbach's alpha = 0.90). We named the measure as 'Zest of Life' (ZoL), emphasising a subdimension of the morale. It should be noted that the PGCMS is a multidimensional scale which contains items we also measured independently such as life satisfaction, loneliness, and happiness, so the variables in the outcome domain are expected to correlate. We operationalised the domains of our variables as follows:

- Sociodemographical factors: age, gender, education, subjective economical situation, and marital status.
- Physical, psychological, and social resources: subjective health, self-reported IADL and ADL-difficulties, subjective self-determination (I can decide over my own issues: 0 = not at all, 5 = fully), satisfaction with social networks and traumatic life-events during the past two years.

- Environmental factors: self-reported barriers for indoor and outdoor mobility, sub-jective access to public transport and other amenities, living alone.
- Care: self-reported intensity of homecare services, use of intramural hospital care during the last 12 months, visiting polyclinic during the last 12 months, subjective access to care (I have access to care when I need), subjective evaluation of posi-tive impact of care (homecare makes it possible for me to stay at home), subjec-tive evaluation of sufficiency of the types and amount of care (the care I get is appropriate and I get enough).
- Outcomes: Life-satisfaction (LS) (I am satisfied with my life), zest of life/morale (ZoL), happiness (H) (I am as happy now as when I was younger), loneliness (I feel lonely: 0 = never, 5 = always).

In health-related QoL research, subjective health has usually been treated as an out-come measure, but here we assumed it to be a condition or determinant of QoL rather than a QoL outcome. We apply Lawton's model in the interpretation of our data and use factor analysis and logistic and linear regression analyses as the main methods of our study.

EMPIRICAL RESULTS

Factor Domains

First, we carried out a varimax-rotated factor analysis to differentiate between dimen-sions within the data. It provided 10 factors, from which the first described subjective QoL, with the others relating to internal and external determinants (Table 2).

The first factor relates to a QoL factor as ZoL, H, LS, and loneliness all load on this factor. The first three contribute positively, while the fourth decreases well-being. These variables, including morale/ZoL, seem to constitute subjective QoL, and are outcome measures rather than conditions, although consideration needs to be given to possible causal relationships between these four elements. A second factor connects subjective health, physical dependency, use of care, and subjec-tive evaluation of care, suggesting that very dependent people have poorer subjective health and rather negative evaluations of the effectiveness of care. The combination also suggests that this factor (including health) indicates conditions of QoL rather than constitutive components. A third differentiates married males, suggesting that gender, marital status, and living alone all impact on well-being. Satisfaction with access to care and satisfaction with the amount of care form a factor together, sup-porting the idea that subjective quality of care contributes to QoL. Economic situ-ation and education level load together as a socio-economic factor of QoL. Use of hospital and polyclinics describe acute illness, suggesting that in addition to dependency, acute illness has its own role for QoL in our study population. Living environment divides into two dimensions – indoor and outdoor – suggesting that both have their own importance for well-being. A ninth factor is somewhat mixed as it combines age, subjective self-determination and satisfaction with social net-works. If we are willing to interpret age to indicate a certain level of maturity and

TABLE 2. Rotated Component Matrix

	Component									
	1	2	3	4	5	6	7	8	9	10
Morale/ZoL	0.749	–	–	–	–	–	–	–	–	–
Happiness	0.693	–	–	–	–	–	–	–	–	–
Life satisfaction	0.660	–	–	–	–	–	–	–	–	–
Loneliness	-0.624	–	–	–	–	–	–	–	–	–
IADL – problems 0–1	–	0.751	–	–	–	–	–	–	–	–
Intensity of home care, mean 0–3	–	0.689	–	–	–	–	–	–	–	–
BADL – problems 0–1	–	0.686	–	–	–	–	–	–	–	–
Subjective care effectiveness	–	-0.500	–	–	–	–	–	–	–	–
Subjective health	–	-0.405	–	–	–	–	–	–	–	–
Married	–	–	0.845	–	–	–	–	–	–	–
Living alone	–	–	-0.818	–	–	–	–	–	–	–
Gender (1 = male, 0 = female)	–	–	0.581	–	–	–	–	–	–	–
Subjective access to care	–	–	–	0.821	–	–	–	–	–	–
Subjective appropriateness of care	–	–	–	0.803	–	–	–	–	–	–
Higher (occupational) education	–	–	–	–	0.767	–	–	–	–	–
Subjective economic status	–	–	–	–	0.710	–	–	–	–	–
Use of hospital care during last 12 months	–	–	–	–	–	0.719	–	–	–	–
Visiting policlinic during last 12 months	–	–	–	–	–	0.682	–	–	–	–
Bad access to amenities	–	–	–	–	–	–	0.739	–	–	–
Hinders for outdoor mobility (dummy)	–	–	–	–	–	–	0.629	–	–	–

(Continued)

TABLE 2. Rotated Component Matrix—Continued

	Component									
	1	2	3	4	5	6	7	8	9	10
Hinders for indoor mobility (dummy)	–	–	–	–	–	–	–	0.771	–	–
Age	–	–	–	–	–	–	–	–	0.683	–
Subjective self-determination	–	–	–	–	–	–	–	–	0.518	–
Satisfaction with social networks	–	–	–	–	–	–	–	–	0.460	–
Traumatic life events during last 2 years	–	–	–	–	–	–	–	–	–	0.834

Note: Extraction method: Principal Component Analysis; Rotation method: Varimax with Kaiser Normalisation. Rotation converged in 12 iterations.

the social network to stand, in this combination, for social competence, then we might describe this factor as representing the factor of sense of control or resilience. But quite obviously this factor seems to be insufficiently captured by the variables in the study. Traumatic life events load on their own factor, suggesting that they have their own role to play within QoL.

The analysis confirms the multidimensionality of the concept of QoL. The rather strong correlations between the variables of well-being were to be expected. Also the importance of subjective quality of care is confirmed. Additionally, the importance of the physical living environment appears high. Situational factors (acute illness as distinct from long-term dependency and traumatic life-events such as becoming seriously ill or widowed) also appear to be important independent factors for QoL in the study population, and they can be seen also as risk-factors for good life quality. Unfortunately our measurement possibilities for client autonomy and control were restricted, and the measurement of self-determination did not adequately represent the dimension.

Relationships between Quality of Life Outcomes and Determinants

To analyse relationships within the QoL model, we carried out several regression analyses. First we analysed life satisfaction and happiness by stepwise logistic regression analyses. Even though the factor analysis suggested that life satisfaction (LS), happiness (H), loneliness, and morale/ZoL are parts of subjective well-being, we used ZoL first as a measure of psychological resources to explain variation in LS (model 1). The rational for treating morale/ZoL as a psychological factor was – as indicated above – that the PGCMS scale does include a sense of control dimension. Morale/ZoL and loneliness were introduced into seperate analyses – as the former includes a factor of the latter – together with variables of the other nine factors (Table 3).

In the first model, the stepwise regression analysis picked up only morale/ZoL and subjective quality of care as important factors explaining the variations in LS. According to the model, morale/ZoL, almost alone explains the variation in LS. The model also demonstrates a direct connection between subjective quality of care and LS, confirming our assumption of the important role of care in production of welfare in care-dependent older people. When we excluded morale/ZoL from the analysis, subjective quality of care and loneliness became the most powerful variables impacting on LS (model 2).

According to this, morale/ZoL alone explains almost all the LS of a frail older person, but subjective quality of care also has an impact on it: clients who are satisfied with their access to care and with the amount they get are more likely to have good LS than the dissatisfied clients. When ZoL is left out, loneliness decreases LS by 83% compared with those not feeling lonely, and people who are satisfied with care show an increase of LS which is considerably higher compared with dissatisfied people. Loneliness also decreases the level of happiness. This means loneliness also contributes to LS and to H. This supports a suggestion by Lawton that loneliness in fact has two dimensions: a psychological dimension which would go together with

TABLE 3. Variation in Life Satisfaction (LS) and Happiness (H). Stepwise Logistic Regression Analysis ($n = 331$)

Model	Model 1, LS	Model 2, LS (ZoL excluded)	Model 3, H (ZoL excluded)
Variable	Exp(β) (Sig β)	Exp(β) (Sig β)	Exp(β) (Sig β)
Constant	0.101 (0.001)	6,005 (0.000)	2,644 (0.002)
Morale/ZoL	372,176 (0.000)	– –	– –
Satisfaction with care	2,933 (0.006)	3,835 (0.000)	– –
Loneliness	– –	0.169 (0.005)	0.194 (0.004)
Barriers for indoor mobility (0.1)	– –	– –	0.472 (0.021)
Bad access to amenities	– –	– –	0.361 (0.055)

emotional feelings such as happiness, and a more social dimension, going together with general life satisfaction. In model 3, it is interesting that barriers for indoor mobility and poor access to public transport and other amenities decrease happiness, indicating some independent influence of objective environment factors on the present well-being of a person.

These analyses suggest that ZoL/morale (positively) and loneliness (negatively) contribute to life satisfaction. The results also confirm the strong connection between subjective quality of care (satisfying access and amount of care) and QoL. This we interpret, following Lawton's model, in a way that subjective quality of care (in terms of access and appropriateness) and subjective qualities of the environment are important dimensions of QoL, and that is why we refer to this as 'care-related quality of life'. The independent influence of mobility and access factors opens an avenue for strategies of improving one's living environment (objective QoL factors) to produce more subjective QoL outcomes.

The results suggest that care has a great possibility to contribute positively to a client's QoL by compensating for the deficits in his/her living environment, by meeting the needs caused by physical dependency, and by decreasing perceived loneliness and supporting the psychological well-being of the client. Thus, the quantity, quality, and appropriateness of care seem to be those care inputs that should be employed in the production of QoL for frail older clients of homecare. In this context, we also see the care system to be a broader concept involving improvements of the living conditions of older clients when needed.

We decided not to include subjective health among the outcome measures but among the determinants of QoL. The factor analysis confirmed our assumption suggesting that morale/ZoL may be more a comprehensive outcome measure of well-being. As these two measures were normally distributed in the sample, it was possible to employ linear regression analyses to analyse the causal relations between these and other variables in the 10 factors (Table 4).

If we interpret model 4 to be a more health-related model of QoL, then we see that morale/ZoL plays an important role, subjective effectiveness of care contribute positively to subjective health, and problems in daily activities, especially in personal care, and barriers for indoor mobility decrease it. If we look at model 5, we see that subjective health plays a less important role for morale/ZoL than morale/ZoL plays for subjective health, and that the model picks up economic safety, satisfaction with the amount of care and loneliness, which we may identify as social variables. The models correspond to earlier results, suggesting that subjective health and morale/ZoL are different phenomena, and that subjective health is important but not the only important element of QoL. It may be better conceived as a subjective measure of health and as a condition of QoL. This interpretation would also correspond

TABLE 4. Variation in Subjective Health and Zest of Life. Stepwise Linear Regression Analysis ($n = 331$)

Model	Model 4, Subjective health	Model 5, ZoL
Variable	B	B
	(Sig.)	(Sig.)
Constant	0.557	0.482
	(0.000)	(0.000)
Morale/ZoL	0.258	–
	(0.000)	–
Loneliness	–	–0.232
	–	(0.000)
Good subjective care effectiveness	0.065	–
	(0.010)	–
Satisfaction with amount of care	–	0.054
	–	(0.014)
Good economic situation	–	0.184
	–	(0.001)
Barriers for indoor mobility (0,1)	–0.069	–
	(0.014)	–
Subjective health	–	0.140
	–	(0.002)
BADL – problems^2	–0.184	–
	(0.000)	–
IADL – problems	–0.138	–
	(0.021)	–
R^2	0.246	0.295
F	17,728	27,593
P	0.000	0.000

better to the finding that poor health and disabilities do not necessarily imply low QoL – at least as perceived by the frail older person (see also Cummins, 1997).

Another interesting result is the independent impact of life events and acute illness on QoL. This demonstrates subjective QoL to be a dynamic phenomenon, which in our study population is vulnerable to situational factors such as becoming seriously ill or widowed. The care system should be sensitive for these factors to be able to identify these as risks for QoL which call for special interventions if care is to support older clients in coping with these issues.

DISCUSSION

We have explored factors determining and describing subjective QoL in care-dependent older people to provide evidence for a model of crQoL that we believe will be useful in evaluating the impact of care on the well-being of frail older people. The theoretical discussion about a possible model of crQoL and the empirical explorations reported in this chapter suggest the following relationships between care and QoL.

All analyses confirm care to be a crucial element of life quality in older people living at home and being dependent on care. The analyses suggest that care can contribute positively to the QoL of a client if it is easily accessible and meets the client's needs. It seems that both accessibility and quantity as well as quality of care are important factors for subjective QoL, indicating that this dimension of environmental support is of special importance to older people dependent on care.

Physical environment had the strongest impact on the dimension of QoL which is called 'happiness', suggesting that a barrier-free home and easy access to public transport and other amenities are important conditions for older people's everyday well-being. Vaarama (2004) found that barriers to outdoor mobility caused premature dependency on care, and that for the oldest and most frail, the barriers for indoor mobility were crucial risks for dependency and even for admission to institutional care. Further, Vaarama et al. (2006) found that physical living environment impacted strongest on subjective QoL for people aged 80 and over, and that a poor living environment decreased their subjective QoL in all dimensions. This makes it clear that the physical environment cannot be excluded if the care is to be effective in provision of QoL for older people.

One interesting finding in this study was the factor named as resilience which is one dimension in the Lawton model (behavioural competence, sense of control, and self-determination). This supports the idea of importance of autonomy and adaptation for good life quality amongst care-dependent older people. Even if we could not really confirm the meaning of this factor, it seems to be an important element in our model. The results also indicate the independent impact of acute illness and traumatic life-events, suggesting that QoL is dynamic and – in care-dependent people – vulnerable for acute illness, personal losses, or other traumatic events. Since they are more adequately construed as situational or risk-factors, rather than as long-term context or person factors, they have to be incorporated separately in future crQoL models.

In this view, we might also expect that different life events will mobilise different coping mechanisms and that the profile of well-being may change depending on the current relevance of a life-event (e.g. physical impairment may include coping by learning new skills and enhance feelings of competence; loss of partner may include coping by searching for a new social identity) (Pieper, 2004). This interdependence of life experiences and QoL profiles has also been suggested by Tester *et al.* (2003) in their research on the experience of a recent relocation from home to a care home.

Finally, we can conclude our results in the following model of crQoL (Figure 2).

The model demonstrates the importance of five domains of factors important for crQoL: person factors, environmental support factors (including care), person–environment fit, subjective evaluation of well-being as QoL factors, and situational factors. The results support the four-dimensional Lawton model, and we provide it with a new interpretation to be applicable as a framework for evaluating QoL among old, frail people who are dependent of external help and support.

The empirical explorations used a model which employed all the factors of our theoretical model of crQoL. However, it is important to emphasise the preliminary nature of the reported research. The data used for the empirical explorations was

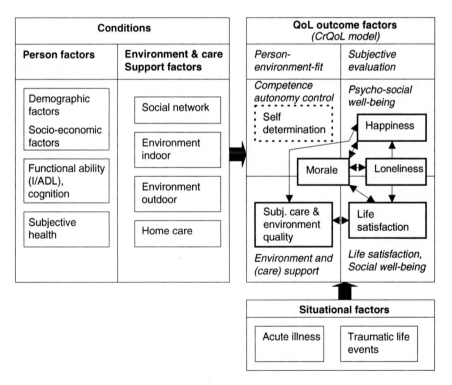

Figure 2. Model of care-related quality of life (crQoL) in old people living at home (enhanced: crQoL variables; italicised: crQoL model)

taken from a previous survey and was not specifically tailored to the requirements of the crQoL model. This meant that not all the dimensions of the model were adequately represented in the analysis, such as cognitive capacities of the client, control, and involvement in the care decision-making process and its impact on QoL.

The fact that the explorations still supported the basic features of the model indicates the generic quality of the conceptual framework. However, the results provided in this chapter should be seen as illustrating and justifying the approach in general, rather than as clear empirical evidence. While our empirical research indicates that care has a role in the production of QoL for frail older people, the issues are when, how and under what conditions? A key challenge will be to utilise measures and instrumentation that are sensitive enough to capture the nuances of care and its quality, and to extract the role of care in diverse client-specific circumstances. A programme of further work is being carried out within the Care Keys project to further elaborate the crQoL model and to carry out more extensive empirical explorations in the five participating countries (Vaarama et al., forthcoming).

To sum up, our results demonstrate that QoL of an older person living at home with lowered functional abilities can be improved by appropriate care interventions and by other inputs, such as improving his or her living environment. The results have important messages for the care sector. They underline the need for a comprehensive needs assessment in planning care, which incorporates all five domains of factors and four dimensions of QoL as described in our model. Both quantity and quality of care matter; it is important that the client gets easy access to care and an appropriate amount of help, and that the care is given in a way which satisfies their needs. The care system should be sensitive to the multiple, complex, and changing needs of older clients, and understand how vulnerable their life quality is to the diverse risks that accompany advanced age. Specific interventions should be tailored to the diverse situational risks to help support the older person cope with them. This study attempted to give a voice to older homecare clients and our experiences and results confirm that very old and frail people do have a voice and will use it if they are empowered to do so. Our belief is that a key factor in the success of long-term care is for this voice to be heard and acted upon.

NOTES

1. Project number QLRT-2002-02525, available at: http://www.carekeys.net

REFERENCES

Abeles, R.P. (1991) 'Sense of Control, Quality of Life and Frail Older People', in J.E. Birren, J.E. Lubben, J.C. Rowe, and D.E. Deutschmann (eds) *The Concept and Measurement of Quality of Life in Frail Elderly*, New York, Academic Press.

Birren, J.E., Lubben, J.E., Rowe, J.C., and Deutschmann, D.E. (eds) (1991) *The Concept and Measurement of Quality of Life in Frail Elderly*, New York, Academic Press.

Bowling, A. (1995) *Measuring Disease: A Review of Disease-specific Quality of Life Measurement Scales*, Buckingham, Open University Press.

Bowling, A. (1997) *Measuring Health: A Review of Quality of Life Measurement Scales*, 2nd edn, Buckingham, Open University Press.

Bowling, A. (2004) *Measuring Health: A Review of Quality of Life Measurement Scales*, 3rd edn, Buckingham, Open University Press.

Brandstädter, J. and Renner, G. (1990) 'Tenacious Goal Pursuit and Flexible Goal Adjustment: Explication and Age-Related Analysis of Assimilative and Accommodative Strategies of Coping', *Psychology and Ageing*, 5, 58–67.

Brown, I. and Brown, R.I. (2003) *Quality of Life and Disability. An Approach for Community Practitioners*, New York, Kingsley.

Cummins, R.A. (1997) 'Assessing Quality of Life', in R.I. Brown (ed.) *Quality of Life for People with Disabilities: Models, Research and Practice*, Cheltenham, Stanley Thornes.

Davies, B. and Knapp, M. (1981) *Old People's Homes and the Production of Welfare*, London, Routledge and Kegan Paul.

Diener, E. (1994) 'Assessing Subjective Well-being: Progress and Opportunities', *Social Indicators Research*, 31, 103–157.

Felce, D. and Perry, J. (1997) 'Quality of Life: The Scope of the Term and its Breadth of Measurement', in R.I. Brown (ed.) *Quality of Life for People with Disabilities: Models, Research and Practice*, Cheltenham, Stanley Thornes.

Freund, A.M., Li, K.Z.H., and Baltes, P.B. (1999) 'Successful Development and Aging: The Role of Selection, Optimization, and Compensation', in Brandstädter and R.M. Lerner (eds) *Action and Self-Development. Theory and Research through the Life Span*, Thousand Oaks, CA, Sage.

Hughes, B. (1990) 'Quality of Life', in S. Peace (ed.) *Researching Social Gerontology*, London, Sage.

Knapp, M. (1984) *The Economics of Social Care. Studies in Social Policy*, Hong Kong, Macmillan Education.

Knapp, M. (1995) *The Economic Evaluation of Mental Health Care*, Aldershot, UK, Personal Social Services Research Unit (PSSRU), Center for the Economics of Mental Health (CEMH).

Lawton, M.P. (1975) 'The Philadelphia Geriatric Centre Morale Scale: A Revision', *Journal of Gerontology*, 30, 85–89.

Lawton, M.P. (1991) 'A Multidimensional View of Quality of Life in Frail Elders', in J. Birren, J. Lubben, J. Rowe, and D. Deutchman (eds) *The Concept of Measurement of Quality of Life in Frail Elders*, San Diego, CA, Academic Press.

Lawton, M.P. (1997) 'Assessing Quality of Life in Alzheimer Disease Research', *Alzheimer Disease and Associate Disorders*, 11, 91–99.

Lawton, M.P. and Nahemow, L. (1973) 'Ecology and the Aging Process', in C. Eisdorfer and M.P. Lawton (eds) *The Psychology of Adult Development and Aging*, Washington, DC, American Journal of Psychology Association.

Pieper, R. (2004) 'Quality of Life of Care-Dependent Older People', *Conference 'Quality of Life in Older Age*, Tartu, Estonia. http://www.carekeys.net

Renwick, R., Brown, I., and Nagler, M. (eds) (1996) *Quality of Life in Health Promotion and Rehabilitation*, Thousand Oaks, CA, Sage.

Rotter, J.B. (1966) 'Generalised Expectancies for the Internal Versus External Control of Reinforcement', *Psychological Monographs*, 90, 1, 1–28.

Schalock, R.L. and Siperstein, G.N. (1996) *Quality of Life: Conceptualisation and Measurement*, Washington, DC, American Association of Mental Retardation.

Sixsmith, A.J. (1994) *Quality of Life: Meanings and Interpretations*. Unit 4. Open University course K256 'An Ageing Society', Buckingham, Open University Press.

Skevington, S.M., Lofty, M., and O'Connell, K.A. (2004) 'The World Health Organization's WHOQoL-BREF Quality of Life Assessment. A Report from the WHOQoL Group', *Quality of Life Research*, 13, 299–310.

Staudinger, U.M., Freund, A.M, Linden, M., and Maas, I., (1999) 'Self, Personality, and Life Regulation: Facets of Psychological Resilience in Old Age', in P.B. Baltes and K.M. Mayer (eds) *The Berlin Aging Study*, Cambridge, Cambridge University Press.

Tester, S., Hubbard, G., Downs, M., MacDonald, C., and Murphy, J. (2003) *Exploring Perceptions of Quality of Life of Frail Older People during and after Their Transition to Institutional Care*. Research Findings 24, Growing Older Project. www.shef.ac.uk/uni/projects/gop/GOFindings24.pdf Accessed 13/5/05.

Vaarama, M. (2004) 'Predictors of Dependency in Old Age and the Demand of Care. The State of the Art and Development up to Year 2015 in Finland', in *Finland for People of all Ages*, Helsinki, Prime Minister's Office, Publications 34/2004.

Vaarama, M., Pieper, R., Hertto, P., and Sixsmith, A. (2004) *Care-Related Quality of Life: Exploring a Model*, Deliverable 9.6, Care Keys project. http://www.carekeys.net, Accessed 13/5/05.

Vaarama, M. and Pieper, R. (2005) *Managing Integrated Care for Older Persons. European Perspectives and Good Practices*, Stakes and the European Health Management Association, Saarijärvi, Gummerrus Printing.

Vaarama, M., Luoma, M-L., and Ylönen, L. (2006) 'Ikääntyneiden elinolot, toimintakyky ja koettu elämänlaatu. (The Living Conditions, Functional Ability and Subjective Quality of Life among older Finns)', in M. Kautto (ed.) *Suomalaisten hyvinvointi 2005 (Well-Being among Finns 2005)*, Jyväskylä, Gummerrus Printing.

Vaarama, M., Pieper, R., and Sixsmith, A. (eds) (forthcoming) *Care-Related Quality of Life in Old Age. Concepts, Models and Measures*. New York, Springer.

Veenhoven, R. (2000) 'The Four Qualities of Life: Ordering Concepts and Measures of the Good Life', *Journal of Happiness Studies*, 1, 1–39.

CONCLUSIONS AND OUTLOOK

14. QUALITY OF LIFE IN OLD AGE

Synthesis and future perspectives

This volume is intended to provide a comprehensive, international, and multidisciplinary perspective on the current state of knowledge on the most important components of quality of life (QoL) in old age. With these concluding remarks we attempt to extract and synthesise the various contributors' main 'messages' and identify the remaining gaps with respect to theory, methodology, and research. With regard to research we can additionally draw on a recently completed European Framework Programme project which, among other things, has been developing recommendations for research on QoL in old age (Walker and Cook, 2004).

THE MULTIFACETED PERSPECTIVE ON QoL IN OLD AGE

At first glance the contributions assembled in this volume on QoL in old age seem rather heterogeneous and put together arbitrarily. Indeed, the issues they raise are seldom related to each other. With a few exceptions, the authors proceed from different theoretical concepts and use different methodologies and instruments to assess their subject. They hardly take note of the other contributors' work nor do they mutually refer to their published findings. Nevertheless, the diverging topics represent the core components which repeatedly, over a long period, have been found to affect older people's QoL, and which were confirmed by Bowling's composite analyses of generic QoL scales and older people's own definitions: person-related psychological variables; health and functional status; social relations, support and activity; economic circumstances and independence; environmental conditions; and leisure activities and mobility. Hence, the chapters mirror simultaneously the amorphous and concrete, abstract and real, elusive and particular character implied in the term 'quality of life'.

Psychological Variables

Psychological variables include, in particular, personal control and mastery (Chapters 3, 7, and 9) but also cognitive adjustment (Chapter 12), social expectations and comparisons, optimism-pessimism (Chapter 2), and resilience (Chapter 13). Mastery was found to be a major predictor for all dimensions of well-being in Daatland and Hansen's Norwegian study (age range 40–79) (see Chapter 3). Older people had nearly the same level of well-being as those in their mid-life even when their sense of mastery was lower. Given the same level of mastery, the older people had in fact higher well-being on all measures (except positive affect). These findings suggest a persistent positive effect of instrumental control. Personal control seems to be an important contributor to well-being even in very old age, when capabilities decrease and environmental constraints weigh heavily (Chapter 3).

H. Mollenkopf and A. Walker (eds.), Quality of Life in Old Age, 235–248.
© 2007 *Springer.*

Findings from the European ENABLE-AGE study on very old people (aged 80–89) living alone in urban districts (Iwarsson *et al.*, 2004), reported on in this volume by Wahl and colleagues (Chapter 7), show that beside meaningful bonding to the home and high usability and accessibility, housing-related control beliefs were linked to the maintenance of independence in daily living and well-being. Conversely, a lack of personal control over the environment constituted a plausible explanation for increased perceived environmental stress (PES) and depression among ethnically diverse urban older adults living in low income subsidised senior housing in a small city in the north-eastern USA (Chapter 9). For some of the residents, unfavourable characteristics of the buildings and the immediate neighbourhood could be linked to a loss of perceived control, connected with decreased physical and mental health status. Resilience, a dimension in Lawton's (1991) model of QoL, comprises behavioural competence, sense of control and self-determination. The impact of these factors on QoL in care-dependent older people in Finland as reported by Vaarama and colleagues shows also the importance of autonomy (Chapter 13).

Taken together, the findings support the notion that a sense of actual control over important life domains, one's environment and/or desirable outcomes adds considerably to well-being among older people. It may be even more important for autonomy and self-respect among very old and vulnerable people when individual competencies decrease and environmental circumstances are or become stressful. In the case of changes which are entirely out of personal control such as the loss of a partner or the occurrence of a severe health impairment, cognitive adjustment may reduce mental incongruence and, by this, help to maintain QoL (Chapter 12).

Health and Functional Status

In a multitude of studies health has proved to be of prime importance for QoL, both as a significant predictor emerging from scientific analyses and as a salient aspect of life according to older people's own definitions. This was confirmed by Bowling's and Deeg's studies reported in Chapters 2 and 12, respectively.

At the same time, Deeg's detailed cross-sectional and longitudinal analyses revealed interesting differentiations and changes. An immediately compelling finding is that QoL in terms of both life satisfaction and positive affect is higher among the healthy than among the ill older people. However, there is a similar change over time in these aspects, and while health is considered as important by most older people, this is less so for the ill than for the healthy people. Moreover, older people rated their QoL higher after making an inventory of aspects of life important to them than was to be expected in view of their score on a standard health-related QoL measurement instrument. Hence, the relationship between health and QoL may not be as strong in older age as has been assumed.

The reason why physical health decreases in its significance in the case of illness might be the psychological process of cognitive dissonance or mental incongruence, an adaptive strategy to losses of important aspects of life that cannot be regained.

Social Relations, Support, and Activity

The fundamental role that the manifold dimensions of social relations play in QoL is widely acknowledged. Older people themselves mentioned social relationships, social roles, and activities in the first place when answering open-ended survey questions and in-depth interviews on the constituents of the 'good things' that gave quality to their lives (Chapter 2). Also the older participants of the LASA study considered a good marriage among the three most important aspects of life (besides physical and mental health) (Chapter 12).

In their overview, Antonucci and Ajrouch (Chapter 4) describe impressively the multidimensionality of social resources – ranging from structural characteristics such as network size, contact frequency, geographic proximity, and network composition, to the more intimate emotional and instrumental aspects of support provided and received, as well as expressions of both positive and negative relationships – and, by this, address the complex ways in which they can influence QoL in old age.

Social support appears to be particularly salient in the case of those experiencing structural disadvantages as they are commonly shared across ethnic subgroups within developed countries (Chapter 11). The family, in particular, seems to enhance QoL under disadvantaged macro-structural conditions. In the face of meso-structural conditions such as deprived neighbourhoods, social support can work like a buffer, as the study by Disch and colleagues showed. Many of the older residents living in low-income housing in environmentally stressful contexts had strong social support systems which many of them derived from their building networks. Those who had the highest social support experienced no PES or depression while those in the PES and depression group had the lowest support (Chapter 9).

Economic Circumstances and Independence

As many previous studies on the general population have shown (for an overview see Diener *et al.*, 2003; Veenhoven, 1996), financial circumstances are the most important predictors for life satisfaction beside health status. Older people seem to be more satisfied with their financial situation than younger ones, though, and the predictive power of the income level, for both satisfaction with the financial situation and the level of the subjective well-being, decreases with advancing age. However, as Weidekamp and Naegele stress, these findings should not be interpreted to the effect that financial resources are losing their importance for a good QoL in old age (Chapter 5). Instead, the objective significance of sufficient earnings increases because the more health impairments arise and the more fragile social support networks become, the more important become other supporting resources. This interpretation is supported by the research they present. It shows that it is not the absolute level of income that plays the most important role for the subjective evaluation of the financial resources but the living standard it enables, that is whether the available economic resources suffice to attain a satisfactory living standard including, for instance, participation in social life despite mobility losses or affording the rising costs of health care or nursing care.

As Weidekamp and Naegele found for older German people, Bowling shows similarly in her multimethodological approach to QoL among older British people that among the main constituents of the 'good things' that gave QoL financial circumstances were only placed seventh (after diverse social aspects, leisure pursuits, health, psychological aspects, and home and neighbourhood). Nevertheless, having health and enough (or more) money were the two most frequently mentioned things older people themselves emphasised would improve their QoL. Likewise, a good income was selected among the three most important aspects of life by less than 25% of the older people who had participated in the LASA study (Chapter 12). However, its importance had increased after six years particularly among older people who reported decreased importance of good physical health – who were also those in poorer health.

The level of income affects many domains of life. As Rojo-Pérez and associates mention in Chapter 8, the heaviest explanatory parameter for overall residential satisfaction of older people living in the Madrid region was the perception of their own economic resources and the household in which they live. This subjective aspect was underpinned by objective conditions in so far as wealthier, higher class people live in relatively more modern and better equipped houses. Probably due to their better economic resources they could afford a house in a neighbourhood according to their wishes – and consequently were more satisfied with their residential environment. Satisfaction with income was also revealed to be an important predictor for QoL in the MOBILATE study (Mollenkopf et al., 2005) presented in this volume by Wahl and colleagues (Chapter 7). Interestingly, this factor affected significantly the cognitive dimension of life quality (satisfaction with life in general) in all European regions under study but was less or not at all important for its emotional dimension (positive affect).

Whether older people have sufficient financial resources at their disposal to enable them to maintain a satisfactory standard of living is largely dependent on the country they live in and the welfare system prevailing in that country. Sheykhi has highlighted, in particular, the differences between European and Asian countries (Chapter 10). Chappell pointed to the relationship between subcultural group status and economic hardship within developed countries (Chapter 11). Motel-Klingebiel demonstrated through his research in three European countries that, to some extent, the levels and, in particular, the variations of QoL among older people are significantly influenced by the countries' differing welfare systems (Chapter 6). In such a comparative perspective England (as an example of a liberal welfare system) showed the lowest level of objective resources. Older people did best in Germany (representing a conservative-corporatist regime) and somewhat less so in Norway (representing the social-democratic welfare regime typical of the Scandinavian countries). Moreover, Germany and Norway revealed the lowest variability in objective income measures while England showed the highest variability on all indicators used for assessing QoL of older people. Nevertheless, England's older people do not seem to feel deprived when asked for a subjective appraisal of their QoL. This raises the question of older people's frames of reference when evaluating their QoL.

Environmental Conditions

Closely connected with financial circumstances are the options for choosing where to live – a connection which has often been neglected in QoL research although older people themselves emphasise the importance of home and neighbourhood (what Bowling termed 'social capital') (Chapter 2).

As Wahl and colleagues underline in their chapter, indoor and outdoor environments can constitute major resources for older people's QoL (see also Chapters 8, 9, and 13). Based on Lawton's (1991) theoretical approach towards person–environment relationships they provide empirical evidence that both objective and subjective components have to be considered to understand better QoL outcomes such as autonomy, life satisfaction and emotional well-being. Likewise, Disch and colleagues show that the structural characteristics of buildings and neighbourhoods affect the residents' QoL (Chapter 9). At the same time their findings indicate that residents with more positive social support experienced higher QoL. If the social environment is compromised, mental health problems and negative perceptions of the built environment increase.

The importance of good housing and an appropriate neighbourhood increases when severe illness or chronic diseases make continued functioning in daily life more critical (Chapter 12). For older people who are receiving care, in particular, the physical environment cannot be excluded if the care is to be effective in promoting their QoL (Chapter 13).

Hence, it is important to consider personal aspects, on the one hand, and to extend individualised person–environment system approaches to the meso and macro levels of analysis, on the other, in order to understand the complex nature of the relationships between environmental factors and QoL.

Leisure Activities and Mobility

Two further components which are only rarely addressed in generic QoL scales and research turned out to be of great importance to older people: activities and mobility. Both aspects are closely interconnected, and their significance for an autonomous and meaningful life in old age becomes obvious in the light of the age-related increase in mobility restrictions. That applies especially to modern societies where mobility is not only a fundamental precondition for overcoming the growing distances between functional areas but also constitutes a highly appreciated societal value.

Objective and subjective aspects of out-of-home activities and mobility were significant predictors of satisfaction with life and emotional well-being among the older adults participating in the MOBILATE study (Mollenkopf et al., 2006, Chapter 7). In Bowling's study (Chapter 2), too, the lay models of QoL emphasised the importance of leisure and social activities, including those enjoyed alone. Furthermore, based on their review of available research findings on material well-being, Weidekamp and Naegele (Chapter 5) state that in very old age satisfaction with mobility (alongside satisfaction with one's state of health and social networks) exceeds income in the subjective evaluation of QoL. And finally, according to analyses performed with the LASA data (Chapter 12) the importance of meaningful spending of time increases for ill older adults. Therefore aspects of activity and mobility need to be included in future QoL approaches.

FUTURE PERSPECTIVES

Even though the chapters of this book cover a wide range of aspects that are salient for QoL in older years, a number of questions remain open. Most of them were raised by the authors themselves. They refer to the various core components of QoL, to methodological issues and theoretical approaches. Therefore, the final part of this chapter turns from a consideration of what we know to an outline of what more we need to know and can be done when QoL in old age is the focus.

Remaining Knowledge Gaps

The chapter authors pointed to a large number of still unexplored findings and open research questions, a small selection of which are repeated here:

Why, for instance, are levels of satisfaction and well-being in general so high among older people? Why is health less important among ill than among healthy older people? Which role does age and ageing play for their subjective QoL? What really matters in very old age or in the case of chronic illness and the need for care?

What exactly is the role of social network resources in relation to QoL in late life? To what extent does intra-group supportiveness – for instance among ethnic minorities – arise out of necessity? What is the significance of the residential context among both older people who live on their own and those who live dependently in a residential institution? Which physical and social conditions are suited to support the person's QoL? What is the interrelationship between the two? While empirical research indicates that care has a role in the production of QoL for frail older people, the issues are when, how, and under what conditions? And, on a more general level: which personal resources and which environmental conditions are contextually most or least important for QoL? More research is needed to determine the strength and contextual salience of each of the variables using a clearly defined QoL model.

Some further knowledge gaps concern cohort-related aspects, for instance: What is the frame of reference of today's older people's evaluations of their lives – and with whom and at what times will future generations compare their situation? What will QoL be for future cohorts of older people in the light of demographic change and structural uncertainty, precarious jobs, long-term unemployment, cuts in pension levels, and reduced welfare provision? Will they be able to cope or compensate for changing environments and resources?

The questions raised here correspond with the research gaps and priorities which emerged in two European Framework Programme projects which have focused specifically on QoL in old age: the FORUM on Population Ageing Research and the European Research Area in Ageing, ERA-AGE. The FORUM project (2002–2004) conducted a series of scientific workshops on three topics – QoL; health, and social care; and genetics, longevity, and demography – aimed at identifying knowledge gaps and prioritising research from a European-wide perspective. The outcomes of the FORUM process command a high level of consensus among both scientists and key research end-user groups. The ERA-AGE project (2004–2008) is designed to promote long-term coordination of national research programmes and to promote interdisciplinary research and international collaboration in the field of ageing across Europe. A wide range of recommendations were made by the FORUM projects and reiterated by ERA-AGE, but here the focus will be only on four sets of them in the QoL field: environmental resources, sociodemographic and economic resources, health resources, and personal resources, social participation, and support networks. (The full set of recommendations including those covering the topics of health, social care, genetics, longevity, and demography, and those intended for national and European research funders, and policymakers can be viewed on the FORUM and ERA-AGE websites (http://www.shef.ac.uk/ageingresearch and http://era-age. group.shef.ac.uk/).

Environmental Resources

The environment (at all levels) is to be treated as a key component and dynamic context of QoL in old age. The scientific discussion about its role envisaged a three-dimensional framework linking together individual factors (from health and personal ability to life story), psychological and social factors (security, loneliness, autonomy, attachment, diversity, cohort, ethnicity, culture, gender, and material resources) and environmental factors (migration, transport, accommodation, technology, neighbourhood, and the natural world). Within this framework the urgent priorities for research in the environmental dimension include a deeper understanding of:

- The spatiality of ageing and the experience of interior and exterior space in later life across different countries and regions (urban vs. rural areas, developed vs. less developed regions).
- The relationship between living arrangements and the community, neighbourhoods and care services. This is closely related to issues of personal mobility, accessibility, urban planning, and transportation systems, as well as those of urban and/or community safety and security.
- The environments that are accessible or inaccessible for older people, and on the intergenerational dimension of integration or segregation within public and private spaces. Many of the problems faced by older people are shared by other social groups, for example poverty, unemployment, poor housing, ill health, and so on. What are the specific risks in late life?
- How older people with learning difficulties or intellectual disabilities and older people with dementia are ageing in place which also calls for imaginative approaches to communication within these groups.

There is also a need to balance knowledge about older people living in 'special' settings such as residential and nursing homes with research on those living in 'ordinary' ones. Moreover, more evaluations of practical environmental interventions are necessary to provide knowledge on how to improve the lives of older people.

This holds especially for technological developments which have an immense impact upon different spheres of the everyday life of older people because with advancing age, the significance of mobility, environmental stimuli and demands, and the potential of technology as prerequisites for a life of quality increases. And, finally, there is a substantial knowledge gap about the impact of major crises related to environmental issues such as climate change, heat waves, power supply shortages on older peoples' QoL.

Sociodemographic and Economic Resources

Four key priority issues were highlighted with regard to sociodemographic and economic resources.

- First, QoL research needs to explore further the question of diversity. There is a need to understand the causal factors behind inequalities between countries and social groups, including the interrelationship with experiences gained earlier in the life course; the extent to which some circumstances and experiences are universal; and how the priority order of factors determining quality varies between different groups of older people. Given the changes in male and female life course trajectories it is important to investigate issues such as gendered changes in working life, the experience of long-term and discontinuous employment, changes in pension policy, the transition to retirement and their impact on QoL.

- Second, it is important to focus research on the economic status of future cohorts of older people and the relationship between ageing, income, and other material resources. New knowledge is required on how the income needs of older people change as they age, their perceptions of income and how these change over time. Too little is known about wealth and inheritance, including the economic power of older people in society and within families, and how wealth is transmitted between generations. What is the impact of inherited wealth within families? What is the impact on potential demand for long-term care services and on financial markets? The absence of reliable data on this topic means that new research is needed to collect comparative information on wealth and goods in kin at both the individual and household levels.

- Third, further research is required on employment in later life and the transition to retirement. For example, what are the economic incentives to continue to work in later life? What is the relationship between work, age of exit from the labour market, pensions, and inheritance? What effect has retirement on QoL and subjective well-being and what is the role of different local, regional and national policies, and welfare systems in shaping the standard of living, social inclusion and QoL in old age? What inequalities exist among older people? Are there new ones or are the classic inequalities persistent? Is there polarisation or convergence within countries and between them?

- Fourth, more knowledge is needed about 'active ageing'. How does active ageing relate to the policies and politics of statutory retirement in Europe? What does active ageing mean beyond working? How is active ageing defined in different countries, what are the differences and what is their relevance to policy? What strategies and policies are needed to promote active ageing – provided it constitutes a substantial component of older people's QoL?

Health Resources

In the field of health resources two different sets of research priorities were identified. First, reviews are needed of the existing conceptual and empirical research relating to the concept of QoL covering not only subjective QoL but also all aspects relevant for individual agency (such as resources and competence). To prepare for comparative research it is also necessary to review analyses of policy, health systems, societal structure, and cultures. In addition to the need for preparatory reviews there were five specific field research priorities:

- Aspects of prevention, rehabilitation, and disease management in health-care systems and their effects on health behaviour and QoL.
- QoL of older people with chronic disease.
- Inequalities in health and QoL, related to structural factors such as income, gender, ethnicity, and age.
- Historical health trends within and between cohorts and generations: comparisons between the young-old who have become healthier over time and the old-old who have developed new forms of frailty such as dementia.
- The relationships between migration, ethnicity, health, and social-care systems. On the one hand there are increasing numbers of migrants in some countries that necessitate research on ethnic and cultural variations in attitudes towards and use of services. On the other hand migrants fulfil different roles in different national health and social-care systems such as, in-house domestic carers in Italy and Greece, and as employees in residential and nursing homes in Germany and the UK.

Personal Resources, Social Participation, and Support Networks

A large number of priorities were highlighted in the field of personal resources, social participation, and support networks, a small selection of which are reported here:

- More focus on individual and societal changes in the second half of the life course, both at the micro and macro levels, and on the changing objective living conditions and how these are subjectively perceived and adapted to. These changes have consequences and impact on dependency, care issues, employment, economic and social resources, retirement, lifelong learning, and other important issues.
- Furthermore, these consequences have implications on inequalities and social exclusion in later life. Therefore there is an urgent need to integrate research with policy and interventions studies.
- The needs, the characteristics, the risk, and marginality among particularly vulnerable groups such as ethnic minorities, the very old and frail, and older people suffering from chronic disease and/or dementia and other intellectual disabilities.

- The impact of factors such as bereavement, retirement, disability, low income, living alone (especially older women), age-friendly or age-unfriendly attitudes towards older people, elder abuse, migration on QoL.
- The interaction between societal modernisation, mechanisation, life course trajectories, family change and intergenerational relationships, including new family forms, on the one hand, and the resources of older people, and personal coping and adaptation to the risks and challenges associated with later life, on the other. Special attention should be paid to long-term effects of societal changes in the framework of global society on future cohorts of older people.

Methodological Issues

With respect to methodological issues, again a large number of recommendations were made by the authors of this book and in the course of the two European coordination projects. The main ones are referred to here.

- First of all, there is a need for further theoretical work on the models of QoL and the instruments used to measure it. In particular the implicit theories held by older people concerning the quality of their lives must be incorporated into a basic definition of QoL. In other words investigators should ensure that their models are grounded in lay perspectives, standards and norms, and not purely in theoretical constructions. Thus a model is required for use in both descriptive and evaluative research that captures individual agency and perspectives, on what constitutes quality and well-being with other relevant factors as preconditions (Chapter 2).
- Such theory development needs to be undertaken by disciplines working in collaboration, moving beyond the common emphasis on health and functioning, which is prevalent in much of the QoL literature, so that the different factors shaping QoL – from genetics to pensions – can be incorporated.
- Comparative QoL research is greatly inhibited by the wide variations in the type and quality of data available on this topic in different countries. Thus, there is an urgent need for comparable approaches and measures to be adopted if the full potential of past, ongoing, and future research is to be realised. Such a harmonisation may consist of both the post-harmonisation of existing data and pre-harmonisation aimed at developing comparable instruments.
- There is a need for coordinated longitudinal and repeated cross-sectional studies on the dynamics of QoL. Such research is required urgently to assess and distinguish cohort effects, effects of ageing, and the impact of changing values and expectations in QoL. Most existing national longitudinal studies concern one historic cohort in which ageing-related changes in QoL are studied (an exception is LASA in the Netherlands, which adds new cohorts at specific time intervals; see Chapter 12). But, given generational and social changes, a cohort-sequential design is necessary to distinguish these from those changes associated with ageing.
- In view of the unique spread of nations and cultures in Europe as well as in other parts of the world it is vital that definitions and methods are cross-cultural and

dynamic. Cross-national studies should include both standardised instruments plus additional culture-specific items considering cultural peculiarities.

* More attention should be given to the heterogeneity of ageing and the aged. Frequently average data conceal differences between specific groups or conditions. The lack of information on ethnic minority elders (Chapter 11) and of people in need of care (Chapter 13) has already been referred to.

* Similarly, many findings may not be generalised for countries with differing living standards and welfare state arrangements. QoL may also vary across cultures. These questions need therefore be explored and validated in various cultures and contexts.

Theoretical Approaches

The most striking observation that is prompted by both reviewing the contributions to this book and participating in European projects such as FORUM and ERA-AGE is how little research and conceptual thinking is shared among this community of scientists. The only theoretical approaches which were used in more than one chapter are Lawton's (1991) and Veenhoven's (2000) multidimensional models of QoL.

Wahl and colleagues (Chapter 7) define QoL in terms of person-environment relations, using – among other concepts – Lawton's proposal that personal life and ageing are always embedded in given environmental conditions able to shape the overall QoL for better or worse. In the empirical illustrations of this approach they demonstrate that a person-physical environment perspective is able to substantially add to the understanding of QoL and stress the importance of considering both objective components and subjective and experiential aspects of the socio-physical environments of ageing.

Vaarama and associates (Chapter 13) build a care-related model of QoL, drawing as well on Lawton's four-dimensional model (Lawton, 1991) and elaborating it further by using Veenhoven's (2000) fourfold taxonomy of QoL. They combine both approaches with a new interpretation to be applicable as a framework for evaluating the QoL among frail old people who are dependent on external help and support. Other authors refer to Lawton's and Veenhoven's work as well, albeit without extensively building on these approaches.

Apart from these common references there is a general agreement that a basic definition and comprehensive model of QoL are urgently needed. Such a model should incorporate different perspectives (individual, societal, and social policy) and conceptualisations (at the societal and objective level to those referring to the individual and subjective levels) and enable research on the societal level as well as on the individual level. For that very reason the basic dimensions included in a model of QoL should reflect science, social policy, and the views of older people. There is also consensus about the process-oriented character of QoL and hence the need to take individual, societal, and historical changes into account. Furthermore, theoretical work is needed to clarify and give reasons for methodological key concepts and operationalisations, indicators, and scales.

TOWARDS A BROADER PERSPECTIVE OF QoL:
COMPARATIVE RESEARCH AND INTERDISCIPLINARY COLLABORATION

Although QoL may be criticised easily for its amorphous nature, there is no doubt that it is a broad-based multidisciplinary concept and one that is the focus of increasing interest among gerontologists. As noted previously an important driver of this interest is the policy making process. Several strands of recent research in this field may be emphasised. Most important of all there is a discernible shift away from the application of health-related proxies for QoL – functional capacity, health status, psychological well-being, social support related to incapacity, morale, dependence, coping with and adjustment to disability – without reference to the ways in which older people in general, or specific groups of older people, or service users, define their own QoL or the value they place on the different components used by the 'experts'. In practice older people are remarkably consistent, across a wide range of studies, in the domains they identify as being important for the quality of their lives: family and other relationships, and contact with others; emotional well-being; religion and spirituality; independence; social activities; finance and standard of living; their own health and the health of others (Brown et al., 2004).

The danger with the previous approach to assessing QoL in old age, that has dominated both scientific and professional worlds, was that it tended to homogenise older people rather than recognising diversity and differences based on age, gender, race and ethnicity, and disability. A key element in this homogenisation is the prevailing use of statistical techniques which focus on means and general coefficients of association rather than on internal sample differentiation (see Singer and Ryff, 2001 for a review of statistical methods addressing diversity). Also inherent in this paradigm was a conception of older age as a distinct phase of the life course, one that is detached from middle age and earlier phases (Gubrium and Lynott, 1983; Bond, 1999). In its place, gradually, are appearing interpretive approaches which aim, among other things, to build on the implicit theories of QoL held by older people themselves. In particular two complementary approaches to assessing QoL from the perspective of older people are, on the one hand, from lifespan development psychology, attempts to understand subjective meanings of QoL within the context of the person's life course and, on the other, the operationalisation of QoL as a multidimensional phenomenon reflecting lay perspectives (Grundy and Bowling, 1999; Bowling et al., 2002; see also Chapter 2). Combining the strengths of these two approaches operationally calls for both quantitative and qualitative research methods.

In Europe, the absence of a common perspective on ageing has begun to be corrected over the last decade with an increasing number of comparative studies. The main driving force has been the scientific communities' recognition of the need for more cross-national research and the sharing of knowledge and expertise. In this context the role of the EU itself has to be acknowledged. For example there was the creation of the European Observatory on Ageing and Older People in 1991 (Walker et al., 1991, 1993) and the highlighting of ageing research in Key Action 6 under the

Fifth Framework Programme (FP5) (1998–2002), though its absence from Framework Programme Six (FP6) was undoubtedly a setback. Many of the European projects include references to QoL outcomes, such as the *Ageing Well* project (Ferring *et al.*, 2003), Care Keys (Chapter 13), ENABLE-AGE (Iwarsson *et al.*, 2004; see also Chapter 7), MOBILATE (Mollenkopf *et al.*, 2005; see also Chapter 7), OASIS (Tesch-Römer *et al.*, 2001; Lowenstein and Ogg, 2003, see also Chapter 6).

The main areas of consensus about QoL in old age are its dynamic multifaceted nature, the combination of life course and immediate influences, the similarities and differences in the factors determining QoL between younger and older people, the most common associations with QoL and the likely variations between groups, and the powerful role of subjective self-assessment. In other words, QoL ...

- should be regarded as a dynamic, multifaceted, and complex concept.
- must reflect the interaction of objective, subjective, macro, micro, positive, and negative influences.
- is the outcome of the interactive combination of life course factors and immediate situational ones. furthermore,
- a model of QoL must include some reference to the individual's scope for action – the various constraints and opportunities that are available in different societies and to different groups.
- Subjective self-assessments of psychological well-being and health are more powerful than objective economic or sociodemographic factors in explaining variations in QoL ratings.
- Sources of QoL in old age often differ between groups of older people. European research also points to different priority orders among older people in different countries with regard to factors that contribute to QoL.

Comparative research is necessary not only to share knowledge and good practice but also to provide a critical perspective on the portability of different models of practice. Comparisons are needed of QoL in old age in different countries, because the existing aggregate data provide only a superficial view, and such studies must relate QoL to the national cultural and to the institutional context. Comparative research will also help to avoid ethnocentric value biases in definitions of the good life.

With regard to interdisciplinary collaboration European scientists in the FORUM and ERA-AGE projects emphasised the importance of disciplinary identities but also stressed the need to integrate knowledge to produce broader models of QoL. The essential point being that the nature of the collaboration should be determined by the specific research question and, therefore, a range of different sorts of interdisciplinary working may be envisaged. If such collaboration is to be successful, it was felt that one discipline should not be dominant.

Finally we hope that this book will succeed in bringing closer together the wide array of scientists, politicians, and practitioners who are concerned with older people and help to develop a deeper understanding of what makes up QoL in old age. This, in our view, should be the essential pre-condition for effective policies to promote quality for the years later in life.

REFERENCES

Bond, J. (1999) 'Quality of Life for People with Dementia: Approaches to the Challenge of Measurement', *Ageing and Society*, 19, 561–579.

Bowling, A., Banister, D., Sutton, S., Evans, O., and Windsor, J. (2002) 'A Multidimensional Model of the Quality of Life in Older Age', *Ageing and Mental Health*, 6, 355–371.

Brown, J., Bowling, A., and Flynn, T. (2004) *Models of Quality of Life: A Taxonomy and Systematic Review of the Literature*, University of Sheffield, FORUM Project (http://www.shef.ac.uk/ageingresearch).

Diener, E., Oishi, S., and Lucas, R.E. (2003) Personality, Culture, and Subjective Well-being: Emotional and Cognitive evaluations of Life, *Annual Review of Psychology*, 54, 403–425.

Ferring, D., Wenger, G.C., and Hoffmann, M. (2003) *Comparative Report on the European Model of Ageing Well*, Bangor, Centre for Social Policy Research and Development, University of Wales.

Grundy, E. and Bowling, A. (1999) 'Enhancing the Quality of Extended Life Years', *Ageing and Mental Health*, 3, 199–212.

Gubrium, J. and Lynott, R. (1983) 'Rebuilding Life Satisfaction', *Human Organisation*, 42, 1, 33–38.

Iwarsson, S., Wahl, H.-W., and Nygren, C. (2004) 'Challenges of Cross-National Housing Research with Older Persons: Lessons from the ENABLE-AGE Project', *European Journal of Ageing*, 1, 1, 79–88. DOI 10.1007/s10433-004-0010-5.

Lawton, M.P. (1991) 'Background: A Multidimensional View of Quality of Life in Frail Elders' in J.E. Birren, J. Lubben, J. Rowe and D. Deutchman (eds) *The Concept and Measurement of Quality of Life in the Frail Elderly*, San Diego, Academic Press.

Lowenstein, A. and Ogg, J. (eds) (2003) *Oasis: Old Age and Autonomy. The Role of Service Systems and Intergenerational Family Solidarity*, Haifa, OASIS.

Mollenkopf, H., Baas, S., Kaspar, R., Oswald, F., and Wahl, H.-W. (2006) 'Outdoor Mobility in Late Life: Persons, Environments and Society' in H.-W. Wahl, H. Brenner, H. Mollenkopf, D. Rothenbacher, and C. Rott (eds) *The Many Faces of Health, Competence and Well-Being in Old Age: Integrating Epidemiological, Psychological and Social Perspectives*, Dordrecht, The Netherlands, Springer, pp.33–46.

Mollenkopf, H., Marcellini, F., Ruoppila, I., Széman, Z., and Tacken, M. (eds) (2005) *Enhancing Mobility in Later Life – Personal Coping, Environmental Resources, and Technical Support: The Out-of-Home Mobility of Older Adults in Urban and Rural Regions of Five European Countries*, Amsterdam, IOS Press.

Singer, B. and Ryff, C. (2001) 'Person-centred Methods for Understanding Aging: The Integration of Numbers and Narratives' in R. Binstock and L. George (eds) *Handbook of Aging and the Social Sciences*, San Diego, CA, Academic Press, pp.44–65.

Tesch-Römer, C., von Kondratowitz, H.J., and Motel-Klingebiel, A. (2001) 'Quality of Life in the Context of Intergenerational Solidarity'; in S.O. Daatland and K. Herlofson (eds) *Ageing, Intergenerational Relations, Care Systems and Quality of Life*, Oslo, Nova, pp.63–73.

Veenhoven, R. (1996) 'Average Level of Satisfaction in 10 European Countries: Explanation of Differences'; in W.E. Saris, Veenhoven, R., Scherpenzeel, A.C., and Bunting, B. (eds) *A Comparative Study of Satisfaction with Life in Europe*, Budapest, Eotvos University Press, pp.243–253.

Veenhoven, R. (2000) 'The Four Qualities of Life. Ordering Concepts and Measures of the Good Life', *Journal of Happiness Studies*, 1, 1–39.

Walker, A. and Cook, J. (2004) European Research Priorities in the Field of Ageing. Summary Paper of the Recommendations of the European Forum on Population Ageing Research. Sheffield. (www.sheffield.ac.uk/ageingresearch)

Walker, A., Guillemard, A.-M., and Alber, J. (1991) *Social and Economic Policies and Older People*, Brussels, European Commission.

Walker, A., Guilllemard, A.-M., and Alber, J. (1993) *Older People in Europe – Social and Economic Policies*, Brussels, European Commission.

(http://www.shef.ac.uk/ageingresearch)
(http://era-age.group.shef.ac.uk/)

Kristine J. Ajrouch, Associate Professor of Sociology and Director of the Gerontology Program, Eastern Michigan University; Adjunct Associate Research Scientist, Life Course Development Program, Institute for Social Research, University of Michigan, USA

Toni C. Antonucci, Elizabeth M. Douvan Collegiate Professor of Psychology, Senior Research Scientist and Program Director, Life Course Development Program, Institute for Social Research, University of Michigan, USA

Ann Bowling, Professor of Health Services Research, Department of Primary Care and Population Sciences, University College London, UK
http://www.ucl.ac.uk/primcare-popsci/aps/

Neena L. Chappell, Professor, Centre on Aging and Department of Sociology, and Canada Research Chair in Social Gerontology, University of Victoria, British Columbia, Canada
http://web.uvic.ca/~nlc/index.htm

Christiane Claus, Research Assistant, Institute of Psychology, University of Halle-Wittenberg, Halle, Germany

Svein Olav Daatland, Senior Researcher, Norwegian Social Research (NOVA), Oslo, Norway
http://www.nova.no/

Dorly J.H. Deeg, Professor of Epidemiology of Ageing and Scientific Director, Longitudinal Aging Study Amsterdam, Vrije Universiteit Medical Centre/LASA, Amsterdam, The Netherlands
http://www.lasa-vu.nl

William B. Disch, Ph.D., Department of Psychology, Central Connecticut State University, USA
www.ccsu.edu

Gloria Fernández-Mayoralas, Scientific Researcher at the Spanish Council for Scientific Research, Madrid, Spain, and Member of the Research Group on Ageing – CSIC
http://www.ieg.csic.es/grupos/gie/

Thomas Hansen, Ph.D. student, Norwegian Social Research (NOVA), Oslo, Norway
http://www.nova.no/

Heidrun Mollenkopf, Senior Research Scientist (retired), former German Centre for Research on Ageing at the University of Heidelberg, Department of Social and Environmental Gerontology, Germany
http://www.psychologie.uni-heidelberg.de; heidrun.mollenkopf@web.de

Andreas Motel-Klingebiel, Deputy Director and Head of Research, German Centre of Gerontology, Berlin, Germany
http://www.dza.de/english/allgemein/mitarbeiter/motel.html
http://www.dza.de

Gerhard Naegele, Professor of Social Gerontology and Social Policy, Department of Sociology, and Director of the Institute for Gerontology, University of Dortmund, Germany
www.fb12.uni-dortmund/gerontologie
www.FFG.uni-dortmund.de

Frank Oswald, Senior Research Scientist and Deputy Chair of the Department of Psychological Ageing Research, Institute of Psychology, University of Heidelberg, Germany
http://www.psychologie.uni-heidelberg.de/ae/apa/

Richard Pieper, Professor of Sociology, Urban Studies and Social Planning, University of Bamberg, Germany
http://www.uni-bamberg.de/index.php?id=7676

Kim E. Radda, MA, RN, Project Director, Institute for Community Research, Hartford, Connecticut, USA
www.incommunityresearch.org

Julie T. Robison, Ph.D., Assistant Professor, Center on Ageing, University of Connecticut Health Center, USA
http://www.uconn-aging.uchc.edu/UCA/Faculty/robison.htm

Vicente Rodríguez-Rodríguez, Research Professor at the Spanish Council for Scientific Research, Madrid, Spain, and member of the Research Group On Ageing – CSIC
http://www.ieg.csic.es/grupos/gie/

José-Manuel Rojo-Abuín, Statistician at the Statistical Analysis Unit, Institute of Economics and Geography, Spanish Council for Scientific Research (CSIC), Madrid, Spain
http://www.ieg.csic.es/laboratorioEstadistica/

Fermina Rojo-Pérez, Scientific Researcher at the Spanish Council for Scientific Research, Madrid, Spain, and member of the Research Group On Ageing – CSIC
http://www.ieg.csic.es/grupos/gie/

Jean J. Schensul, Ph.D., Senior Scientist and Founding Director, Institute for Community Research, Hartford, Connecticut, USA
www.incommunityresearch.org

Mohammad Taghi Sheykhi, Senior Associate Professor of Sociology, Department of Social Science, Al-Zahra University, Tehran, Iran

Andrew Sixsmith, Lecturer in Social Gerontology, University of Liverpool, UK
http://www.liv.ac.uk/primarycare/staff/asixsmith.htm

Marja Vaarama, Professor of Social Gerontology and Social Work, University of Lapland, Department of Social Work, Rovaniemi, Finland
http://www.ulapland.fi/?deptid=8769

Hans-Werner Wahl, Professor of Psychological Ageing Research, Chair, Department of Psychological Ageing Research, Institute of Psychology, University of Heidelberg, Germany
http://www.psychologie.uni-heidelberg.de

Alan Walker, Professor of Social Policy and Social Gerontology, Department of Sociological Studies, University of Sheffield, UK, and Director of the New Dynamics of Ageing Programme and the European Research Area on Ageing
http://www.shef.ac.uk/socstudies/staff/staff-profiles/walker.html
http://www.newdynamics.group.shef.ac.uk/
http://era-age.group.shef.ac.uk/

Manuela Weidekamp-Maicher, Research Associate and Lecturer in Social Gerontology, Department of Sociology, University of Dortmund, Germany
www.fb12.uni-dortmund.de/gerontologie

Accessibility, 101, 103, 108, 109, 124, 132, 136, 138, 162, 228, 236, 241
Achievements, 27, 35, 127
Active ageing, 5, 242, 243
Activity/activities, 21, 23, 24, 34–44, 86, 109, 116–118, 144, 158–160, 162, 235, 237, 240
Adaptation, 6, 7, 34, 43, 50, 60, 102, 128, 161, 205, 209, 219, 220, 228, 244
Adaptive, 34, 35, 43, 44, 205, 209, 236
Affluence, 60, 61, 65
Africa, 56, 59, 168
Appraisal, 39, 54, 114, 118, 167, 200, 238
Asia, 4, 59, 167–177
Aspirations, 35, 169
Australia, 185
Autonomy, 6, 9, 17, 23, 35, 44, 52, 89, 101, 104, 106, 108–113, 124, 171, 179, 199, 201, 206, 216, 220, 225, 228, 229, 236, 239, 241

Barrier-free, 102, 106, 228
Bereavement, 9, 27, 244
Bottom-up, 37, 38, 40, 77

Canada, 179–181, 183, 185–188
Capabilities, 34, 44, 235
Capacity, 6, 8, 9, 35, 37, 44, 51, 68, 70, 146, 169, 171, 217, 218, 246
Care, 5, 9, 15, 17, 22, 24, 26–27, 55, 58, 60, 66, 70, 81, 86, 104, 105, 122, 124, 146, 154, 155, 167, 169, 174–177, 182–184, 186, 188, 215–230, 236, 237, 239–241, 243, 245, 247
Chronic disease, 55, 196, 203, 204, 239, 243
Chronic illness, 240
Civil status, 38, 41, 131, 132, 156
Cognitive adjustment, 199, 235, 236
Cognitive competence, 6, 179
Cohesion, 8, 16, 85, 153
Cohort/cohorts, 7, 17, 24, 38, 40, 50, 56, 72, 80, 81, 85–88, 110, 146, 156, 199, 200, 204, 241–244

Communication, 8, 136, 152, 210, 241
Community, 6, 9, 22, 27, 60, 89, 102, 112, 113, 125, 127, 129, 136, 138, 143, 145, 146, 154, 157, 162, 173, 176, 185–188, 220, 241, 245
Competence, 6, 15, 43, 44, 57, 102, 104, 106, 108, 123, 146, 179, 201, 217–219, 220, 225, 228, 229, 243
Competence-Press-Model, 103–104, 118
Control, 3, 6, 9, 15, 17, 20, 33–45, 95, 103, 108, 132, 153, 161, 162, 172, 181, 216, 220, 225, 230, 235, 236
Control beliefs, 34, 36–38, 40, 41, 108, 109, 236
Control, internal, 34, 37
Control, primary, 33, 35, 37
Convey model,
Coping, 6, 26, 34, 43, 52, 190, 219, 220, 228, 229, 244, 246
Crime, 6, 28, 153, 162
Cross-cultural, 189, 244
Cross-national, 6, 7, 57, 184, 190, 208, 245, 246
Cross-sectional, 17, 40, 88, 96, 201, 205, 236, 244
Cultural context, 16, 57, 59, 61, 182
Culture, 45, 50–53, 57–59, 90, 156, 167, 172, 173, 182–190, 241, 243–245

Dementia, 102, 180, 197, 198, 217–219, 241, 243
Depression, 6, 37, 49, 52, 53, 55, 56, 59, 108, 151–163, 184–186, 197, 198, 201, 207, 236, 237
Disability, 9, 24, 101, 131, 180, 186, 195–198, 208, 244, 246
Domain-specific, 36, 67, 77–80, 82, 123, 209

Ecological resources, 6, 179
Economic resources, 3, 4, 65–81, 94, 113, 125–128, 140, 145, 170, 171, 237, 238, 241, 242

Education, 5, 33, 38, 41, 54–56, 65, 88, 90, 92–97, 126, 128, 131, 134, 135, 139, 155, 156, 170, 171, 175, 181, 183, 196, 201, 220–222

Emotions, 15, 78, 154

Employment, 5, 8, 66, 68, 70, 73, 86, 131, 173, 174, 181, 183, 190, 242, 243

Empowerment, 8, 59

Environment, 3, 4, 9, 15, 22, 34, 44, 51, 57, 60–61, 86, 90, 97, 101–105, 107, 108, 110–113, 118, 123–126, 130, 136, 138–140, 143–145, 151, 152, 157, 161, 162, 169, 176, 187, 216, 217, 219, 220, 222, 228, 230, 236, 238, 239, 241, 245

Environmental circumstances, 5, 104, 236

Environmental conditions, 4, 102, 103, 111, 113, 171, 235, 239, 240, 245

Equity, 85, 86–88, 96, 167, 217

Ethnic groups, 4, 179–184, 186, 191

Ethnic minorities, 181–183, 185, 240, 243

Ethnicity, 151, 155, 157, 179–191, 216, 241, 243, 246

Europe, 4, 5, 9, 59, 88, 123, 167–177, 180, 196, 197, 241, 242, 244, 246

Evaluations, 7, 34, 74, 75, 77, 79, 80, 87, 90, 91, 113, 126, 180, 186, 198, 199, 208, 209, 218, 222, 240, 241

Exclusion, 175, 243

Expectations, 5, 7, 15, 20, 21, 28, 34, 35, 44, 53, 76, 87, 90, 95, 124, 127, 143, 172, 188, 216, 220, 235, 244

Expert/experts, 7, 15, 154, 173, 216, 246

Extraversion, 37, 41

Family, 4, 9, 16, 17, 23, 25–27, 36, 43, 51, 53, 57–59, 65, 89, 106, 111, 123–125, 144, 146, 153, 155, 156, 163, 170–177, 180, 182–189, 196, 197, 201, 203–205, 209, 216, 237, 244

Financial circumstances, 19, 20, 21, 27, 237, 238, 239

Financial situation, 61, 66, 67, 74–76, 79, 90, 91, 92, 125, 140, 142, 237

Frame of reference, 4, 240

Functional capacity, 8, 146, 246

Functional health, 53

Gender, 3, 17, 36–39, 41, 51, 53–55, 88, 92, 94–97, 113, 115–117, 125, 131, 132,

135, 137, 151, 153, 155–161, 171, 176, 181, 197, 198, 200–202, 221, 222, 241, 243, 246

Generation/generations, 7, 36, 54, 65, 66, 85, 86, 132, 170, 173, 183, 189, 240, 242, 243

Good life, 15, 16, 33, 44, 45, 65, 87, 216, 225, 228, 247

Government, 15, 28, 51, 57, 59–60, 172, 173, 191

Guidance, 50, 52

Happiness, 5, 6, 16, 33, 49, 51, 77–79, 169, 182, 186, 190, 200, 216, 218, 219, 221, 222, 225, 226, 229

Hermeneutic approaches, 6

Heterogeneity, 8, 245

History, 49, 50, 51, 56, 57, 59, 167, 180, 182

Home, 17, 19–22, 26–28, 35, 36, 44, 53, 55, 57, 101, 102, 104–114, 118, 124–127, 129–136, 138–141, 143, 145, 146, 180, 182, 185, 189, 206, 217, 220, 223, 228–230, 236, 238, 239, 241, 243

Housing, 5–7, 9, 15, 16, 27, 33, 101–110, 113, 123–126, 133, 134, 138, 140, 141, 144, 146, 151–154, 162, 163, 167, 169, 180, 185, 201, 203–206, 209, 216, 236, 237, 239, 241

Human needs, 5, 6, 16, 179

Identity, 58, 105, 143, 184, 187, 229

Illness, 55, 57, 152, 153, 157, 175, 215, 222, 225, 228, 236, 239, 240

Immigrants, 56, 58, 59, 138, 182–185, 187–189

Immigration, 58, 59, 187

Inclusion, 8, 143, 242

Income, 5, 6, 9, 20, 38, 43, 55, 56, 65–80, 90–97, 116, 118, 123, 125, 131, 132, 134–137, 140, 142, 151, 152, 154–156, 162, 168, 181, 182, 184, 186, 201, 203–206, 209, 216, 236–239, 242–244

Incongruence, 199, 203, 205, 206, 209, 236

Independence, 5, 9, 15, 16, 19–23, 28, 57, 66, 101–103, 108, 109, 132, 146, 170, 180, 235–237, 246

In-depth interviews, 17, 18, 20, 24, 25, 180, 237

Inequality/inequalities, 72, 85–97, 105, 118, 171, 181, 182

Institution/institutions, 68, 124, 146, 217, 240

Insurance, 68–73, 81, 86, 96, 174, 181

Integration, 6, 9, 50, 136, 172, 185, 199, 241

Interactive, 8, 49, 60, 154, 247

Interdisciplinary/interdisciplinarity, 3, 4, 9, 82, 154, 175, 199, 241, 246, 247

Intervention, 60, 88, 162, 198, 217, 228, 230, 241, 243

Involvement, 26, 103, 153, 162, 184, 186–188, 220, 230

Japan, 53, 55, 58, 168, 175, 176

Lay people, 15, 21

Leisure, 17, 20, 21, 22, 103, 112, 114, 116–118, 124–126, 142, 153, 157–159, 161, 162, 174, 180, 201, 216, 235, 238, 239

Level of living, 35, 80, 87

Life course, 3, 8, 36, 44, 49–51, 54, 59, 86, 87, 89, 105, 106, 172, 195, 242–244, 246, 247

Life domains, 66, 77, 79, 87, 90, 216, 218, 220, 236

Life satisfaction, 3, 5, 6, 8, 9, 15, 16, 23, 35–43, 49, 53, 65, 77–80, 82, 87, 90–92, 94, 108, 109, 114, 117, 118, 169, 179, 180, 185, 186, 190, 200–202, 206, 209, 216, 218, 219, 221–223, 225, 226, 236, 237, 239

Lifespan, 49, 50, 52, 61, 246

Living conditions, 4, 9, 33–35, 37, 38, 40, 77, 86, 87, 97, 113, 123, 136, 138, 151, 180, 199, 200, 220, 226, 243

Living standard, 33, 42, 45, 81, 174, 237, 245

Loneliness, 20, 24, 52, 54, 136, 172, 185, 216, 218, 221–223, 225–227, 241

Longevity, 6, 54, 167, 171, 177, 179, 241

Longitudinal, 17, 24, 40, 75, 87, 146, 190, 199, 202, 204, 205, 206, 236, 244

Macro, 4, 15, 86, 105, 106, 114, 118, 179, 190, 237, 239, 243, 247

Marital status, 51, 54, 56–57, 131, 139, 155, 186, 221, 222

Marriage, 18, 201–205, 237

Mastery, 34, 37–45, 52, 108, 109, 218, 235

Material resources, 77, 219, 241, 242

Mental health, 15, 16, 38, 104, 151–155, 161–163, 169, 176, 188, 195, 197–198, 201–205, 207–209, 236, 237, 239

Meso, 4, 237, 239

Micro, 4, 5, 8, 15, 86, 105, 118, 179, 243, 247

Midlife, 8, 43

Migration, 58, 173, 183, 241, 243, 244

Minority, 9, 56, 151, 154, 179–183, 196, 205, 245

Minority groups, 181, 182

Mobility, 5, 8, 53, 58, 77, 81, 101, 104, 106, 107, 112–114, 116–118, 124, 127, 134, 138, 144, 146, 169, 171, 174, 207, 222–224, 226–228, 237, 239, 240–242

Model/models, 3, 5–7, 15–17, 20, 38, 41, 49, 50, 53, 61, 68, 76, 77, 79, 80, 92, 94, 96, 104, 109, 116, 117, 123, 130, 140, 143–145, 162, 179, 182, 195, 199, 215–220, 222, 225–230, 236, 239, 240, 244, 245, 247

Modern/Modernisation, 58, 66, 85, 88, 96, 112, 114, 118, 134, 135, 169, 170, 174, 175, 188, 238, 239, 244

Morale, 5, 6, 16, 169, 179, 186, 216, 221–223, 225–227, 246

Morbidity, 6, 23, 195–196, 199, 208

Mortality, 6, 151, 167, 174, 196–198, 200, 208

Motivation, 35, 104, 110, 215

Multidimensional, 3, 16, 17, 24, 49, 77, 86, 102, 103, 145, 175, 207, 209, 218, 221, 245, 246

Multiple discrepancies theory, 35

Multiple-generation families, 54

Negative affect, 37–43, 54, 108, 109, 114

Neighbourhood, 6, 19–22, 25, 27, 60, 61, 102, 104–106, 114, 124, 125, 127, 129–131, 136–141, 143, 146, 147, 153, 156, 162, 179, 182, 216, 236–239, 241

Network, 4, 6, 8, 51, 52, 55, 59, 61, 81, 86, 110, 116, 117, 123–126, 130, 142, 143, 145, 161, 169, 170, 172, 181, 183, 188, 216, 221, 222, 224, 225, 229, 237, 239–241, 243

Neuroticism, 37, 38, 41, 43
Nursing homes, 241, 243

Objective, 3–8, 15, 16, 33, 34, 37, 38, 49,
    56, 67, 75, 77, 78, 81, 82, 87, 88, 94,
    95, 96, 102–104, 106–110, 113, 117,
    118, 123, 125, 126, 135, 140, 143, 145,
    167, 179, 180, 182, 185, 190, 195, 197,
    210, 217, 218, 226, 237–239, 245, 247
Organization for Economic Cooperation and
    Development (OECD), 65, 90
Outdoor mobility, 101, 113, 222, 228

Paradox, 7, 35, 36, 41, 43, 44, 76, 77, 81,
    102, 180
Participation, 6, 9, 22, 81, 101, 103, 112,
    136, 157, 158, 168, 171, 181, 185, 187,
    216, 237, 241, 243
Pension, 28, 66, 68–73, 81, 86, 131, 132,
    169, 175, 240, 242, 244
Perceptions, 5, 6, 9, 15, 16, 21, 33, 34, 126,
    179, 180, 197, 207, 239, 242
Personality, 9, 23, 24, 37, 38
Person-environment relationships, 101, 104,
    105, 239
Physical decline, 53
Physical health, 9, 15, 21, 22, 38, 49, 54, 56,
    90, 91, 151–153, 157, 158, 160–162,
    180, 195, 198, 201–207, 236, 238
Physical strength,
Place attachment, 105, 109
Policy, 5, 7, 15, 17, 24, 59, 68, 69, 81, 85,
    86, 97, 102, 144–147, 163, 175, 208,
    217, 242, 243, 245, 246
Policymakers, 5, 49, 163, 241
Positive affect, 37–44, 80, 108, 109,
    114–118, 143, 199–201, 205, 206, 209,
    235, 236, 239
Poverty, 7, 61, 65–67, 73, 78, 97, 136, 151,
    171, 173, 175, 180, 183, 189, 241
Privacy, 17, 103, 104, 106, 110, 132
Psychological aspects, 4, 238
Psychological variables, 21, 235–236

Race, 51, 54, 56, 172, 180, 246
Reciprocity, 52, 58, 60
Regional conditions, 4, 112, 114
Relationships, 8, 9, 16, 17, 19, 20, 22, 25,
    37, 40, 49, 50, 52–55, 61, 79, 80, 90,

101, 102, 104, 105, 107, 112, 123, 126,
    130, 136, 139, 145, 151–153, 161, 170,
    180, 182, 187, 191, 203, 205, 215, 219,
    222, 225, 237, 239, 243, 244, 246
Religion, 23, 180, 201, 209, 246
Residence, 6, 17, 20, 112, 124, 130, 136,
    140, 144, 152, 156, 186
Resident/residents, 9, 126, 130, 143, 146,
    151–155, 161–163, 185, 236, 237, 239
Residential environment, 123–126, 130,
    136, 138–140, 144–147, 238
Resignation, 35, 44
Resilience, 35, 220, 225, 228, 235, 236
Resources, 4, 6, 8, 9, 16, 21, 34, 35, 38, 41,
    43, 45, 49–82, 86–91, 94, 96, 97, 101,
    105–107, 110, 113, 118, 125–127, 136,
    145, 168, 170, 171, 173, 174, 181, 189,
    191, 217, 219, 220, 225, 237–240,
    243, 244
Retirement, 28, 66, 68, 70, 73, 89, 123, 131,
    132, 153, 156, 157, 169, 172–174, 183,
    200, 242, 244
Rural, 36, 60, 61, 106, 112–117, 136, 139,
    145, 173, 241

Safety, 6, 16, 20, 61, 103, 106, 136, 152,
    155, 156, 161, 162, 206, 227, 241
Satisfaction paradox, 7, 76, 77, 81
Scales, 6, 8, 15, 16, 18, 20, 21, 22, 186,
    235, 239, 245
Security, 8, 16, 27, 35, 53, 60, 65–70, 72,
    73, 76, 79–82, 85, 86, 88, 96, 97,
    132, 136, 152, 153, 162, 169, 170,
    173, 241
Segregation, 241
Self-assessment, 3, 9, 247
Self-efficacy, 20, 34, 37, 179
Senior housing, 151–154, 162, 236
Services, 5, 6, 16, 21, 24, 27, 42, 60, 61, 74,
    90, 112, 113, 124, 126, 127, 129, 130,
    132, 136, 138, 143, 145–147, 156, 163,
    171, 177, 182, 183, 186, 188, 191, 215,
    216, 218, 220, 242, 243
Sexual activity, 23, 24
Situational characteristics, 49, 51, 53, 57
Social capital, 6, 16, 21, 51, 179, 239
Social care, 5, 15, 17, 24, 215, 241, 243
Social cohesion, 8, 16, 153
Social gerontology, 7, 65, 85, 215

Social indicators, 3, 6, 33, 106, 179, 182, 184, 190
Social policy, 17, 24, 81, 85, 86, 97, 102, 175, 217, 245
Social relations, 4, 49, 50–54, 56–59, 61, 62, 77, 235, 237
Social roles, 19, 20, 25, 172, 237
Social structure, 33, 88, 92, 94, 96, 97, 172, 181, 184, 190
Sociodemographic characteristics, 17, 126
Sociodemographic factors, 9, 247
Socio-economic, 8, 9, 20, 60, 61, 75, 79, 82, 90, 126, 134, 139, 151, 168, 170, 171, 173, 176, 179, 182, 184, 186, 187, 196, 216, 222
Socio-emotional selectivity theory, 35
Sociology, 33, 85
Spirituality, 23, 180, 187, 201, 246
Standard of living, 6, 16, 28, 73–77, 79, 80, 82, 179, 180, 238, 242, 246
Stimulation, 103, 106, 110–112
Stress, 51–53, 103, 151–153, 156–161, 236, 237, 245
Structural conditions, 33, 237
Support, 6, 9, 15, 20, 21, 23–25, 27, 33, 35, 38, 43, 50–53, 56–61, 76, 79, 81, 89, 90, 101, 102, 104, 105, 109, 111, 112, 118, 126, 145, 152–154, 157–163, 169–171, 174, 182, 183–188, 191, 205, 206, 215–217, 220, 225, 226, 228–230, 235–237, 239–241, 243, 245, 246
Survey, 17–21, 23, 25, 75, 76, 79, 89, 108, 126, 154, 157, 230, 237

Technological development, 242
Technology, 8, 241, 242
Tests, 6, 130, 158, 159, 161

Theory, 3, 5, 10, 34, 35, 38, 49, 50, 86, 103, 105, 173, 216, 217, 235
Third age, 8
Top-down, 37, 38, 40
Traits, 37, 38, 40
Transport/transportation, 6, 28, 52, 58, 104, 105, 112, 114, 116, 117, 138, 185, 222, 226, 228, 241

Urban, 36, 89, 106, 108, 112–117, 125, 132, 138, 143, 145, 151, 152, 154, 157, 161, 169, 171, 236, 241
USA, 55, 56, 58–60, 75, 76, 78, 112, 113, 151, 152, 156, 162, 180, 181, 184–186, 190, 196, 236

Values, 3, 5, 6, 16, 28, 44, 59, 73, 75, 78, 81, 90–92, 94, 105, 112, 114, 126, 127, 131, 144, 173, 179, 182, 187, 188, 244

Wealth, 6, 9, 61, 65, 67, 73, 74, 81, 103, 190, 217, 242
Welfare, 4, 45, 68, 73–75, 82, 85–97, 174, 217, 225, 238, 240, 242, 245
Welfare State, 45, 85, 86, 88, 96, 245
Welfare system, 4, 86, 96, 97, 238, 242
Well-being, 3–6, 9, 15, 16, 19–22, 24, 27, 28, 33–45, 51–53, 55, 57, 58, 65–67, 74, 76–80, 86, 87, 92, 101, 104–106, 108, 109, 112–118, 126, 151, 153, 175, 179, 180, 185, 186, 190, 191, 215–220, 222, 225–229, 235, 236, 239, 240, 242, 244, 246, 247
WHOQoL/WHOQoL-Bref, 6, 16, 21, 23, 24, 90–92, 123, 216
Widowhood, 52, 55, 131, 132, 172, 203
World Health Organization (WHO), 3, 16, 90

Social Indicators Research Series

1. V. Møller (ed.): *Quality of Life in South Africa.* 1997   ISBN 0-7923-4797-8
2. G. Baechler: *Violence Through Environmental Discrimination.* Causes, Rwanda Arena, and Conflict Model. 1999   ISBN 0-7923-5495-8
3. P. Bowles and L.T. Woods (eds.): *Japan after the Economic Miracle.* In Search of New Directories. 1999   ISBN 0-7923-6031-1
4. E. Diener and D.R. Rahtz (eds.): *Advances in Quality of Life Theory and Research.* Volume I. 1999   ISBN 0-7923-6060-5
5. Kwong-leung Tang (ed.): *Social Development in Asia.* 2000
   ISBN 0-7923-6256-X
6. M.M. Beyerlein (ed.): *Work Teams: Past, Present and Future.* 2000
   ISBN 0-7923-6699-9
7. A. Ben-Arieh, N.H. Kaufman, A.B. Andrews, R. Goerge, B.J. Lee, J.L. Aber (eds.): *Measuring and Monitoring Children's Well-Being.* 2001
   ISBN 0-7923-6789-8
8. M.J. Sirgy: *Handbook of Quality-of-Life Research.* An Ethical Marketing Perspective. 2001   ISBN 1-4020-0172-X
9. G. Preyer and M. Bös (eds.): *Borderlines in a Globalized World.* New Perspectives in a Sociology of the World-System. 2002
   ISBN 1-4020-0515-6
10. V. Nikolic-Ristanovic: *Social Change, Gender and Violence: Post-communist and war affected societies.* 2002   ISBN 1-4020-0726-4
11. M.R. Hagerty, J. Vogel and V. Møller: *Assessing Quality of Life and Living Conditions to Guide National Policy.* 2002   ISBN 1-4020-0727-2
12. M.J. Sirgy: *The Psychology of Quality of Life.* 2002   ISBN 1-4020-0800-7
13. S. McBride, L. Dobuzinskis, M. Griffin Cohen and J. Busumtwi-Sam (eds.): *Global Instability.* Uncertainty and new visions in political economy. 2002
   ISBN 1-4020-0946-1
14. Doh. Chull Shin, C.P. Rutkowski and Chong-Min Park (eds.): *The Quality of Life in Korea.* Comparative and Dynamic Perspectives. 2003
   ISBN 1-4020-0947-X
15. W. Glatzer: *Rich and Poor.* Disparities, Perceptions, Concomitants. 2002
   ISBN 1-4020-1012-5
16. E. Gullone and R.A. Cummins (eds.): *The Universality of Subjective Wellbeing Indicators.* A Multi-disciplinary and Multi-national Perspective. 2002
   ISBN 1-4020-1044-3
17. B.D. Zumbo (ed.): *Advances in Quality of Life Research 2001.* 2003
   ISBN 1-4020-1100-8

18. J. Vogel, T. Theorell, S. Svallfors, H.-H. Noll and B. Christoph: *European Welfare Production*. Institutional Configuration and Distributional Outcome. 2003 ISBN 1-4020-1149-0

19. A.C. Michalos: *Essays on the Quality of Life*. 2003 ISBN 1-4020-1342-6

20. M.J. Sirgy, D. Rahtz and A.C. Samli (eds.): *Advances in Quality-of-Life Theory and Research*. 2003 ISBN 1-4020-1474-0

21. M. Fine-Davis, J. Fagnani, D. Giovannini, L. Højgaard and H. Clarke: *Fathers and Mothers: Dilemmas of the Work-Life Balance*. 2004

ISBN 1-4020-1807-X

22. M.J. Sirgy, D.R. Rahtz and D.J. Lee (eds.): *Community Quality-of-Life Indicators*. Best Cases. 2004 ISBN 1-4020-2201-8

23. A. Dannerbeck, F. Casas, M. Sadurni and G. Coenders (eds.): *Quality-of-Life Research on Children and Adolescents*. 2004 ISBN 1-4020-2311-1

24. W. Glatzer, S. von Below and M. Stoffregen (eds.): *Challenges for Quality of Life in the Contemporary World*. 2004 ISBN 1-4020-2890-3

25. D.T.L. Shek, Y. Chan and P.S.N. Lee (eds.): *Quality of Life Research in Chinese, Western and Global Contexts*. 2005 ISBN 1-4020-3601-9

26. A.C. Michalos (ed.): *Citation Classics from Social Indicators Research*. The Most Cited Articles Edited and Introduced by Alex C. Michalos. 2005

ISBN 1-4020-3722-8

27. A. Ben-Arieh and R.M. Goerge (eds.): *Indicators of Children's Well Being*. Understanding Their Role, Usage and Policy Influence. 2006

ISBN 1-4020-4237-X

28. M.J. Sirgy, D. Rahtz and D. Swain (eds.): *Community Quality-of-Life Indicators*. Best Cases II. 2006 ISBN 1-4020-4624-3

29. R.J. Estes (ed.): *Advancing Quality of Life in a Turbulent World*. 2006

ISBN 1-4020-5099-2

30. David Kucera (ed.): *Qualitative Indicators of Labour Standards*. Comparative Methods and Applications. 2007 ISBN 1-4020-5200-6

31. Heidrun Mollenkopf and Alan Walker (eds.): *Quality of Life in Old Age*. International and Multi-Disciplinary Perspectives. 2007

ISBN 978-1-4020-5681-9